VOLUME 160

International Review of Cytology
A Survey of Cell Biology

Edited by

Kwang W. Jeon
Department of Zoology
University of Tennessee
Knoxville, Tennessee

Jonathan Jarvik
Department of Biological Sciences
Carnegie Mellon University
Pittsburgh, Pennsylvania

VOLUME 160

ACADEMIC PRESS
San Diego   New York   Boston   London   Sydney   Tokyo   Toronto

This book is printed on acid-free paper. ⊚

Copyright © 1995 by ACADEMIC PRESS, INC.

Academic Press, Inc.
A Division of Harcourt Brace & Company
525 B Street, Suite 1900, San Diego, California 92101-4495

United Kingdom Edition published by
Academic Press Limited
24-28 Oval Road, London NW1 7DX

International Standard Serial Number: 0074-7696

International Standard Book Number: 0-12-364563-8

PRINTED IN THE UNITED STATES OF AMERICA
95 96 97 98 99 00 EB 9 8 7 6 5 4 3 2 1

CONTENTS

Molecular Mechanisms for Passive and Active Transport of Water

T. Zeuthen

The Comparative Cell Biology of Accessory Somatic (or Sertoli) Cells in the Animal Testis

Sardul S. Guraya

Factors Controlling Growth, Motility, and Morphogenesis of Normal and Malignant Epithelial Cells

Carmen Birchmeier, Dirk Meyer, and Dieter Riethmacher

Dynamics of the Seminal Vesicle Epithelium

Lucinda R. Mata

Molecular Organization of Hepatocyte Peroxisomes

Takashi Makita

CONTRIBUTORS

Numbers in parentheses indicate the pages on which the authors' contributions begin.

Carmen Birchmeier (221), *Max Delbrück Center for Molecular Medicine, D-13122 Berlin, Germany*

Sardul S. Guraya (163), *Department of Zoology, Punjab Agricultural University, Ludhiana-141004, India*

Keith E. Latham (53), *Fels Institute for Cancer Research and Molecular Biology and Department of Biochemistry, Temple University School of Medicine, Philadelphia, Pennsylvania 19140*

Takashi Makita (303), *Department of Veterinary Anatomy, Yamaguchi University, Yoshida, Yamaguchi City 753, Japan*

Lucinda R. Mata (267), *Department of Cell Biology, Gulbenkian Institute of Science, 2781 Oeiras Codex, Portugal*

James McGrath (53), *Department of Genetics and Pediatrics, Yale University Medical Center, New Haven, Connecticut 06511*

Dirk Meyer (221), *Max Delbrück Center for Molecular Medicine, D-13122 Berlin, Germany*

Yasuhiko Ohta (1), *Laboratory of Animal Science, Department of Veterinary Science, Faculty of Agriculture, Tottori University, Tottori 680, Japan*

Dieter Riethmacher (221), *Max Delbrück Center for Molecular Medicine, D-13122 Berlin, Germany*

Davor Solter (53), *Max Planck Institute for Immunobiology, D-79108 Freiburg, Germany*

T. Zeuthen (99), *Department of Medical Physiology, The Panum Institute, University of Copenhagen, DK-2200 Copenhagen N, Denmark*

Sterility in Neonatally Androgenized Female Rats and the Decidual Cell Reaction

Yasuhiko Ohta

Laboratory of Animal Science, Department of Veterinary Science, Faculty of Agriculture, Tottori University, Tottori 680, Japan

In female rats, administration of androgen during early postnatal life results in anovulatory sterility characterized by polyfollicular ovaries and persistent vaginal cornification in adulthood. In these androgen-sterilized rats, the acyclic male pattern of gonadotropin secretion results from permanent damage to hypothalamic centers normally responsive to steroid feedback. The capacity of the endometrium to differentiate into deciduoma in response to endometrial stimulation is markedly reduced in the uterus of the androgenized rats receiving an appropriate regimen of progesterone and estrogen injections. The hypothesis is presented that the lowered uterine responsiveness in the androgenized rats is largely ascribable to the effect of androgen given neonatally rather than to the influence of continued exposure to ovarian estrogen. This review deals with the nature of the uterine response to ovarian hormones and the deciduogenic stimulus in androgenized rats in order to verify the mechanism involved in androgen action on the neonatal uterus.

KEY WORDS: Androgenization, Persistent estrus, Decidual cell reaction, Uterus, Female rat.

I. Introduction

Treatment of female rats with androgen or estrogen during a critical neonatal period induces permanent anovulation, resulting in persistent vaginal cornification in the adult. This syndrome reflects the masculinization of the secretion pattern of gonadotropins owing to permanent alterations in the hypothalamic centers normally involved in the ovulatory discharge of

1

gonadotropins from the pituitary. Many studies have analyzed the sexual differentiation of the central nervous system (CNS) controlling reproductive functions and behavior (Takewaki, 1962; Barraclough, 1966; Flerkó, 1971; Gorski, 1971, 1979; MacLusky and Naftolin, 1981; Arnold and Schlinger, 1993). However, permanent disruption of normal function is also observed in organs other than the brain in the androgen- or estrogen-sterilized rats.

Short-term administration of large doses of sex steroids to laboratory rodents during a critical period results in persistent changes in reproductive organs, which often develop precancerous or cancerous lesions with age, hyperplastic and/or neoplastic changes in the genital tracts and mammary glands (Takasugi, 1976; Iguchi, 1992; Bern, 1992a,b), and polycystic changes in ovaries (Reiter, 1969) comparable to those seen in the Stein-Leventhal syndrome in domestic animals and humans (Goldzeiher, 1981). Thus, neonatally androgenized rats, in addition to their use as laboratory model for neurological studies, also provide a model for oncological and gerontological investigations of age-related changes in reproduction. Little attention has been paid to the effect of androgen on the neonatal uterus, with reference to fertility in the androgenized rats. The infertility in these rats is mainly accounted for by the absence of ovulation caused by the lack of ovulatory surge of luteinizing hormone (LH) (Gorski, 1971) or by inadequate hormonal secretion from the induced corpora lutea (CL) (Barraclough and Fajer, 1968; Hahn and McGuire, 1978). The sensitivity of ovaries to gonadotropins and the ovarian binding of human chorionic gonadotropin (HCG) are known to be lower in androgenized females than in controls (Uilenbroek and van der Werff ten Bosch, 1972; Kolena *et al.*, 1977). During the past few decades, however, a number of studies have shown that altered uterine reactivity to ovarian hormones, in addition to hormonal deficiency of the CL, may be the cause, at least in part, of implantation failure in the androgenized rats, and that the uterus as well as the anterior pituitary is less responsive to estrogen in such rats (Wrenn *et al.*, 1969). Unusual reactivity of uterine mucosa is also shown by the metaplasia and hypertrophy of the epithelium following estrogen injections (Takewaki, 1968). Moreover, it has been reported that estrogen-binding sites are reduced in the uterus, pituitary, and hypothalamus of sterilized rats (Heffner and van Tienhoven, 1979; Flerkó *et al.*, 1969), which may account for the decreased estrogen sensitivity of these organs.

In many mammalian species, the endometrium differentiates into a decidua in response to blastocyst implantation. The main event in this differentiation is the transformation of stromal cells into polyploid cells, the decidual cells (Finn and Porter, 1975b; Parr and Parr, 1989). It may be thought, therefore, that a failure in the decidual reaction is due to a disturbance of the implantation process in the androgenized rats. On the other hand, similar endometrial differentiation has been reported in animals subjected to an experimental condition: in pseudopregnant rodents and in those adequately sensitized with ovarian steroids, decidual cell reaction (DCR) can also be induced by an artificial endometrial stimulus mimicking blastocyst stimulation, such as intrauterine instillation of oil (Finn and Porter, 1975b; Parr and Parr, 1989; Abrahamsohn and Zorn, 1993). This decidual response has been used in the study of hormonal control of implantation (Finn and Porter, 1975b) and employed as a criterion for luteal function (Gibori et al., 1984). In neonatally androgenized rats, however, little is known of the mechanism of endometrial differentiation into deciduoma in response to uterine stimulation after exposure to ovarian steroids following ovariectomy. This chapter reviews the decidual response in androgen-sterilized rats, with attention to the cytological basis for the mechanism of androgen action on the neonatal uterus.

II. Fertility in Neonatally Androgenized Female Rats

The anovulatory sterility induced by androgen given in early postnatal life varies with the initiation day and the dose (Barraclough, 1966; Rudel and Kincl, 1966; Kincl, 1990). Neonatal or earlier treatment induced sterility with a lower dose of androgen than did later treatment. Barraclough and Gorski (1962) and Rudel and Kincl (1966) have reported that persistent estrus is indicative of sterility, and that a daily dose of 10 μg of testosterone propionate (TP) given in early postnatal life is the minimally effective dose for producing sterility in adulthood, whereas several authors have shown that androgen treatment with lower doses of TP, 5 μg or less, is effective in inducing anovulatory sterility after 10 weeks (Barraclough, 1966; Gorski, 1968; van der Werff ten Bosch et al., 1971). Androgenization with small doses, 5–10 μg TP, beginning at 3 or 5 days of age, did not impede mating behavior (Barraclough and Gorski, 1962; Uilenbroek and van der Werff ten Bosch, 1972; Hahn and McGuire, 1978; Shanbhag and Maqueo, 1986) and allowed the treated rats to attain full-term pregnancy early in postpuberal life (Swanson and van der Werff ten Bosch, 1964; Shanbhag and Maqueo, 1986); later these rats became anovulatory, although their sexual receptivity was retained (Swanson and van der Werff ten Bosch, 1964).

The lightly androgenized rats entered into anovulation during their lactation period (Shanbhag et al., 1987). Given 20 μg of TP at 3 days of age, a number of females became pregnant after mating, notwithstanding irregular persistent estrus, but most of them showed resorption of fetuses by midpregnancy (Dorner and Fatschel, 1970). Adult rats androgenized neonatally with 50–250 μg TP still retained mating ability despite having entered into persistent estrus, although increasing doses decreased the percentage of mated animals, and suppressed ovulation and subsequent pregnancy (Kramen and Johnson, 1971; Hahn and McGuire, 1978; Johnson, 1979; Shanbhag and Maqueo, 1986). Postnatal treatment with large doses, more than 1 mg TP, almost abolished mating behavior (Dorner and Fatschel, 1970; Kramen and Johnson, 1971). However, administration of LH increased the percentage of mating in rats given 1.25 mg TP neonatally (Kramen and Johnson, 1971). Thus, varying degrees of infertility in androgenized rats appear to depend on both TP dose and age of animals exposed to androgen.

Ovulation and CL formation occured in androgen-sterilized immature and adult rats when gonadotropins or their releasing hormones were administered (Kincl, 1990). Ovaries of androgenized rats are responsive to exogenous and endogenous gonadotropins, but require more hormone than those of normal cycling rats (Ying, 1973). Kramen (1974) reported successful implantation with a subnormal number of embryos in LH-treated, androgen-sterilized rats bearing pituitary grafts under the kidney capsule. He suggests that the failure to obtain the normal number of embryos results from a deficit of the hormones maintaining pregnancy rather than from depression of uterine sensitivity.

Hahn and McGuire (1978) have confirmed this hypothesis in a study of rats exposed to 100 μg TP on the third day after birth, indicating that these animals failed to conceive normally even when ovulation and fertilization were induced by LH-releasing hormone, mating and progesterone (P) or prolactin (PRL), or by transplantation of normal pituitary glands under renal capsules. Furthermore, transfer studies revealed that embryos recovered from androgenized rats showed normal competence. This is in good agreement with the findings by Kramen and Johnson (1971) that fertilized eggs ovulated with LH and recovered from persistent estrous (PE) females androgenized with 50 μg of TP were morphologically normal. They developed normally to day 18 of gestation when they were transferred on day 1 into normal pseudopregnant recipients.

In androgenized rats, luteal function following gonadotropin-induced ovulation was assessed by several authors. Using criteria such as sudanophilic staining characteristics, Zeilmaker (1964) revealed that in the PE rats treated with 1 mg TP at 5 days of age, cervical stimulation was incapable of inducing functional CL, whereas reserpine treatment or isotransplants of pituitary glands beneath the kidney capsule resulted in functional CL and extensive mucification of the vaginal epithelium, suggesting the secretion of progestins. Administration of reserpine or PRL to the females androgenized with 1 mg TP at 5 days resulted in secretion of P in high concentrations into ovarian vein blood, whereas HCG barely stimulated P secretion (Barraclough and Fajer, 1968). When the dose of TP given neonatally was decreased to 50 or 100 μg, ovaries of androgenized rats were capable of producing and secreting P in response to the administration of LH (Cortes *et al.*, 1971) or pregnant mare's serum gonadotropin (PMSG) (Johnson, 1979).

Serum P was subnormal in mated androgenized rats aborted at early pregnancy in which the endometrium showed minimal progestational changes, although CL exhibited normal structure (Hahn and McGuire, 1978). These findings may indicate a permanent modification in the secretion pattern of PRL from the anterior pituitary of the androgenized rats.

Van der Schoot and de Greef (1983) have reported that treating adult rats sterilized by 1.25 mg TP at 4 or 5 days of age with HCG induced ovulation followed by a period with high levels of plasma P. In contrast to normal cyclic female rats, stimulation of the uterine cervix was not necessary for a raised level of P after HCG-induced ovulation in these androgenized rats. The two daily surges of PRL secretion following ovulation (secretion pattern of normal females) (de Greef and Zellmaker, 1978) were not observed in androgenized female rats, which showed a high basal level of PRL secretion (Johnson, 1979; Gala, 1981; Watanobe *et al.*, 1991b; Collado and Aguilar, 1993), although in androgenized females, pituitary content of PRL was approximately the same as or lower than in the control females (Mallampati and Johnson, 1974). Thus, the control of luteal activity in androgenized females may reflect the male characteristics (Neil, 1972) brought about by the neonatal androgen treatment and subsequently cause continuous secretion of estrogen (Gala, 1981). Hormonal imbalance eliciting an uncongenial uterine milieu is one cause of fertility failure in spite of ovulation and frequent matings in androgen-sterilized rats.

III. Decidual Response in Androgen-Sterilized Female Rats

A. Androgenized Pseudopregnant Rats

Van der Schoot and de Greef (1983) have investigated the uterine capacity for deciduoma formation in neonatally androgenized rats with HCG-induced pseudopregnancy. Endometrial scratching on the fifth day after ovulation did not result in the formation of deciduomata. Furthermore, treatment of the androgenized females with P did not lead to higher uterine weights than those seen in untreated androgenized females. These authors concluded, therefore, that the inability of the uterus to respond to either a deciduogenic stimulus or P treatment might explain the failure of pregnancy even though high levels of P were measured in the plasma. This is in contrast to an earlier study reporting that a deficit in progestins interfered with normal pregnancy in TP-treated females (Hahn and McGuire, 1978). Treatment with P or PRL resulted in normal implantation and maintenance of pregnancy. The difference between these two studies may be ascribable to the doses of androgen injected neonatally. Hahn and McGuire (1978) employed a lower dose of TP than that used by van der Schoot and de Greef (1983). Failure of the decidual response has been also reported in pseudopregnant rats given 1 mg TP at 5 days of age (Matton and Maurasse, 1985). In pseudopregnancy induced by cervical stimulation or administration of PMSG and LH, massive deciduomata were produced by uterine traumatization in controls, but no response was observed in androgenized rats, although in immature rats with gonadotropin-induced pseudopregnancy, trauma elicited a greater response in androgenized animals than in controls.

B. Androgenized Rats Given Ovarian Hormones

The findings in neonatally androgenized pseudopregnant rats indicate that the failure of the uterus to show a decidual reaction is involved in infertility. Deciduoma formation in response to trauma has been also investigated in ovariectomized (OX) rats treated with androgen neonatally after exposure to progestins alone or in combination with a small amount of estrogen. In earlier reports, Burin *et al.* (1963) found that deciduomata were elicited in androgenized rats following a trauma applied on the fifth day of 8 daily injections of 5 mg P. In contrast, in a group of neonatally androgenized rats that were ovariectomized after HCG injections, Zeilmaker (1964) could not induce deciduoma formation by traumatizing the endometrium on the fourth day of an 8-day period of injections of 5 mg P and 0.5 μg estrone (E1). Although the latter observation appears to sustain the conclusion of van der Schoot and de Greef (1983), the exact data on the decidual response were not provided.

Takewaki and Ohta (1974), Kramen and Johnson (1975), and Matton and Maurasse (1985) have confirmed and extended the study done by Zeilmaker (1964) by a more detailed analysis of uterine ability to form deciduomata in androgenized, OX rats given ovarian hormones. Takewaki and Ohta (1974) reported that PE rats androgenized with 1.25 mg of TP neonatally invariably failed to form deciduomata in response to endometrial scratching, even when they were given 7 daily injections of 2 mg P together with 0.2 μg estradiol (E2) starting on the day following ovariectomy as adults. These hormone doses were sufficient for the development of deciduomata in reaction to trauma in nonandrogenized rats (Fig. 1).

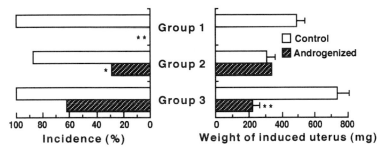

FIG. 1 Deciduoma formation in response to uterine trauma in neonatally androgenized (1.25 mg TP) and control rats. Groups 1 and 3 were given simultaneous injections of 2 mg P and 0.2 μg E2 for 7 days after ovariectomy as adults, while group 2 received 7 daily injections of 5 mg P and 0.2 μg E2. Group 3 underwent hormone treatment about 1 month after ovariectomy as adults. The weight of the induced uterus is the mean weight of uteri bearing deciduomata. *$P < .05$, **$P < .01$ versus control.

Kramen and Johnson (1975) showed a reduction in decidual response after trauma in ovary-bearing PE rats androgenized with 1.25 mg TP at 5 days of age and given 9 daily injections of 2 mg P alone as adults. In contrast, PE rats, androgenized with 1 mg of TP at 5 days, ovariectomized as adults, and given 8 daily injections of 2 mg P, formed deciduomata in response to trauma, although the magnitude of response estimated by the weight of the traumatized uterus was decreased by 50% compared with controls (Matton and Maurasse, 1985). The incidence of deciduomata was markedly elevated by increasing the daily dose of P from 2 to 5 mg, given together with estrogen, although the response was always less in androgenized females than in controls given similar injections (Takewaki and Ohta, 1974). The incidence of positive response was unchanged by a single injection of 0.1 μg estradiol benzoate (EB) given on the fourth or fifth day of 9-day treatment with P in the androgenized rats (Kramen and Johnson, 1975).

Takewaki and Ohta (1974) examined the decidual response in androgenized rats having a resting period between ovariectomy and commencement of treatment with ovarian steroids to determine whether the uterus of the androgenized rat recovers normal sensitivity to ovarian hormones after blockade of persistent estrus. Interruption of persistent estrus by ovariectomy for as long as 2 months in such rats prior to the start of treatment with ovarian steroids did not restore uterine responsiveness, indicating that neonatal androgenization results in a permanent or at least a long-lasting reduction of uterine sensitivity to ovarian steroids. This finding again suggests that lowered uterine sensitivity was brought about by neonatal injection of androgen rather than by continued exposure to endogenous estrogen. In a similar experiment using PE rats androgenized with 1 mg TP neonatally, the uterine response had not fully recovered to the normal level 12 to 15 days after ovariectomy (Matton and Maurasse, 1985), confirming the previous suggestion (Ohta and Takewaki, 1974).

The decline of the decidual response in androgenized females could be related to tonic maximal stimulation of the uterus (Lobl et al., 1974; Lobl and Maenza, 1975) or to the modified capacity of the uterus to bind ovarian hormones. Many authors have indicated that uptake of radioactive estrogen into the uterus is decreased in androgenized rats (Flerkó et al., 1969; McGuire and Lisk, 1969; Tuohimaa and Johansson, 1971; Lobl, 1975a; Maurer and Wooley, 1975; Heffner and van Tienhoven, 1979, among others) despite normal concentrations of cytosolic estrogen receptor (ER) (Gellert et al., 1977; Thrower et al., 1978; Campbell, 1980; Morikawa et al., 1982; Campbell and Modlin, 1987).

Neonatal androgenization also reduced specific (5S) nuclear binding of estrogen and the formation of estrogen-induced proteins (Lobl, 1975a). In addition, Lobl (1975a,b) and Thrower et al., (1978) have suggested that androgenization increases the rat uterine cytosol proteins which inhibit translocation of estrogen to the nucleus, thus reducing specific nuclear binding. P binding in the rat uterus is decreased by ovariectomy and can be restored to intact levels following estrogen stimulation (Milgrom and Baulieu, 1970; McGuire and DeDella, 1971; Feil et al., 1972). Accordingly, a decrease in estrogen binding would probably be followed by a reduction in P binding. White et al., (1981) have shown that the concentration of progesterone receptor (PR) was low in the uterus of the androgenized rats, and administration of estrogen did not increase this concentration, although PR in the normal rat uterus is under estrogenic control and increases following estrogen injection (Kurl and Borthwick, 1979; Castellano-Diaz et al., 1987). A marked decrease in uterine response to trauma in androgenized rats may be explained by these findings.

In contrast, on the basis of findings that the addition of 0.1 μg E2 to daily injections of P completely abolished decidualization in androgenized rats ovariectomized as adults but enhanced decidualization in controls, Matton and Maurasse (1985) have suggested that, rather than being insensitive to estrogen, the uteri of androgenized adult rats are more sensitive to estrogen than those of control rats. Furthermore, the decidual response obtained in the resting period is significantly lowered by increasing the amount of estrogen given during the steroid injection period, whereas in the control rats, such treatment had either a beneficial effect or none, indicating that the response in androgenized rats is somewhat similar to what occurs in normal rats when the dose of estrogen is raised above a certain threshold (Yochim and DeFeo, 1963). These authors concluded, therefore, that the uteri of androgenized females differ from those of controls in two ways: increased sensitivity to estrogen and inability to achieve maximal growth in response to a deciduogenic stimulus.

In addition, Matton and Maurasse (1985) suggested that reduced weight of the deciduoma could be attributed to the reduced number of stromal cells present in the uterus of the androgenized rats (Lobl and Maenza, 1975; White *et al.,* 1981). In the view of these findings, it is concluded that neonatal androgenization results in persistent changes in reactivity of the uterus to the ovarian steroids, which are responsible for the lowered decidual response.

C. Effect of Dose of Androgen and Age

The dose and duration of neonatal exposure of laboratory rodents to gonadal steroids can vary to induce different reproductive alterations. Permanent changes in peripheral target organs are induced by neonatal treatment with steroids in a dose-dependent manner. Treatment of female mice with estrogen and androgen during neonatal life induces two different types of persistent proliferation and cornification or parakeratosis of the vaginal epithelium in adulthood. One type of persistent vaginal change, brought about by low doses of estrogen or androgen, is ascribable to a permanent alteration of the hypothalamo-hypophysial system and is directly related to a continued secretion of ovarian estrogen, whereas the other, elicited by high doses of the steroids, is due to permanent changes in the vaginal epithelium itself and is estrogen-independent (Takasugi, 1976; Bern and Talamantes, 1981).

The incidence of sterility in androgenized female rats is also dependent upon the dose of androgen given neonatally and the age of animals (see Section II). Administration of low doses of androgen to neonatal female rats results in a delayed anovulatory syndrome, which is characterized by maintenance of ovulatory cyclicity for some weeks to months after puberty (Swanson and van der Werff ten Bosch, 1964; Gorski, 1968; Harlan and Gorski, 1977). The fertility of such animals is maintained until establishment of vaginal acyclicity (Swanson and van der Werff ten Bosch, 1964; Shanbhag and Maqueo, 1986; Shanbhag *et al.,* 1987). It would be interesting to know the uterine responsiveness to deciduogenic stimulus in the delayed anovulatory syndrome, but no data are available.

Kramen and Johnson (1975) have investigated the decidual response in the uteri of adult females of different ages given various doses of TP at 5 days of age. Female rats were given doses of 50, 250, or 1250 μg of TP, which could induce a full anovulatory syndrome until about 2 months of age. A decidual response was induced in adult females with ovaries by uterine trauma on the sixth day of a 9-day period of injection of P alone or in combination with EB. In 2-month-old rats given a single injection of 250 μg TP neonatally and sacrificed after the last injection of P in doses varying from 2 mg for the first 5 days to 1 or 4 mg for the following 4 days, uterine trauma elicited a response in all rats. In 6- to 8-month-old rats, however, females receiving a single injection of 250 μg or less of TP failed to give a positive response, even when a single injection of 0.1 μg EB was added on the day before traumatization. The incidence of deciduomata was markedly reduced in 3- to 4-month-old females given 1250 μg TP neonatally.

Female rats androgenized with 1 mg TP neonatally and receiving P plus E2 following ovariectomy as adults, as well as the controls, responded to the trauma postpuberally (1 month of age), but failed to respond at 3 months (Matton and Maurasse, 1985). Although females aged 2 months given 1.25 mg TP neonatally and daily injections of P plus E2 after ovariectomy failed to form deciduomata in response to trauma (Takewaki and Ohta, 1974), Ohta (1983) showed that if the dose of TP was reduced to 100 μg, trauma was effective in all females at 2 months of age. These findings indicate that the uterus of young females treated neonatally with the lower dose of TP responds as well as that of normal females, which is in good agreement with successful implantation in young animals given less than 50 μg TP, and the fact that it takes a long time to elicit total failures in these females (see Section II). With the higher dose of TP, the uterine capacity to form deciduomata is lost earlier than with the lower dose. Thus, the general lack of success in obtaining a response in females treated with TP is probably related to the dose used neonatally, and to the age of the animals when challenged.

The mechanism of the aging effect involved in the failure of the uterine response in the androgenized rats is unclear, but may be accounted for by the continued secretion of estrogen from their ovaries. Kramen and Johnson (1975) have pointed out that the uterine histology provides some clues in this matter: before 3 months of age, the uterine epithelium of females treated with TP is simple, tall columnar with basal nuclei, and sparse uterine glands are lined by low columnar to cuboidal epithelium, whereas with increasing age, the uterine epithelium undergoes squamous metaplasia. Hyperplasia and/or metaplasia of the uterine epithelium has been noted in rodents treated with androgens (Takewaki, 1965; Hayashi, 1968; Reiter, 1969; White et al., 1981). Since in some cases, the transformation occurs even in the absence of the ovaries (Takewaki, 1965), the possibility of a direct effect of androgen upon the uterine response to subsequent hormonal stimulation, in addition to the effect of continued exposure to estrogen, cannot be ruled out. Moreover, Kramen and Johnson (1975) suggested that changing endocrine levels in aging females treated with androgen may also be involved in the failure to elicit a decidual reaction. PRL levels increase with age and with the dose of TP given at 5 days of age (Mallampati and Johnson, 1974). The role of prolactin in uterine physiology is uncertain, although the hyperprolactinemia induced by pituitary grafts results in adenomyosis in the mouse uterus (Mori *et al.*, 1987).

D. Critical Period for Induction of Permanent Loss of Responsiveness

Induction of anovulatory sterility by neonatal treatment with androgen depends on the age of the animal. In rats, there is the androgen-sensitive period between 1 and 10 days after birth, although the incidence of the sterility is dependent upon the dose of androgen administered in addition to the time of treatment (Barraclough, 1966; Kincl, 1990). A critical period for steroid-induced abnormalities also occurs in peripheral target organs: the critical period for induction of estrogen-induced testicular damage in male mice is limited to the first 10 days after birth (Takasugi, 1970) and for induction of estrogen independent vaginal cornification in female mice, it is the first 3 days of age (Takasugi, 1966). A critical period may also exist during postnatal uterine development (Clark and Gorski, 1970; Takewaki and Ohta, 1975a; Sananès *et al.*, 1976; Ohta *et al.*, 1993a). Takewaki and Ohta (1975a) gave a single injection of 1.25 mg TP to female rats at 8 or 10 days of age in an attempt to determine the androgen-sensitive period for inducing the permanent loss of uterine responsiveness to a deciduogenic stimulus. Adult rats given TP at 8 days invariably exhibited persistent estrus, whereas most of the females androgenized at 10 days showed estrous cycles, with only some animals showing persistent estrus. These two groups of androgenized rats invariably formed deciduomata in response to trauma in the presence of an appropriate hormone supply. However, the uterine response was significantly less in rats injected with TP at 8 days. Since there is an excellent correlation between the weight and the histology of the uteri bearing deciduomata (Velardo *et al.*, 1953), it is evident that deciduomata were better developed in rats given TP at 10 days than in those androgenized at 8 days. Thus, it is apparent that treatment of female rats with TP at 10 days in no way affected uterine responsiveness in adulthood.

To study the possible effects of a continued exposure to estrogen on uterine response to the trauma in rats treated with TP at 10 days, female rats ovariectomized at 9 days and treated with 1.25 mg TP on the next day were given a series of 30 daily injections of 0.1 μg E2 from 30 days of age onward before the rats were placed on the standard injection schedule of P plus E2 for determining decidual response. The uterine response in the rats ovariectomized prior to treatment with TP at 10 days was not significantly different from that in ovariectomized rats given no TP, whether given alone or in combination with subsequent injections of estrogen for 30 days. An androgen-sensitive period appears to exist in the female rat between birth and the eighth day of age, during which administration of TP results in a long-lasting decrease in uterine reactivity. However, although treatment with TP at 5 days induces a marked reduction of both incidence and magnitude of deciduomata, treatment at 8 days has a less drastic effect, causing a decrease in size of deciduomata but not in their incidence. This critical period appears to be a little shorter than that for the induction of anovulatory sterility, since some rats treated with TP at 10 days were acyclic, showing that the central mechanism controlling gonadotropin secretion was masculinized in some animals.

To characterize the critical period for the neonatal steroid effects on the uterus, Ohta (1981) investigated the capacity for deciduoma formation in infantile rats. Infantile rats given 7 daily injections of 1 mg P together with 0.1 μg E2 starting at 10 days of age invariably responded to trauma by forming deciduomata. In contrast, animals receiving injections of the ovarian steroids commencing at 3 days of age always failed to form deciduomata. Injections of steroids starting at 5 or 7 days allowed a positive response to the trauma in increasing percentages of the animals. An electron microscopic study was carried out on endometrial stromal cells from rats given P plus E2 for a 7-day period from 3 or 10 days (Ohta, 1981). In traumatized uteri of rats receiving injections of steroids from 3 days of age, endometrial stromal cells invariably failed to differentiate into decidual cells, being similar in structure to those of intact uteri, although their nuclear and cytoplasmic volumes were increased in the traumatized uteri as well as in intact uteri compared with those in age-matched animals given no steroids.

In contrast, in rats receiving a similar schedule of steroid treatment beginning at 10 days, stromal cells in the antimesometrial side of the endometrium responded to the trauma by transforming into large cells. These cells usually possessed two large nuclei and numerous tonofilaments in the peripheral regions of the cytoplasm, which is characteristic of antimesometrial decidual cells as described by Sananès and Le Goascogne (1976) and Parr et al. (1986). In the intact uteri, stromal cells were approximately similar in structure to those in rats given P plus E2 injections from 3 days, containing more numerous cytoplasm organelles than in the age-matched controls.

The structure of stromal cells in intact horns is in good agreement with that reported by Tachi and Tachi (1974) in ovariectomized and adrenalectomized (AX) adult rats given P injections. However, since cellular stimulation was less marked in the rats treated with the steroids from 3 days of age than in animals so treated from 10 days, the failure of decidual reaction in the former might be ascribable to a lower sensitivity of stromal cells to the injected hormones. These findings suggest that the capacity of endometrial stromal cells to differentiate into decidual cells in response to trauma is not fully developed within about 10 days after birth. Consistent with this, Sananès and Le Goascogne (1976) concluded that the period from 7 to 10 days of age is of particular significance in the maturation of the rat uterus.

In rats, the uterine ER increased in number after birth and reached a maximum by 10 days of age (Clark and Gorski, 1970), whereas the P binding sites were markedly lower in prepuberal rats than in adult animals (Milgrom and Baulieu, 1970), indicating that the ability of endometrial cells to synthesize receptor proteins for ovarian hormones may be involved in uterine maturation. The distribution of ER was investigated immunohistochemically in the genital tract of C57BL/Tw female mice from the day of birth to 50 days of age (Sato et al., 1992). Uterine stromal cell nuclei showed positive ER immunoreaction at birth, whereas nuclei of the epithelial cells were negative until 4 days after birth, as has been reported in other strains of mice by Yamashita et al. (1989) and Bigsby et al. (1990). The number of positive epithelial cells and their staining intensity gradually increased until 10 days. On the other hand, immunohistochemical study of the distribution of PR in rat uterus from the day of birth to 30 days revealed that epithelial and stromal cells showed a negative PR immunoreaction on the day of birth (Ohta et al., 1993b); PR did not appear in the epithelium until day 5. The number of stained epithelial cells and the staining intensity gradually increased from 7 to 15 days and remained unchanged until day 30. In contrast, stromal cells still showed a negative PR reaction at 10 days. The staining of the stromal cells appeared on day 15 and increased steadily thereafter until day 30.

P or diethylstilbestrol (DES) given during a period extending from 7 to 9 days is reported to alter the decidual response in both immature and adult rats (Sananès and Le Goascogne, 1976; Sananès et al., 1980), indicating that this period is particularly significant in uterine maturation. Treatment of rats with P at 7–9 days reduced the number of nuclear bodies normally appearing in uterine epithelial cells during this period, and prevented deciduoma formation during prepubertal life, whereas treatment with DES at the same age increased nuclear bodies and facilitated the decidual response (Sananès and Le Goascogne, 1976; Le Goascogne and Baulieu, 1977). Sananès et al. (1980) suggested that the antidecidual effects of neonatal P may be due to antagonism of endogenous estrogen by progestin during the critical period associated with nuclear changes in epithelial cells. Thus, it seems likely that endogenous estrogen may play a key role in uterine development during this stage.

In both epithelial and stromal cells, however, the initiation of appearance of PR was not affected by ovariectomy performed on the day of birth or 2 to 3 days prior to the appearance of PR (Ohta *et al.*, 1993b). This is in good agreement with the findings that the uterine growth as estimated by DNA content, protein content, and weight is not affected by ovariectomy for 10–20 days, but thereafter the absence of the ovaries greatly inhibits uterine development (Clark and Gorski, 1970; Döcke *et al.*, 1981). Two or 3 daily injections of 0.1, 1, or 10 μg E2 induced PR staining in the stromal cells, regardless of ages at start of the treatment, days 1, 5, or 10, whereas staining in the epithelial cells was not induced by any estrogen treatment but rather was inhibited by estrogen (Ohta *et al.*, 1993b). In any dose, however, estrogen injection beginning on day 10 failed to increase the staining intensity of the stromal cells over that seen in intact rats at day 15. Daily injection of 0.1 μg E2 from days 10 to 14 suppressed PR appearance in epithelial cells for 10 to 15 days, when PR in the stromal cells normally appeared independently of PR in the epithelium. The staining of PR in the stromal cells induced by 3 daily injections of 10 μg E2 from the day of birth disappeared by day 6. In rats ovariectomized as adults, the PR reaction in the epithelial cells is decreased by estrogen and increased by progestin (Ohta *et al.*, 1993a).

The appearance of PR in rat uterus, at least in the epithelial cells, is programmed by estrogen before birth and/or determined by factors other than estrogen, in striking contrast to the appearance of ER in mouse uterine epithelial cells, which could be induced by a single injection of estrogen on the day of birth (Sato *et al.*, 1992). Moreover, 5 daily injections with 10 μg of T, 5α-dihydrotestosterone (DHT), and P did not induce PR in uterine epithelial and stromal cells (Y. Ohta, unpublished data). In regard to PR, therefore, it is not likely that gonadal steroids play any important role in its ontogenic expression in association with the critical period for postnatal maturation of the uterus.

Moreover, Ohta (1982) investigated the effects of ovariectomy carried out before and after the critical period in the development of the decidualizing ability of the rat uterus. Ovariectomy performed within 10 days of age, even on the day of birth, did not inhibit deciduoma formation in adult rats. The presence of the ovary for more than 10 days after birth was incapable of exerting any facilitatory effect on the decidual response. These findings indicate that hormones secreted from the neonatal ovary have little effect on the completion of uterine maturation. This conclusion is substantiated by the results of the experiments of ovarian transplantation showing that the presence or absence of the ovary during the first 10 days after birth produced no difference in decidualization as adults.

Nevertheless, since there is a critical period in rat uterine development during neonatal life, extraovarian estrogen may play some role. Secretion of estrogen by the adrenal in neonatal rats has been reported by several workers (Weisz and Gunsalus, 1973; Rabii and Ganong, 1976). The capacity of the uterus to form deciduomata in response to intraluminal instillation of oil has been investigated in prepuberal rats adrenalectomized and ovariectomized at 5, 10, and 19 days of age, and in rats given injections of an antiestrogen, MER-25, or estrogen after operation (Ohta, 1985a). The AX-OX rats formed deciduomata in response to endometrial stimulation, regardless of their age at the time of operation; the weight of the instilled uterus was approximately twice as heavy as that in OX rats with their adrenals intact. The decidual response in AX adult rats is normal when rats are given corticosterone (Ohta, 1984). Five daily injections of MER-25 following adrenalectomy and ovariectomy at day 5 failed to produce any significant decrease in the decidual response, and a similar treatment with estrogen during this period did not enhance responsiveness. Administration of MER-25 for 5 postnatal days has little effect on decidualization in adults (Ohta *et al.*, 1989).

From these findings, it is apparent that endogenous estrogen from any source is not needed for the development of the uterine capacity to form deciduomata during the critical period. This conclusion, however, seems to be at variance with the reports of Sananès and Le Goascogne (1976) and Sananès *et al.* (1980), but to be in harmony with the findings of Ogasawara *et al.* (1983) that sex steroids secreted from ovaries and adrenals of neonatal mice play no significant role in proliferation of uterine cells.

Uterine growth was impaired by adrenalectomy and ovariectomy at 5 or 10 days, but this suppressive effect could be overcome to some extent by later hormone treatments. The reduction in weight of the untreated uteri in the AX-OX rats may be attributable to the shortage of circulating estrogen after 10 postnatal days. Injections of MER-25 evoked a further reduction in weight in AX-OX rats, suggesting the existence of endogenous estrogen secreted from some unknown source(s) in neonatal rats. Although the involvement of endogenous estrogen in uterine development remains undefined at present, the critical period for the action of neonatal androgen administration on the rat uterus appears to end with completion of functional maturation of the endometrium. Androgen injection induced a permanent modification of uterine responsiveness in rats only when it acted on differentiating endometrial cells.

E. Responses to Various Stimuli of DCR

The DCR can be initiated by a variety of artificial stimuli in the absence of blastocysts. Mechanical stimulation such as endometrial scratching was widely adopted as the effective method in pseudopregnant rodents and in those adequately sensitized with ovarian steroids for decidualization (Finn and Porter, 1975b) and was applied to rats sterilized by neonatal androgen. Intrauterine instillation of oil acts as a decidual stimulus in rats, and Finn and Hinchliffe (1964) concluded that oil may imitate the action of the blastocyst at the surface of uterine epithelial cells rather than act traumatically.

Intrauterine instillation of various chemicals as well as oil has also been shown to be deciduogenic in hormone-primed animals (DeFeo, 1967; Finn and Porter, 1975b; Kennedy, 1990). Among them, prostaglandin (PG) has been studied extensively as a mediator of DCR because its concentration is elevated at implantation sites and in uteri after the application of artificial deciduogenic stimuli (Kennedy, 1986b, 1990; Leavitt, 1989; Yee et al., 1993). Administration of PGE2 or PGF2α, or their precursor, arachidonic acid (AA), induced a decidual response in the uteri of hormone-primed animals (Tachi and Tachi, 1974; Tobert, 1976; Sananès et al., 1981; Kennedy, 1985). In rabbits and rats, PGE2 is reported to be more effective than PGF2α (Hoffman et al., 1977; Kennedy, 1986a; Hamilton and Kennedy, 1994). Endometrial binding of E-series PGs has been reported in rats (Kennedy et al., 1983) and rabbits (Cao et al., 1984) during early pregnancy.

Uterine responsiveness to different kinds of deciduogenic stimuli has been studied in PE rats produced by a single injection of 100 μg TP 4 days after birth (Ohta, 1983) (Fig. 2). The nonandrogenized OX control rats given 7 daily injections of 3 mg P and a single injection of 0.1 μg E2 on the third day of the injection period, mimicking a "nidatory surge of estrogen," invariably formed deciduomata in response to any endometrial stimulus, including traumatization and intraluminal instillation of sesame oil, PGE2, or AA. Trauma elicited a greater response than did intrauterine

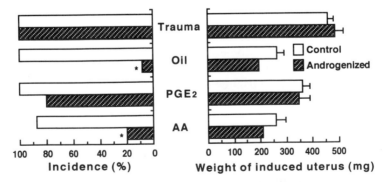

FIG. 2   Deciduoma formation in neonatally androgenized (100 μg TP) and control rats. The rats were given 7 daily injections of 3 mg P, including a single injection of 0.1 μg E2 together with the third P injection after ovariectomy as adults. The weight of the induced uterus is the mean weight of uteri bearing deciduomata. *$P < .05$ versus control.

instillation (DeFeo, 1963). If rats were given 7 daily P injections and treated no further, the incidence of deciduomata after the AA instillation was markedly lower than in rats receiving an E2 injection with P treatment.

In contrast, in androgenized OX rats similarly treated with ovarian steroids, the response varied according to the type of endometrial stimulation. Trauma invariably elicited deciduomata in androgenized rats, and the response was similar to that in the control females. The incidence of deciduomata in response to oil instillation, however, was much lower in androgenized rats than in the controls. However, the uteri of the androgenized rats were sensitive to oil instillation if the occurrence of persistent estrus was inhibited by ovariectomy on day 20. Accordingly, the effect of continued exposure to endogenous estrogen, but not the influence of a single dose of androgen given neonatally, is likely to be responsible for the reduced response to oil instillation in androgenized rats given 100 μg TP neonatally.

The instillation of PGE2 produced deciduomata in androgenized rats, whereas instillation of its precursor, AA, was rarely effective, although the two substances equally produced deciduomata in the control rats (Tobert, 1976; Sananès et al., 1981). As evidence has accumulated suggesting the involvement of PGs in implantation and in artificially induced DCR, it seems highly probable that the failure of the oil instillation to induce decidualization in androgenized rats is the result of reduced PG biosynthesis owing to blockade of cyclooxygenase-mediated conversion of AA to PG in the endometrium.

Hoffman et al. (1984) demonstrated elevation of PG concentration at implantation sites in the rabbit uterus after removing the blastocysts and the myometrium. PGs are produced by endometrial cells from pregnant or pseudopregnant rats in vivo and in vitro (Phillips and Poyser, 1981; Moulton, 1984; Parr et al., 1988). PG synthase was localized histochemically in the luminal epithelium in OX rats given hormone treatment for the development of deciduomata (Ohta, 1985b). In contrast, this enzyme was detected in both luminal epithelial and adjacent stromal cells of the rat endometrium on days 4 and 5 of pregnancy and its activity declined in the epithelial cells from days 5 to 7 (Parr et al., 1988), during which the uterus becomes refractory to implantation (Psychoyos, 1973a).

Moulton and Russell (1989) reported that the AA contents were decreased at the implantation site of rat uterus on day 6 of pregnancy. In the luminal epithelium of rat uterus, furthermore, mobilization of AA from phospholipids and phospholipid methylation appear to initiate the response to artificial stimuli (Moulton and Koenig, 1986; Moulton et al., 1987). It has been suggested, therefore, that PG production by the epithelial cells may play a role primarily during early stages of implantation and decidualization, whereas PGs produced by the stromal cells may be necessary during the later stages (Parr et al., 1988).

The uterine epithelium is of particular interest as a source of PGs because the deciduogenic stimulus appears to be transmitted through the luminal epithelium to the stroma in the induction of decidualization (Rankin et al., 1979; Lejeune et al., 1981; Kennedy, 1983a). In mice, Rankin et al. (1979) suggested that deciduogenic substances, such as cyclooxygenase products, produced from the endometrial epithelium are necessary for transmission of a signal through oil-stimulated epithelial cells to stromal cells. The intervention of such transmitter substances is not necessary, however, if the endometrium is subjected to trauma because trauma can act directly on the stroma (Finn, 1965). In contrast, Lejeune et al. (1981) reported the inability of deepitheliated horns to decidualize, even following application of traumatic stimuli, suggesting a transmitter role for the epithelium in decidualization. The results from androgenized rats appear to show that the mechanisms involved in decidualization induced by oil instillation and by trauma are different, at least at the initial step (Finn and Martin, 1974; Rankin et al., 1979), probably because of a difference in their effects on PG synthesis.

A short surge of estrogen is known to occur on day 4 of pregnancy in rats and mice (McCormack and Greenwald, 1974a; Watson et al., 1975). This is called "nidatory" or luteal phase estrogen; its involvement in implantation is well documented in these animals (Finn et al., 1992; Finn and Porter, 1975b; Psychoyos, 1986; Yoshinaga, 1989). Ovariectomy before this estrogen surge inhibited blastocyst implantation, resulting in delayed implantation (McLaren, 1973; McCormack and Greenwald, 1974b). A single injection of estrogen followed by daily injections of P brought about implantation of blastocysts in the females with delayed implantation caused either by lactation or by ovariectomy (McLaren, 1973). The timing of uterine sensitivity to blastocyst implantation has been reported to be influenced by the estrogen secreted at proestrus, whereas estrogen secretion after mating seems to be important in limiting the duration of sensitivity (DeFeo, 1967; Finn and Pollard, 1973). The time and dose of the estrogen injection are critical in decidualization induced artificially in the OX animals given ovarian hormones (Finn, 1966a; Finn and Porter, 1975b; Ohta, 1985b). The dose of P, however, is not critical if it is provided sufficiently.

At the molecular level, P priming for 3 days following ovariectomy increased the number of RNA polymerase binding sites in rat endometrium available for the initiation of RNA synthesis, whereas a single injection of estrogen given after priming with progestin resulted in a rapid decrease in the number of RNA initiation sites on uterine chromatin (Glasser and McCormack, 1979), indicating that the estrogen surge appears to alter the P-induced transcription. In the uterus of delayed-implanting mice, RNA synthesis was activated by estrogen injection (Yamada and Nagata, 1993).

Progestin priming increased ER concentration of the stromal cells of rat endometrium, which might lead to enhancement of their response to estrogen (Martel and Psychoyos, 1982).

A single injection of estrogen in association with continued injection of progestin is essential for the induction of decidualization by oil instillation, but is not required for response to trauma or PG instillation (Finn, 1965, 1966a; McLaren, 1969; Sananès *et al.*, 1981). The hormonal requirements for the uterine reaction induced by oil instillation are similar to those of natural implantation, whereas trauma seems to be initiating the decidual reaction by bypassing the need for an initial reaction, e.g., estrogen sensitization of the uterine epithelium (Pollard and Finn, 1972). The instillation of AA also requires the nidatory estrogen for inducing decidual response. Moreover, the amount of the nidatory estrogen is critically concerned in the induction of deciduomata by instillation of oil or phosphate-buffered saline gelatin, but not by trauma (Finn, 1966a; Kennedy, 1980; Ohta, 1985b). On the other hand, an impairment of the estrogen-binding capacity of the uterus in androgenized rats has already been noted. Therefore, the limited effect of deciduogenic stimuli requiring nidatory estrogen in PE rats androgenized with 100 μg TP could be interpreted as the consequence of a decrease in estrogen-binding capacity caused by continued exposure to estrogen from polyfollicular ovaries.

Although there is evidence suggesting that nidatory estrogen plays an important role in blastocyst implantation, the mechanism of its action during implantation or decidualization remains undefined. It is possible that the estrogen surge may act in some way on the receptivity of the uterus to the stimulus provided by the blastocyst: the estrogen primarily acts through an effect on the endometrium, probably on the luminal epithelium. In OX mice, following administration of progestin and estrogen, the uterus showed only a narrow lumen (Finn and Martin, 1969). Two stages of luminal closure have been recognized by electron microscopy (Pollard and Finn, 1972): (1) interlocking of the microvilli from apposed epithelial cells and (2) subsequent reorganization of the contacting surfaces of the epithelial cells similar to the surface reaction occurring in the epithelium at the time of implantation (Nilsson, 1966; Enders and Schlafke, 1967; Pollard and Finn, 1972). A small amount of estrogen is necessary, in addition to progestin, to achieve the second stage of closure (Pollard and Finn, 1972; Finn and Pollard, 1973). In addition, since estrogen is known to stimulate PG production *in vitro* through activating fatty acid cyclooxygenase in rat aortic smooth muscle cells and platelets (Chang *et al.*, 1981), it seems likely that the nidatory estrogen induces prostaglandin synthesis in the luminal epithelial cells during decidualization.

The hypothesis appears to be substantiated by the results in PE rats given 100 μg TP showing that inducing stimuli requiring the estrogen surge in addition to progestin always failed to produce deciduomata. In this regard, Ohta (1985b) has reported effects of estrogen on PG synthase in hormonally sensitized rat endometrium with reference to DCR. The PG synthase demonstrated by histochemical methods using AA as substrate (Janszen and Nugteren, 1971) was restricted to the luminal and glandular epithelium. The OX rats given no progestin showed no PG synthase. The intensity of the reaction was dependent upon the dose of progestin. In OX rats given progestin injections, a single estrogen injection acted neither synergistically nor antagonistically. Thus, progestin may be involved in endometrial PG synthesis by regulating cyclooxygenase activity, whereas estrogen seems to have little effect on this activity.

This conclusion is not in agreement with the results of biochemical studies by Naylor and Poyser (1975) and Ham et al. (1975) indicating that estrogen, but not progestin, was capable of altering uterine PG synthesis by modulating PG synthase activity. Since Downing and Poyser (1983) have demonstrated that estrogen initiates PG synthesis by activating phospholipase A2, which is responsible for releasing AA from phospholipids in the progestin-primed guinea pig uterus, the nidatory surge of estrogen may prove to be involved in the production of PGs during decidualization.

Although PGs appear to mediate the DCR, it is unknown whether they act directly on endometrial cells inducing decidualization or indirectly by some other mechanism. In some laboratory rodents, cyclic adenosine monophosphate (cAMP) may be a mediator of some of the effects of PGs at implantation or during decidualization (Leroy et al., 1974; Rankin et al., 1981; Kennedy, 1983b; Jonston and Kennedy, 1985; Yee and Kennedy, 1993). A histochemical analysis of localization of adenylate cyclase (AC) (Sananès and Psychoyos, 1974) and a possible effect of deciduogenic stimulus on AC have been reported in hormonally sensitized rat endometrium (Ohta and Takewaki, 1987). A positive AC reaction was found in the luminal and glandular epithelia of the endometrium sensitized by injections of ovarian steroids. The AC reaction occurred mainly along the plasma membrane of epithelial cells, whereas the reaction was negative in endometrial stromal cells and smooth muscle cells. In uterine blood vessels, endothelial cells showed a positive reaction. In OX rats given deciduogenic stimulation following hormone treatment, the AC reaction was similar to the reaction seen in rats given no deciduogenic stimuli. In endometrial stromal cells of rats sacrificed 5 min or 5 hr after application of the stimulus, the AC reaction was negative. Thus, deciduogenic stimulation failed to provide any definite histochemical evidence for an increase in AC activity.

Although an increase in cAMP concentration following deciduogenic stimulation has been demonstrated biochemically (Rankin et al., 1977, 1979; Kennedy, 1983b; Sanders et al., 1986), the endometrial stromal cells do not seem to be involved in the increase. In uteri bearing deciduomata, decidual cells and vascular endothelial cells showed a positive AC reaction, whereas luminal and glandular epithelial cells were negative.

On the basis of several lines of evidence, a hypothesis has been proposed that deciduogenic stimuli increase uterine concentration of PGs, particularly of the E-series, and this in turn raises intracellular cAMP concentration by activating AC, initiating decidualization (Rankin et al., 1981; Kennedy, 1983a,b). However, PGE2 added to the incubation medium brought about no distinct changes in localization and magnitude of the AC reaction in sections of uteri removed from OX rats given progestin and estrogen. Accordingly, it appears likely that the effect of PGE2 on endometrial stromal cells is not mediated by cAMP. This may account for the lack of deciduomal response to the intrauterine administration of cAMP or its analog (Leroy et al., 1974; Hoffman et al., 1977; Rankin et al., 1979). Adenylate cyclase present in endothelial cells appears to indicate that the increase in endometrial vascular permeability observed prior to decidualization (Kennedy, 1983b; Rogers, 1992) is mediated by cAMP. The uterine behavior of PGs and cAMP in androgen-sterilized rats invites investigation.

F. Neonatal Administration of Nonaromatizable Androgen

Neonatal treatment of female rats with androgen induces a permanent masculinization of the neuroendocrine mechanism controlling reproductive function, resulting in anovulatory sterility. Increasing evidence, however, suggests that the initial mode of androgen action on the neuroendocrine organization may involve its intraneuronal aromatization to estrogen (Naftolin and Brawer, 1978; Shinoda, 1994). A non-aromatizable androgen, 5α-dihydrotestosterone, unlike the aromatizable androgen, testosterone (T), is incapable of masculinizing the brain of female or neonatally castrated male rats (McDonald and Doughty, 1972; Korenbrot et al., 1975).

On the other hand, some authors have demonstrated in mice that neonatal treatment with relatively high doses of nonaromatizable androgens as well as aromatizable androgens for 5 to 10 postnatal days resulted in estrogen-independent vaginal changes (Iguchi and Takasugi, 1976; Ohta and Iguchi, 1976). A high incidence of estrogen-independent persistent stratification and/or squamous metaplasia also took place in the uterine epithelia of OX adult mice given neonatal injections of TP or of DHT or its propionate (DHTP) (Iguchi and Takasugi, 1976).

Takewaki and Ohta (1975c) have investigated whether deciduoma formation in response to uterine trauma would be affected by neonatal administration of nonaromatizable androgens. Adult female rats given a single injection of 1.25 mg DHT or DHTP at 5 days of age showed almost regular

4-day cycles and contained CL in their ovaries. Uterine trauma applied on the fourth day of the treatment with P plus E2 resulted in well-developed deciduomata in all animals by the day after the last injection. This was in sharp contrast to the failure of female rats receiving TP neonatally to give a positive response under similar experimental conditions (Takewaki and Ohta, 1974). However, the incidence of deciduomata was considerably lower only in animals treated with DHTP neonatally, if they received estrogen for a 30-day period prior to the treatment with P plus E2. At present, it is not known whether this inability of DHT or DHTP to prevent uterine decidualization is due to the failure of these steroids to be converted to estrogens in neonatal uteri. Taking these findings together with the results of experiments with DHT benzoate (DHTB) (Takewaki and Ohta, 1976) into account, the effects of neonatal steroid treatment upon uterine responsiveness in adulthood appear to be specific to the steroid and independent of its androgenicity. Although it is well established that the action of androgen on the male accessories is considerably enhanced by its esterification, the effects of free and esterified androgens on organs other than male accessories have not been extensively compared.

IV. Effect of Persistent Estrus on Decidual Response

As androgen-sterilized female rats invariably show persistent estrus when adult, it is not clear whether alterations in the decidual response are due to a direct action of androgen given neonatally or to the ovarian hormones secreted following neonatal exposure of androgen. Takewaki and Ohta (1974) put forward a hypothesis that the lowered DCR observed in the androgenized adult rats results from the injection of androgen rather than from the continued exposure to ovarian estrogen. In contrast, Kramen and Johnson (1975) and Matton and Maurasse (1985) have shown that the reduction of DCR is age dependent in androgenized rats, and that the decidual reaction is affected by modification of endocrine secretions by neonatal androgen treatment, as observed by Uilenbroek et al. (1976). Altered titers of circulating ovarian hormones similar to those in androgenized animals can be induced by treating neonates with estrogen. In female rodents, moreover, this phenomenon is one of the age-related neuroendocrine impairments associated with reproductive senescence. Studies using these animals seem to provide a favorable basis for understanding the influence of sustained exposure of the uterus to ovarian hormones during the postpubertal period on uterine structure and function in androgenized rats.

A. Neonatal Administration of Estrogen

Administration of estrogen by single or repeated injections during the first few days after birth induces a sterility syndrome characterized by acyclicity and anovulation in female adult rats, which is superficially indistinguishable from that induced by neonatal administration of androgen (Rodriguez *et al.*, 1993; Takewaki, 1962; Barraclough, 1966; Flerkó, 1971; Gorski, 1971). Early exposure to estrogens, as well as androgens, induces hyperplastic or neoplastic changes in the female reproductive tracts (Takasugi, 1976, 1979; Bern and Talamantes, 1981; Iguchi, 1992; Bern, 1992a,b). In neonatally estrogenized, unlike androgenized, rats, however, there are few reports dealing with the decidual response.

Ohta and Takewaki (1974) investigated the uterine response in female rats showing anovulatory and persistent estrus resulting from a single injection of 12.5 μg E1 neonatally, and compared their response with that in adult rats androgenized with 1.25 mg TP neonatally (Takewaki and Ohta, 1974). The rats were ovariectomized when adult and were given daily injections of 2 or 5 mg P in combination with 0.2 μg E2 and subjected to uterine trauma on the fourth day of injection. The estrogenized rats invariably showed continued vaginal cornification until ovariectomy and had no CL. After ovariectomy, estrogenized and androgenized rats were both diestrous during the entire period of hormone administration. Thus, the estrogen-sterilized rats were not distinguishable from the androgen-sterilized females in either ovarian structure or vaginal response. However, uterine responses to traumatic stimuli were considerably different in these two types of animal. In androgenized rats, uterine trauma invariably failed to induce a positive response, whereas half of the estrogenized rats reacted to the trauma by forming deciduomata. The deciduomata produced in the estrogenized rats were generally small nodules. When the daily dose of P was increased to 5 mg, there was no significant difference between the estrogenized and androgenized animals in mean weight of the traumatized uteri bearing deciduomata. However, the incidence of deciduomata was again higher in the former than in the latter. Since the androgen- and estrogen-sterilized females were similarly anovulatory and showed persistent estrus, the difference in direct effect upon the uterus between neonatal administration of androgen and estrogen may be due largely to a difference in reaction pattern to the trauma as adults.

An impairment of estrogen binding capacity of the uterus in adult rats treated neonatally with androgen has already been noted (see Section II, B). In mice, Terenius *et al.* (1969) reported that the uterine estrogen binding was slightly reduced by neonatal treatment with estrogen but not with androgen, whereas Shyamala *et al.* (1974) reported that the specific nuclear binding in the uteri of neonatally estrogenized mice was not significantly different from that in control animals. Female rats treated neonatally with 100 μg EB exhibited reduced uterine growth responses to exogenous estrogen in comparison with controls at 23 days of age but not at 31 days, whereas neonatal exposure to 100 μg TP did not impair prepubertal uterine responses (Gellert *et al.*, 1977).

By contrast, Johnson and Witschi (1963), and Campbell (1980) indicated that neonatal androgen treatment reduced uterine responsiveness in the immature animal. In prepubertal rats given neonatal injections of estrogen, the uterine cytoplasmic ER was significantly reduced, although neonatally administered androgen had no effect on the concentration of cytoplasmic estrogen binding sites (Gellert *et al.*, 1977; Campbell, 1980; Campbell and Modlin, 1987). Uterine tissue of neonatally estrogenized mice showed a marked reduction in the response to estrogen and in the concentration of cytoplasmic ER in adults (Aihara *et al.*, 1980). Neonatal treatment with DES decreased the number of Type II binding sites for estrogen without altering receptor affinity in adult rats (Csaba *et al.*, 1986).

Nuclear translocation of ER complexes was not affected in either estrogenized immature rats (Gellert *et al.*, 1977) or in similarly treated adult mice (Aihara *et al.*, 1980). In estrogenized rats, accordingly, reduction of estrogen binding also results in a decline of progestin binding. In view of these findings, neonatal injection of estrogen appears to have a direct effect upon concentrations and binding capacities of receptors for estrogen and progestin in the uterus, which is different from that exerted by neonatal administration of androgen. In this regard, McGuire and Lisk (1969) have recorded different effects of early treatment with androgen and estrogen on the CNS. After neonatal treatment with estrogen, there is a significant disruption of ER function throughout the hypothalamus and pituitary, whereas after treatment with androgen, extensive disruption of ER function occurs only at the level of the pituitary. Moreover, neonatal estrogenization and androgenization of female rats also increase pituitary and plasma PRL concentrations in adults (Vaticón *et al.*, 1985; Watanobe *et al.*, 1991a; Collado and Aguilar, 1993). However, the hyperprolactinemia induced by estrogenization in females did not disappear following ovariectomy (Vaticón *et al.*, 1985), whereas ovariectomy of androgenized animals reduced prolactin levels to values found in controls. Thus, androgenization and estrogenization may differ in their ability to alter hypothalamic function, since, according to Mallampati and Johnson (1974), plasma PRL levels can be used to detect hypothalamic damage.

B. Neonatal Injection of Nonaromatizable Androgen Simultaneously with Estrogen

A number of early studies demonstrated an interaction between estrogen and androgen given neonatally (Takasugi, 1954; Velardo et al., 1956; Takasugi and Furukawa, 1972). In female rats, daily injections of estriol (E3) for 20 days from the day of birth induced persistent diestrus in adults, whereas rats receiving TP concurrently with E3 showed persistent estrus (Takasugi, 1954), indicating that permanent suppression of gonadotropin secretion caused by neonatal estrogen is counteracted by simultaneously injected androgen. Van der Schoot et al. (1976) found that neonatal injections of DHTP in male rats decreased the concentration of testicular testosterone that masculinized the brain of male neonates (Price and Ortiz, 1965), resulting in inhibition of masculinization of male rat brain. However, DHTP injected simultaneously with TP into neonatal female rats failed to prevent TP-induced anovulation and decrease of lordosis behavior in response to appropriate stimuli (van der Schoot et al., 1976). In addition, nonaromatizable androgen may act competitively with estrogen and thus reverse the estrogen-induced failure of decidualization.

Neonatal female rats were given a single injection of 60 μg EB, 1 mg DHTB, or the two steroids in combination. The ovaries removed around the age of 60 days from the rats given EB or EB plus DHTB neonatally were significantly smaller than those from females given no steroids neonatally, whereas those from animals given DHTB alone neonatally were not significantly different from controls. The ovaries from rats given EB alone or in combination with DHTB were polyfollicular and lacked CL, in contrast to those from the DHTB-treated rats, which invariably contained some CL. The DHTB-treated rats formed well-developed deciduomata in response to trauma on the fourth day of treatment with ovarian steroids following ovariectomy as adults. About half of the rats given EB neonatally formed deciduomata, although in most of them the deciduomata were small and nodular, as has already been mentioned for PE rats given estrogen neonatally (Ohta and Takewaki, 1974). The rats given EB plus DHTB failed to react to trauma. It is evident that the uterus was strongly affected by neonatal administration of 60 μg EB and that as much as 1 mg DHTB given simultaneously failed to eliminate the effect of estrogen.

C. Decidual Response in Aging Rats

The reproductive malfunction associated with advancing maternal age arises from a variety of abnormalities in the reproductive system (Finch,

1976; Soriero, 1978; Mulholland and Jones, 1993). An age-related decline in uterine response to decidualizing stimuli has already been shown in some species of rodents. In old OX mice and rats given injections of ovarian steroids on a schedule mimicking the endocrine condition of early pregnancy, an artificial stimulus, either endometrial traumatization or intraluminal infusion of air, arachis, or sesame oil, failed to produce a DCR comparable to that seen in young mice treated similarly (Finn, 1966b; Chatterjee and Mukherjee, 1974; Holinka and Finch, 1977; Hsueh et al., 1979; Saiduddin and Zassenhaus, 1979; Craig, 1981).

In pseudopregnant golden hamsters, the DCR to chemical or mechanical stimuli was less in older animals than in the young, and increased ovarian activity stimulated with PMSG can compensate for the response to only a limited extent (Blaha, 1967). Thus, the reduced uterine ability to form decidual tissue may be a significant factor in the reduction of litter size in senescence (Matt et al., 1986; Day et al., 1989). Age-related changes in DCR in some laboratory rodents are due to impairment of uterine responsiveness to ovarian hormones (Larson et al., 1972; Peng and Peng, 1973; Blaha and Leavitt, 1974; Saiduddin and Zassenhaus, 1979). In mice, however, Pollard and Finn (1974) have shown that this decrease in DCR is not the result of an inability of the stromal cells to undergo division in response to hormonal stimulation. In addition, Finn and Martin (1969) showed that high doses of ovarian hormones in old mice produced more stromal mitoses than in young mice.

On the other hand, there is a decrease in epithelial corrugation in uterine tissue from older mice as well as failure of luminal closure in response to progestin in aged golden hamsters (Thorpe et al., 1974). Factors causing age-related changes in uterine function have not yet been thoroughly worked out, although there is convincing evidence for a decrease in the uptake of estrogen and progestin and a consequent reduction in the sensitivity to circulating progestin in the uteri of aged animals (Larson et al., 1972; Peng and Peng, 1973; Blaha and Leavitt, 1974). ER was greatly reduced in aged rats and mice (Holinka et al., 1975; Saiduddin and Zassenhaus, 1979; Han et al., 1989), whereas PR was not significantly different between young and senescent animals (Blaha and Leavitt, 1978; Saiduddin and Zassenhaus, 1979). However, the uterine receptors (ER and PR) are reported to be qualitatively different between young and old animals: the young rats had a greater proportion of PR with a higher sedimentation coefficient (S) than the receptors from aging animals, whereas ER in the aged rats was about 4.5 S in contrast to 7.5 and 4.5 S in the young (Saiduddin and Zassenhaus, 1979). ER complexes from the aged mouse uterus were less able to bind to nuclei from the young uterus (Belise et al., 1989). It seems highly probable, therefore, that reproductive failure in aging animals, as demonstrated by the absence of DCR, is accompanied by changes in the receptors for estrogen and progestin. In contrast, Maibenco and Krehbiel (1973) and Finch and Holinka (1982) found in mice and rats, respectively, that although the ability of the uterus of these animals to form deciduomata was minimal, an abrupt loss of ova occurred well after implantation, indicating that the reproductive system fosters normal development through the implantation period in some pregnancies of older females. In rats, furthermore, the decline in rate of implantation with advancing age did not parallel the decline in DCR (Shapiro and Talbert, 1974). These findings suggest that in addition to the decline in uterine sensitivity, other factors may contribute to the reproductive decline associated with increasing age.

Age-related changes are also observed in the CNS which controls the female gonad: irregular estrous cycles leading to persistent estrus, repetitive pseudopregnancies, and finally anestrus are observed widely in aged laboratory rodents (Meites, 1982; Matt *et al.*, 1993; Finch *et al.*, 1984). Since a decline in DCR occurs in PE rats produced by neonatal administration of androgen or estrogen, alterations in titers of circulating ovarian hormones owing to the decline in CNS activity with increasing age might also be involved in the decreased DCR-inducing ability. However, there are few reports attempting to relate irregular estrous cycles to the decline in DCR in aged animals. Saiduddin and Zassenhaus (1979) reported that young rats showing normal estrous cycles formed deciduomata in response to uterine trauma, while older rats (more than 20 weeks of age) exhibiting either pseudopregnancies, characterized by prolonged diestrus with copious vaginal mucus and activated CL, or by anestrus failed to show a DCR.

Ohta (1987) has also investigated in rats whether age changes in estrous cycles participate in the age-related decline of DCR. The rats showing regular 4-day estrous cycles before ovariectomy at 4 months of age invariably formed deciduomata in response to any type of stimulation. At 8 months of age, DCR began to decrease, completely disappearing by 12 months. The uterus of 8-month-old rats showing persistent estrus until ovariectomy formed only small deciduomata in response to trauma. The instillation of oil or PGE2 resulted in deciduomata in 8-month-old females only when they had been cycling regularly at the time of ovariectomy. It appears likely, therefore, that continued secretion of ovarian estrogen is involved in the reduction of uterine responsiveness in aged rats as well as in neonatally androgenized or estrogenized animals. When ovariectomy was postponed until 10 months of age, instillation of either oil or PGE2 did not elicit deciduoma formation, regardless of whether the rats had been in state of cyclic estrus or persistent estrus at the time of ovariectomy. This is in good agreement with the finding that DCR was impaired in aging mice even though the mice were still cycling regularly (Holinka and Finch, 1977).

It seems highly probable that the decline in DCR ability of the rat uterus occurs without reference to age-related changes in the CNS, although altered secretion of ovarian hormones associated with aging of the brain appears to accelerate uterine aging. The absence of the ovaries for a period of 5 or 7 months starting with ovariectomy at 4 months and subcutaneous transplantation of an ovary from an immature rat 5 or 7 months later, prevented the occurrence of age-related changes in the uterus as well as in the CNS. Moreover, a 2-month interval between ovariectomy at 8 months and the beginning of the standard treatment schedule restored or maintained uterine ability to form deciduomata in 10-month-old rats, regardless of whether the rats had been in persistent estrus or were cycling regularly at the time of ovariectomy. These findings suggest that the amount of ovarian hormones, estrogen in particular, arriving at the uterus may be intimately related to the induction of age-related changes in DCR.

An increased rate of DNA synthesis and mitosis has been observed in uterine endometrial cells during decidualization (Moulton and Koenig, 1984). Cell proliferation is significantly reduced in aged mice following deciduogenic stimulation (Holinka et al., 1975). In aged rats, Saiduddin and Zassenhaus (1979) demonstrated that uterine response to estrogen following ovariectomy was greater than in young rats despite lower DNA/protein and soluble/total protein ratios in the uteri of aged animals and indicated that much of the increase in weight may be due to fibrous tissue, as occurs in the aged rat and hamster uterus (Rahima and Soderwall, 1977; Craig, 1981). Increased deposition of collagen in the uterus of aged animals may interfere with uterine vascularization and, consequently, with uterine functions (Biggers et al., 1962). Also, reduction in uterine blood supply may be a factor in decreasing reproductive activity in aged rabbits (Larson and Foote, 1972).

The increase in vascular permeability in endometrium exposed to a deciduogenic stimulus seems to play an important role in subsequent decidualization (Kennedy, 1979; Rogers, 1992). In this regard, histological studies reveal that endometrial fibrosis advanced between 4 and 8 months of age, whether the females were in persistent estrus or cycling regularly (Ohta, 1987). However, the fibrosis index remained almost unchanged during the next 4 months between 8 and 12 months, when DCR ability continued to decrease. Therefore, accumulation of collagen in the uterus may not be primary in the increasing failure of DCR in aging animals. This decrease may be accounted for by a reduction in the number of stromal cells and/or in uterine responsiveness to ovarian hormones. On the other hand, neonatal androgenization increases uterine collagen and inhibits the uterine growth response to estrogen (Lobl and Maenza, 1975; Lobl and Mohammed, 1975).

The uterine response in aged rats differed according to the type of endometrial stimulation. This difference may be related to the difference in the mechanisms involved in decidualization evoked by different stimuli (Finn and Martin, 1974; Rankin *et al.,* 1979). Although it has been reported that a reduction in the production of PGE2 by the aging uterus may be responsible for the lower implantation rate and the reduction in fertility of aging rats (Brown *et al.,* 1984), instillation of PGE2 failed to elicit decidual response in aging rats. Furthermore, the uterus responded in a similar manner to instillation of PGE2 and oil in aged rats in spontaneous persistent estrus, whereas in young adult rats showing persistent estrus induced by neonatal treatment with a relatively low dose of androgen, DCR induced by PGE2 was almost the same as that evoked by trauma and much stronger than that induced by oil instillation (Ohta, 1983). The difference in PGE2-induced response between the two types of PE rats may be ascribable to a difference in absorption rate of instilled PGE2 through the luminal epithelium of the uterus. Aged rats in repetitive pseudopregnancies, as well as PE rats, show reduced DCR (Saiduddin and Zassenhaus, 1979).

D. Chronic Injections of Estrogen in Androgenized Rats

The failure to restore uterine responsiveness to a normal level by interrupting persistent estrus with an ovariectomy for as long as 2 months in androgenized rats prior to the start of the treatment with ovarian hormones led Takewaki and Ohta (1974) to conclude that the lowered sensitivity was largely ascribable to the single neonatal injection of androgen rather than to continued exposure to ovarian estrogen. They also pointed out that female rats given injections of estrogen for the first few days after birth showed, as adults, a syndrome of sterility characterized by acyclicity and anovulation which was macroscopically indistinguishable from that in animals receiving androgen during neonatal life, but that the capacity to form deciduomata in response to uterine trauma was different in the two groups.

PE rats at 8 months of age formed deciduomata after trauma (Ohta, 1987). Takewaki and Ohta (1975b) examined DCR in female rats ovariectomized at 3 days of age and given a single injection of 1.25 mg TP on the next day. The incidence of deciduomata in response to trauma applied on the fourth day of the P plus E2 injections was almost as high as that in neonatally ovariectomized, nonandrogenized rats, but the response was significantly less in androgenized rats than in controls. When the daily dose of P was increased to 5 mg, the effect of neonatal TP was no longer evident. If similarly operated females were given injections of 0.1 μg E2 for 30 days prior to treatment with ovarian steroids, deciduoma formation in androgenized rats was markedly reduced in both incidence and size. In contrast, in OX rats receiving no neonatal TP, continued administration of estrogen exerted no significant effect on either size or incidence of deciduomata. If rats were ovariectomized prior to treatment with TP, a single injection of 1.25 mg TP at 10 days, whether given alone or in combination with subsequent injections of estrogen for 30 days, did not bring about any alteration in uterine responsiveness (Takewaki and Ohta, 1975a).

Thus, exposure of rats to estrogen for a prolonged postpubertal period is without effect, unless the animals have received enough androgen neonatally. Accordingly, it is highly probable that in androgenized PE rats, the androgen administered during neonatal life is primarily responsible for the decrease in uterine reactivity to trauma (Ohta and Takewaki, 1974; Takewaki and Ohta, 1974).

This conclusion is substantiated by the fact that if the dose of androgen administered neonatally was reduced to 100 μg, uterine trauma elicited DCR similar in both size and incidence to that in nonandrogenized females despite the appearance of persistent estrus. However, since deciduoma formation in the androgenized rats was further lowered by chronic administration of estrogen (Takewaki and Ohta, 1975b,c), continued secretion of ovarian estrogen appears to cooperate with the androgen to reduce responsiveness further. In contrast, the uterine response to oil instillation, unlike that to trauma, was markedly affected in both lightly androgenized and aged rats showing persistent estrus (Ohta, 1983, 1987), suggesting that the response to this stimulus is reduced by continued exposure to endogenous estrogen. The difference in the effect of androgen given neonatally on the response to these two types of stimulus may arise from the different mechanisms involved in the decidualization induced (Finn and Martin, 1974; Rankin et al., 1979; Sananès et al., 1981).

V. Uterine Responses to Ovarian Hormones

From the findings noted in earlier sections, it is assumed that the changed uterine response to ovarian steroids is responsible for the permanent reduction of DCR in androgenized rats. In this section, uterine response to ovarian hormones is discussed to verify the mechanism involved in androgen action on the neonatal uterus.

A. Uterine Growth and Cell Proliferation

Uterine weights in androgenized adult rats were similar to those in normal diestrous rats (Wrenn *et al.,* 1969; Lobl *et al.,* 1974; Lobl and Maenza, 1975). The uterus was no longer responsive to estrogen in androgenized rats with intact ovaries (Wrenn *et al.,* 1969), indicating that the uterus in the androgenized rats is tonically maximally stimulated (Lobl *et al.,* 1974; Lobl and Maenza, 1975). Although uteri in the androgenized rats, as well as in the nonandrogenized controls, decreased in weight in response to ovariectomy, growth induced by exogenous estrogen was markedly less in androgenized OX rats (Lobl *et al.,* 1974; Lobl and Maenza, 1975; Gala, 1981; Morikawa *et al.,* 1982).

In contrast, Takewaki (1966) found in androgenized adult rats that uterine growth is stimulated by estrogen, but ovariectomy causes no significant decrease in uterine weight. In immature rats, other evidence indicates that neonatal androgenization has no effects on the uterine response to estrogen (Gellert *et al.,* 1977; Campbell, 1980), although Johnson and Witschi (1963) showed a definite reduction in uterine response in androgenized animals. There are many differences in the experimental conditions in these studies. Although there are few studies concerning uterine response to P injection, examination of the intact uterus in rats given a deciduogenic stimulus in the contralateral uterus reveals that uterine growth following progestin treatment is similar to that seen in normal and androgenized rats (Matton and Maurasse, 1985).

In OX rats, progestin injected simultaneously with estrogen had both an agonistic and an antagonistic effect on uterine growth, depending upon the dose of estrogen (Medlock *et al.,* 1994). Suzuki *et al.,* (1994) have investigated the uterine response to progestin and estrogen injections that supports the development of endometrial sensitivity to deciduogenic stimulus in PE rats neonatally androgenized with 100 μg or 1.25 mg TP at 5 days. Uteri from both control and androgenized rats responded to ovariectomy with a decrease in weight. In the control OX rats, 3 daily injections of 3 mg P increased uterine weight, whereas in the androgenized OX rats, progestin failed to increase uterine weight significantly, regardless of the dose of TP given neonatally. However, if a single injection of 0.2 μg E2 mimicking the brief surge of estrogen during early pregnancy was given on the third day of the P treatment, the uterine weight increased greatly, even in rats given 1.25 mg TP, as in the controls.

Uterine epithelial cells underwent mitosis in response to estrogen given alone to OX animals, whereas they failed to response to progestin (Tachi and Tachi, 1974; Finn and Porter, 1975a). Progestin inhibited the estrogen-induced mitotic response when given shortly after estrogen injection (Mar-

tin and Finn, 1971). Estrogen given together with progestin had no effect
on mitosis in nonandrogenized rats (Suzuki *et al.*, 1994) (Fig. 3). This is in
accord with results from AX-OX rats, where pretreatment with progestin
suppressed the mitogenic action of estrogen on the epithelial cells (Tachi
and Tachi, 1974). In androgenized rats, ovariectomy induced a marked
increase in mitosis (Suzuki *et al.*, 1994). The 3-day treatment with P alone
or in combination with the single injection of E2 inhibited postovariectomy
proliferation of the epithelial cells in rats androgenized with 100 μg TP
neonatally, but not in those given 1.25 mg TP. Administration of estrogen
alone had little effect on epithelial proliferation in these androgenized
rats. Since vaginal epithelium showing estrogen-independent metaplasia
induced by neonatal treatment with estrogen or androgen retained high
mitotic activity after ovariectomy (Takasugi, 1976), the proliferative
changes following ovariectomy may be related to metaplastic changes
encountered in some androgenized rats given no further treatment. Uterine
epithelium in the 100-μg TP-treated rats given P and E2 was different in
structure from that in the controls similarly treated: the uterine lumen was
lined by a high columnar epithelium, in contrast to a cuboidal epithelium
in the controls. In these animals, moreover, vaginal epithelium became
cornified despite progestin administered together with estrogen. The single
estrogen injection did not result in cornified vaginal epithelium in either
control or androgenized rats, indicating that, under the progestational
condition, reproductive tracts of androgenized rats may be more sensitive

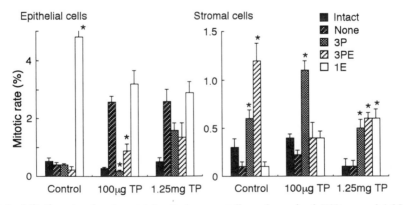

FIG. 3   Mitotic rate of endometrial cells in neonatally androgenized (100 μg and 1.25 mg
TP) and control rats ovariectomized and given ovarian hormones when adult. Intact, sacrifice
at the time of ovariectomy; None, 3 daily injections of vehicle only after ovariectomy; 3P
and 3PE, 3 daily injections of 3 mg P alone or in combination with a single injection of
0.2 μg E2 after ovariectomy; 1E, a single injection of 0.2 μg E2 following 2 daily injections
of vehicle after ovariectomy. *$P < .05$ versus None in each group of androgenized and
control rats.

to estrogen than those in normal rats (Matton and Maurasse, 1985). Female rats given DHT neonatally show vaginal cornification under hormonal conditions that support the development of uterine deciduomata (Takewaki and Ohta, 1976).

Unlike the epithelial cells, uterine stromal cells of OX and AX-OX animals did not respond to either estrogen or progestin alone by increasing mitosis(Tachi and Tachi, 1974; Finn and Porter, 1975a). Progestin stimulates mitosis in stromal cells, if given after priming with estrogen. The 3-day injection of P increased the mitotic rate of stromal cells in the normal rats ovariectomized at estrus (Suzuki *et al.*, 1994). A single injection of estrogen in association with P treatment resulted in a further increase in mitosis as it did in AX-OX rats similarly treated (Tachi *et al.*, 1972). Tachi *et al.*, (1972) put forward a hypothesis that progestin switches the mitogenic action of estrogen from the epithelial to the stromal component of the endometrium.

In androgenized rats, the mitotic response to hormones is different from that in controls (Suzuki *et al.*, 1994). In androgenized rats, administration of P increased the mitotic rate of stromal cells, the rate being twice as high in rats androgenized with 100 μg TP as in controls. However, the synergistic effect of a single injection of estrogen given together with progestin on stromal cell proliferation was no longer evident in androgenized rats. The estrogen injection suppressed the P-induced stromal cell division in the 100-μg TP-treated rats. The single injection of estrogen, given alone, stimulated the mitotic activity of stromal cells in androgenized rats but not in controls. Thus, neonatal androgenization modifies uterine growth and cell proliferation in response to ovarian hormones.

B. Progesterone Receptor Levels

Although much research has been done on uterine ER content in the androgenized rats (see Section III, B), little has been reported on uterine PR content in these animals. Ohta *et al.*, (1993a) demonstrated immunocytochemically the distribution of PR in normal female rat reproductive tracts. The uterine cells of OX rats given hormone treatment to promote DCR showed PR expression similar to that seen in the pregnant females. The PR staining in the epithelial and stromal cells increased shortly after a single injection of 3 mg P, reaching a maximum after 2 daily injections but decreasing after 3. A single injection of 0.2 μg E2 on the third day of the P injection period inhibited this P-induced reduction. The pattern of PR expression in OX rats androgenized with 100 μg or 1.25 mg TP neonatally and given P and E2 injections as adults differed from that in the controls (Suzuki *et al.*, 1994) (Table I).

TABLE I

Effect of Ovarian Hormones on PR Expression in the Uterus of Neonatally Androgenized (100 μg and 1.25 mg TP) and Control Rats Ovariectomized[a]

Neonatal treatment	Cell type	Intact	P and E2 treatment					
			1P	2P	3P	3PE	1E	None
Control	Ep	–	+	+ + +	+	+ +	– / +	+ + / + + +
	St	+ +	– / +	+ +	+	+ +	+ / + +	+ / + +
100 μg TP	Ep	–	–	+ / + +	+ +	–	+ / + +	+ / + +
	St	+	–	– / +	+	+	+	+
1.25 mg TP	Ep	–	–	–	+	–	–	–
	St	+	–	– / +	+	+	– / +	–

[a] Ep, epithelial cells; St, stromal cells; 1P and 2P, a single injection of 3 mg P and 2 daily injections of 3 mg P after ovariectomy; Intact, 3P, 3PE, 1E and None, refer to Fig. 3. – , +, + +, and + + + indicate negative, weak, moderate, and strong staining, respectively.

In the androgenized rats, PR staining in the epithelial and stromal cells increased after 2 daily injections of 3 mg P but did not decrease following 3-day treatment with P. The single injection of estrogen given together with the progestin injections completely suppressed PR staining, regardless of the dose of androgen given neonatally. The injection of estrogen alone had little effect on PR in androgenized rats, although the injection reduced epithelial PR in control rats (Ohta *et al.*, 1993a). PR staining in the stromal cells was weaker in androgenized rats sacrificed at 60 days than in the controls at estrus. In the control rats, moreover, PR increased after ovariectomy, the staining reaching maximum within 3 days after ovariectomy, whereas there was almost no PR in androgenized rats except for the 100-μg TP-treated rats in which PR reappeared 3 days after the operation. This is in good agreement with the finding that PR concentration is abnormally low in the uterus of androgenized rats (White *et al.*, 1981). It is evident that androgenization during neonatal life interferes with the normal responsiveness of uterine cells to ovarian hormones in producing PR in addition to ER; this may account, at least in part, for the altered cell proliferation in the uterus of androgenized rats.

C. Surface Changes in Endometrial Epithelial Cells

Morphological changes that allow the adhesion of the trophoblast to uterine epithelium occur on the apical surface of epithelial cells just prior to implantation. Since there are many reviews of this phenomenon (Finn, 1980; Kennedy, 1983a; Chávez, 1984; Moulton, 1984; Parr and Parr, 1989; Murphy, 1992, among others), only a summary is presented here. Scanning electron microscopy reveals that bulbous cytoplasmic protrusions called pinopods (Enders and Nelson, 1973) or "mushroomlike" structure (Psychoyos, 1973b) appear for a limited period just before implantation, i.e., on day 5 of pregnancy in rats (Psychoyos, 1973b). With the loss of uterine sensitivity, these structures disappear, and the epithelial cells once again show definite microvilli (Moulton, 1984). Tachi *et al.,* (1970) and Psychoyos (1973b) have found the close contact of implanting blastocyst on the surface of the protrusions, and have suggested that these changes in the epithelial surface may contribute to the adhesion of the blastocyst to the uterine epithelium. Furthermore, the protrusions have been shown to mediate endocytosis, which may be involved in the replacement of membrane components needed for adhesion (Parr and Parr, 1974, 1989). Cytoplasmic protrusions were observed in the epithelium of OX rats treated with ovarian hormones (Parr, 1983), as well as during delayed implantation (Given and Enders, 1989). OX rats given 3 daily injections of 5 mg P alone or after 2-day priming with 0.5 μg E2 showed many protrusions on the epithelial surface (Parr, 1983). In contrast to these findings, Ohta and Iguchi (1994) have demonstrated that in OX rats, 3-day treatment with 3 mg P failed to induce morphological changes associated with implantation, although this treatment is adequate to sensitize the endometrium for trauma-induced but not for oil-induced decidualization, which requires a single injection of estrogen in addition to progestin (Ohta, 1983). Numerous protrusions appeared in the epithelial surface in response to a single injection of 0.2 μg E2 given on the third day of the P treatment but not to the injection of progestin or estrogen alone.

Enders and Schlafke (1974) have shown that one of the main changes occurring at implantation sites is a modification of a fine layer of glycoprotein (the glycocalyx) present on both blastocyst and luminal epithelial outer surfaces. The electrical negativity of the glycocalyx of epithelial cells has been demonstrated ultrahistochemically using cytochemical markers such as ruthenium red (Nilsson, 1974), lanthanum (Salazar-Rubio *et al.,* 1980), polycationic ferritin (Hewitt *et al.,* 1979), colloidal thorium (Enders and Schlafke, 1974), or iron (Tachi and Tachi, 1976).

Since negative charge repulsion seems to play an important role in regulating cell surface interactions between the trophoblast and uterine epithelium, the distribution of cell surface anionic sites in the epithelial cells has been studied (Murphy, 1992). In rats, there is a reduction in the thickness of the glycocalyx of the epithelium as implantation approaches (Hewitt *et al.,* 1979; Salazar-Rubio *et al.,* 1980; Murphy and Rogers, 1981), suggesting that the initial adhesion of blastocysts is accompanied by a reduction in negative charge on the uterine epithelial surface. In contrast, some workers have found little ultrastructural evidence for the changes in the glycocalyx at the time of implantation (Enders and Schlafke, 1974; Nilsson, 1974; Enders et al., 1980).

Tachi and Tachi (1976) also failed to demonstrate any conspicuous changes in the distribution of colloidal iron staining in the luminal epithelium either before or after onset of implantation. The discrepancy between these results may be due to differences in staining reaction. Parr and Parr (1989) have mentioned in their review that the role of changes in the surface charge or surface coat in causing blastocyst adherence to the uterine epithelium remains speculative. Murphy and Rogers (1981) have argued that the reduction of surface charge alone is not sufficient for adhesion, as estrogen treatment decreases surface charge without leading to blastocyst adhesion.

Colloidal iron staining in the epithelial surface was approximately the same among the OX rats given one of three different hormone treatments: 3-day treatment with 3 mg P alone, or P in combination with a single injection of $0.2\ \mu g$ E2 given on the third day of the P treatment, or a single injection of estrogen alone (Ohta and Iguchi, 1994). However, if the tissues were subjected to pretreatment with HCl–methanol solution, which was known to be effective in eliminating the reaction to carboxylic glycoprotein among acidic glycoproteins (Spicer, 1960), iron staining was greatly reduced only in those rats given progestin plus estrogen. On day 5 of pregnancy, the epithelial cells exhibited a similar reduction in staining of the surface coat by methylation the day of blastocyst implantation, although the staining on day 5 without methylation was similar to that on the other days of pregnancy, indicating an increase of carboxylic glycoprotein content in the glycocalyx of the epithelial cells on the day of implantation.

The cell surface glycoproteins have been found to vary with hormonal state during early pregnancy (Parr and Parr, 1989). Chávez (1990) demonstrated a decrease in sialoglycoproteins of the uterine cells during early pregnancy, concluding that desialylation resulted in a decrease in thickness of the glycocalyx and permitted cell surface interaction. In rats undergoing delayed implantation, however, Tachi and Tachi (1976) showed that colloidal iron staining is sensitive to neither neuraminidase nor hyaluronidase before or after onset of implantation. Treatment of the endometrium with neuraminidase reduced colloidal iron staining of the epithelial surface in the OX rats given a single injection of estrogen together with a 3-day injection of progestin, but not in those given either hormone alone (Ohta and Iguchi, 1994), indicating that sialoglycoproteins in the glycocalyx were increased by estrogen associated with progestin. The increase in sialoglycoproteins is responsible, at least in part, for the estrogen-dependent increase of carboxylic glycoprotein content in the surface coat.

Although the results obtained by Ohta and Iguchi (1994) in OX rats subjected to a hormonal milieu mimicking early pregnancy do not fully agree with those reported earlier, these authors found differences between nonandrogenized and androgenized rats, in the morphological changes in the surface structure of the epithelial cells. The epithelial cells did not show increased protrusions in response to progestin alone or in combination with estrogen in the androgenized rats, regardless of dose of TP used (100 μg or 1.25 mg) (Ohta and Iguchi, 1994). The number of protrusions increased following treatment with estrogen alone rather than in combination with progestin. In addition, the epithelial cells in androgenized rats failed to respond to estrogen together with progestin, exhibiting no significant reduction of the colloidal iron reaction after methylation of their luminal surface. Pretreatment of the tissue with neuraminidase had little effect on the iron reaction in androgenized rats given P and E2 injections. These findings also suggest that the response of uterine epithelial cells to ovarian hormones, estrogen associated with progestin in particular, is markedly affected by neonatal androgenization.

D. Protein Synthesis

Although the brief surge of estrogen which occurs between days 3 and 4 of pregnancy is necessary for implantation (Psychoyos, 1973b), the mechanism involved in the estrogen-mediated process remains unknown. One possible effect of estrogen in the initiation of implantation is on RNA and protein synthesis. It has been suggested that implantation is prevented by inhibitory proteins produced in the progestational period of early pregnancy, which may act by being secreted into the lumen (Watson *et al.,* 1975). Leroy *et al.,* (1980) suggested a role for nidatory estrogen in inhibition of the progestin-dependent transcription of endometrial RNA. Lejeune *et al.,* (1985) have investigated the pattern of endometrial protein synthesis in relation to implantation in OX rats given hormonal treatment mimicking the progestational period. Three-day treatment with 5 mg P following 2-day priming with 1 μg E2 induced the synthesis of 14 new polypeptides in the epithelium and 6 in the stroma.

A single injection of 0.05 μg E2 given on the third day of P treatment suppressed the production of 10 epithelial proteins and 1 stromal one. This is in good agreement with the findings that in the endometrium of pregnant rats sacrificed on day 5 of pregnancy, no unique proteins appeared, but several peptides produced on days 3 and 4 were no longer synthesized (Mulholland and Villee, 1984), suggesting that the inhibitory protein controlled by progestin can be suppressed by the estrogen surge. In response to the 3-day treatment with 3 mg P given after ovariectomy, the androgenized rats produced fewer kinds of proteins in the endometrium than the nonandrogenized rats ovariectomized at estrus and treated similarly, and synthesized only 3 or 6 of 13 proteins produced in the controls, depending upon the dose of TP (100 μg or 1.25 mg) given neonatally (Suzuki *et al.*, 1994) (Table II). In the controls, a single injection of estrogen in addition to the progestin treatment suppressed synthesis of the progestin-induced proteins except for 4 proteins of which 2 were progestin-specific.

In androgenized rats, such treatment caused the disappearance of 1 or 5 of the progestin-stimulated proteins. Protein synthesis after a single estrogen injection in androgenized rats was also different from that in the controls. Estrogen treatment induced the synthesis of some progestin-specific proteins in androgenized rats: 3 of 9 proteins in rats treated with 100 μg TP and 1 of 3 proteins in those given 1.25 mg. The synthesis of proteins in the uterus evidently affected the androgenized rats. At present, however, it is uncertain how the modified uterine synthesis of proteins may be involved in the reduction of DCR in androgenized rats.

TABLE II

Analysis of Effect of Ovarian Hormones on Uterine Protein Synthesis in Neonatally Androgenized (100 μg and 1.25 mg TP) and Control Rats by Two-Dimensional Electrophoresis[a]

Protein number	Molecular weight(kDa)	pI	Control			100 μg TP			1.25 mg TP		
			3P	3PE	1E	3P	3PE	1E	3P	3PE	1E
1	52–51	6.7–6.85	+	−	+	−	−	+	−	−	−
2	51–49	6.7–6.85	+	−	+	−	−	−	−	−	−
3	45	6.5–6.7	+	−	+	−	−	+	−	−	−
4	44	6.9	+	−	+	−	−	−	−	−	−
5	44	7.1	+	−	+	−	−	−	−	−	−
6	42	6.85	+	−	−	−	−	−	−	−	−
7	40.5	6.85	+	+	−	−	−	+	+	−	−
8	39	6.6	+	−	+	−	−	+	+	−	−
9	38	6.5	+	−	−	+	+	+	−	−	+
10	36	6.7	+	+	−	+	−	+	+	−	−
11	32.5	6.8	+	+	+	−	−	+	+	−	+
12	32	6.5	+	+	+	+	+	+	+	+	−
13	31.5	6.75	+	−	+	−	+	+	+	−	+

[a] 3P, 3PE, and 1E refer to Fig. 3. + and − indicate detectable and undetectable, respectively.

VI. Comments and Conclusions

Administration of androgen to the female rat during a limited period of neonatal life interferes with uterine functions in the adult, as well as with the CNS controlling ovarian function. The decidual response is permanently reduced in such animals. However, the reduction varies with the dose of androgen given neonatally and with the type of endometrial stimulus. If the amount of androgen given neonatally is relatively low (i.e., less than 100 μg TP), the uterus is capable of forming deciduomata in response to a strong stimulus, such as traumatization, but not to a weaker one, such as intrauterine instillation of oil. When larger doses of androgen (more than 1 mg TP) are injected neonatally, the uteri become no longer responsive to any deciduogenic stimulus. Since neonatally androgenized rats show persistent estrus as adults, two aspects of effects on the uterus should be considered: (1) direct action of androgen on the neonatal uterus and (2) indirect action via the CNS, which results in sustained secretion of ovarian estrogen. Several lines of evidence from female rats given the higher dose of androgen indicate that lowered uterine responsiveness in androgenized rats is largely ascribable to the direct effect of neonatal exposure to androgen, although continued exposure of the uterus to endogenous estrogen may play a co-operative role.

In contrast, the uterus of androgenized rats given the lower dose of androgen may be affected by continued exposure to estrogen as adults because the response in these animals is similar to that seen in aged rats showing persistent estrus and is nullified by prepubertal ovariectomy. Furthermore, the difference in response to an endometrial stimulus in the two types of androgen-sterilized rats may arise from the mechanism involved in the action of androgen on the neonatal uterus: the lower dose of androgen appears to have less effect on the endometrial cells but has significant negative effects, mainly on the epithelial cells with the continued secretion of ovarian estrogen at later ages, whereas the higher dose of androgen may bring about a total loss of uterine response to endometrial stimulus in the adult by acting directly on both the epithelial and stromal cells of the developing uterus. This hypothesis is substantiated by the results of morphological and biochemical studies in androgenized rats showing that the uterine response to ovarian steroids is affected to varying degrees, depending upon the dose of androgen given neonatally. The continued secretion of estrogen also impairs the endometrial cells: shorter exposure of the uterus to estrogen affects only epithelial cells and longer exposure affects both epithelial and stromal cells.

Morphological and biochemical analyses of the different uterine responses to ovarian steroids in androgenized and normal rats contribute to an understanding of the mechanism involved in implantation and placentation. Moreover, although age-related reproductive changes in the CNS and peripheral organs exhibit a remarkable variability, most of them occur earlier in androgenized rats than in normal animals. In general, the physiological condition of the CNS controlling the female reproductive system in neonatally androgenized animals is similar to that encountered in normal aged animals. This chapter also reveals the similarity of the uterine function in the androgenized rat to that seen in the aged rat: the uterine response of androgenized rats to steroid hormones and to deciduogenic stimuli is comparable to that observed in normal aging rats. Thus, neonatally androgenized rats, although they have been used for some investigations, including sex differentiation of the CNS in the control of reproduction, could also be used to study fertility, including implantation, and as a profitable model for aging in female reproduction.

Acknowledgments

The author dedicates this article to the late Professor Kiyoshi Takewaki of the University of Tokyo in gratitude for his valuable advice and suggestions, and thanks Emeritus Professor Noboru Takasugi of Yokohama City University and Emeritus Professor Howard A. Bern of University of California at Berkeley for their critical reading of this review. Some studies and unpublished investigations described here were supported by a Grant-in-Aid from the Ministry of Education, Science and Culture of Japan (05640750).

References

Abrahamsohn, P. A., and Zorn, T. M. (1993). Implantation and decidualization in rodents. *J. Exp. Zool.* **266,** 603–628.

Aihara, M., Kimura, T., and Kato, J. (1980). Dynamics of the estrogen receptor in the uterus of mice treated neonatally with estrogen. *Endocrinology (Baltimore)* **107,** 224–230.

Arnold, A. P., and Schlinger, B. A. (1993). Sexual differentiation of brain and behavior: The zebra finch is not a flying rat. *Brain Behav. Evol.* **42,** 231–241.

Barraclough, C. A. (1966). Modification in reproductive function after exposure to hormones during the prenatal and early postnatal period. *In* "Neuroendocrinolgy" (L. Martini and W. F. Ganong, eds.), pp. 61–99. Academic Press, New York.

Barraclough, C. A., and Fajer, A. B. (1968). Progestin secretion by gonadotropin-induced corpora lutea in ovaries of androgen-sterilized rats. *Proc. Soc. Exp. Biol. Med.* **128,** 781–785.

Barraclough, C. A., and Gorski, R. A. (1962). Studies on mating behaviour in the androgen-sterilized female rat in relation to the hypothalamic regulation of sexual behaviour. *J. Endocrinol.* **25,** 175–182.

Belise, S., Bellabarba, D., and Lehoux, J.-G. (1989). Estradiol receptor-nuclear interactions in aging mouse uteri: The role of DNA and nuclear matrix. *J. Steroid Biochem. Mol. Biol.* **32,** 41–43.

Bern, H. A. (1992a). The fragile fetus. *In* "Chemically-Induced Alterations in Sexual and

Functional Development" (T. Colbon and C. Clement, eds.), pp. 9–15. Princeton Sci. Publ., Princeton, NJ.

Bern, A. H. (1992b). Diethylstilbestrol (DES) syndrome: Present status of animal and human studies. In "Hormonal Carcinogenesis" (J. Li, S. Nandi, and S. A. Li, eds.), pp. 1–8. Springer-Verlag, New York.

Bern, H. A., and Talamantes, F., Jr. (1981). Neonatal mouse models and their relation to disease in the human females. In "Developmental Effects of Diethylstilbestrol (DES) in Pregnancy" (A. L. Herbst and H. A. Bern, eds.), pp.129–147. Thieme-Stratton, New York.

Biggers, J. D., Finn, C. A., and McLaren, A. (1962). Long-term reproductive performance of female mice. J. Reprod. Fertil. 3, 313–330.

Bigsby, R. M., Aixin, L., Luo, K., and Cunha, G. R. (1990). Strain differences in the ontogeny of estrogen receptors in murine uterine epithelium. Endocrinology (Baltimore) 126, 2592–2596.

Blaha, G. C. (1967). Effects of age, treatment, and method of induction on deciduomata in the golden hamster. Fertil. Steril. 18, 477–476.

Blaha, G. C., and Leavitt, W. W. (1974). Ovarian steroid dehydrogenase histochemistry and circulating progesterone in aged golden hamsters during the estrous cycle and pregnancy. Biol. Reprod. 11, 153–161.

Blaha, G. C., and Leavitt, W. W. (1978). Uterine progesterone receptors in the aged golden hamster. J. Gerontol. 33, 810–814.

Brown, C., Gosden, R. G., and Poyser, N. L. (1984). Effects of age and steroid treatment on prostaglandin production by the rat uterus in relation to implantation. J. Reprod. Fertil. 70, 649–656.

Burin, P., Thevenot-Dulug, A. J., and Mayer, G. (1963). Exploration des potentialités de l'hypophyse et des effecteurs des hormones genitales chez les Rattes en oestrus permanent provoqué par une injection postnatale de testostérone. C.R. Hebd. Seances Acad. Sci. 157, 1258–1260.

Campbell, P. S. (1980). Impaired prepubertal uterine capacity of response after neonatal exposure to steroid hormone esters. J. Exp. Zool. 214, 345–353.

Campbell, P. S., and Modlin, P. S. (1987). Uterine glucose metabolism in the prepubertal rat treated neonatally with androgen, estrogen and antihormones. Experientia 43, 309–310.

Cao, Z.-D., Jones, M. A., and Harper, J. J. K. (1984). Prostaglandin translocation from the lumen of the rabbit uterus in vitro in relation to day of pregnancy or pseudopregnancy. Biol. Reprod. 31, 505–519.

Castellano-Diaz, E., Gonzalez, M. I., and Daiz-Chico, B. N. (1987). Relationship between occupied form of nuclear estrogen receptor and cytosolic progesterone receptor or DNA synthesis in uteri of estradiol implanted rats. Rev. Esp. Fisiol. 43, 401–406.

Chang, W.-C., Nakano, J., Neichi, T., and Orimo, H. (1981). Effects of estradiol on the metabolism of arachidonic acid by aortas and platelets in rats. Biochim. Biophys. Acta 664, 292–297.

Chatterjee, A., and Mukherjee, A. P. (1974). The aging uterus and its refractoriness of appropriate sex steroid in the development of deciduoma. Endokrinologie 63, 166–168.

Chávez, D. J. (1984). Cellular aspects of implantation. In "Ultrastructure of Reproduction" (V. J. Blerkman and P. M. Motta, eds.), pp. 247–259. Martinus Nijhoff, The Hague, Boston.

Chávez, D. J. (1990). Possible involvement of D-galactose in the implantation process. Trophoblast Res. 4, 259–272.

Clark, J. H., and Gorski, J. (1970). Ontogeny of the estrogen receptor during early uterine development. Science 169, 76.

Collado, D., and Aguilar, E. (1993). Further evidence that prolactin secretion in adult female rats is differently modified after neonatal estrogenization or androgenization: Responses to methysergide, quipazine, and pizotifen. *Physiol. Behav.* **53**, 161–165.

Cortes, V., McCracken, J. A., Lloyd, C. W., and Weisz, J. (1971). Progestin production by the ovary of a the testosterone-sterilized rat treated with an ovulatory dose of LH, and the normal, proestrous rat. *Endocrinology (Baltimore)* **89**, 878–885.

Craig, S. S. (1981). Effect of age upon uterine response to deciduogenic stimulus. *Acta Anat.* **110**, 146–158.

Csaba, G., Inczefi-Gonda, A., and Dobozy, O. (1986). Hormonal imprinting steroids: A single neonatal treatment with diethylstilbestrol or allylestrenol gives rise to a lasting decrease in the number of rat uterine receptor. *Acta Physiol. Hung.* **67**, 207–212.

Day, J. R., LaPolt, P. S., Morales, T. H., and Lu, J. K. H. (1989). An abnormal pattern of embryonic development during early pregnancy in aging rats. *Biol. Reprod.* **41**, 933–939.

DeFeo, V. J. (1963). Determination of the sensitive period for the induction of deciduomata in the rat by different inducing procedures. *Endocrinology (Baltimore)* **73**, 488–499.

DeFeo, V. J. (1967). Decidualization. *In* "Cellular Biology of the Uterus" (R. M. Wynn, ed.), pp. 191–291. Appleton-Century-Crofts, New York.

de Greef, W. J., and Zeilmaker, G. H. (1978). Regulation of prolactin secretion during the luteal phase in the rat. *Endocrinology (Baltimore)* **102**, 1190–1198.

Döcke, F., Rohde, W., Badell, J., Geier, T., and Dörner, G. (1981). Influence of neonatal ovariectomy on the developmental patterns of serum gonadotropins, hypothalamic LH-RH concentration and organ weights in female rats. *Endokrinologie* **77**, 273–285.

Dorner, G. V., and Fatschel, J. (1970). Wirkungen neonatal verabreichter Androgene und Antiandrogene auf Sexualabverhalten und Fertilität von Rattenweibchen. *Endokrinologie* **56**, 29–48.

Downing, I., and Poyser, N. L. (1983). Estimation of phospholipase A2 activity in guinea-pig endometrium on days 7 and 16 of the estrous cycle. *Prostaglandins Leukotrienes Med.* **12**, 107–117.

Enders, A. C., and Nelson, M. D. (1973). Pinocytotic activity of the uterus of the rat. *Am. J. Anat.* **138**, 277–300.

Enders, A. C., and Schlafke, S. (1967). A morphological analysis of the early implantation stages in the rat. *Am. J. Anat.* **120**, 185–226.

Enders, A. C., and Schlafke, S. (1974). Surface coats of the mouse blastocyst and uterus during the preimplantation period. *Anat. Rec.* **180**, 31–46.

Enders, C. A., Schlafke, S., and Welsh, A. O. (1980). Trophoblastic and uterine luminal epithelial surfaces at the time of blastocyst adhesion in the rat. *Am. J. Anat.* **159**, 59–72.

Feil, P. D., Glasser, S. R., Toft, D. O., and O'Malley, B. W. (1972). Progesterone binding in the mouse and rat uterus. *Endocrinology (Baltimore)* **91**, 738–746.

Finch, C. E. (1976). The regulation of physiological changes during mammalian aging. *Q. Rev. Biol.* **51**, 49–83.

Finch, C. E., and Holinka, C. F. (1982). Aging and uterine growth during implantation in C57BL/6J mice. *Exp. Gerontol.* **17**, 235–241.

Finch, C. E., Felicio, L. S., Mobbs, C. V., and Nelson, J. F. (1984). Ovarian and steroidal influences on neuroendocrine aging processes in female rodents. *Endocr. Rev.* **5**, 467–497.

Finn, C. A. (1965). Oestrogen and the decidual cell reaction of implantation in mice. *J. Endocrinol.* **32**, 223–229.

Finn, C. A. (1966a). Endocrine control of endometrial sensitivity during the induction of the decidual cell reaction in the mouse. *J. Endocrinol.* **36**, 239–248.

Finn, C. A. (1966b). The initiation of the decidual cell reaction in the uterus of the aged mouse. *J. Reprod. Fertil.* **11**, 423–428.

Finn, C. A. (1980). The endometrium during implantation. *In* "The Endometrium" (F. A. Kimball, ed.), pp. 43–56. Spectrum, New York.

Finn, C. A., and Hinchliffe, J. R. (1964). The reaction of the mouse uterus during implantation and deciduoma formation as demonstrated by changes in the distribution of alkaline phosphatase. *J. Reprod. Fertil.* **8**, 331–338.

Finn, C. A., and Martin, L. (1969). The cellular response of the uterus of the aged mouse to oestrogen and progesterone. *J. Reprod. Fertil.* **20**, 545–547.

Finn, C. A., and Martin, L. (1974). The control of implantation. *J. Reprod. Fertil.* **39**, 195–206.

Finn, C. A., and Pollard, R. M. (1973). The influence of the oestrogen secreted before oestrus on the timing of endometrial sensitivity and insensitivity during implantation. *J. Reprod. Fertil.* **56**, 619–620.

Finn, C. A., and Porter, D. G. (1975a). The action of ovarian hormones on the endometrium. *In* "The Uterus" pp. 42–56. Publishing Science Groups, London.

Finn, C. A., and Porter, D. G. (1975b). The decidual cell reaction. *In* "The Uterus," pp. 239–248. Publishing Science Groups, London.

Finn, C. A., Pope, M. D., and Milligan, S. R. (1992). Timing of the window of uterine sensitivity to decidual stimuli in mice. *Reprod. Fertil. Dev.* **4**, 565–571.

Flerkó, B. (1971). Steroid hormones and the central nervous system. Curr. Top. Exp. Endocrinol. **1**, 42–80.

Flerkó, B., Mess, B., and Illei-Donhoffer, A. (1969). On the mechanism of androgen sterilization. *Neuroendocrinology* **4**, 164–169.

Gala, R. R. (1981). The influence of estrogen administration on plasma prolactin levels in the neonatally androgenized (NA) female rat. *Proc. Soc. Exp. Biol. Med.* **166**, 216–220.

Gellert, R. J., Lewis, J., and Pétra, P. H. (1977). Neonatal treatment with sex steroids: Relationship between the uterotropic response and the estrogen 'receptor' in prepubertal rats. *Endocrinology (Baltimore)* **100**, 520–528.

Gibori, G., Kalison, B., Basuray, R., Rao, M. C., and Hunzicker-Dunn, M. (1984). Endocrine role of the decidual tissue: Decidual luteotropin regulation of luteal adenylyl cyclase activity, luteinizing hormone receptors, and steroidogenesis. *Endocrinology (Baltimore)* **115**, 1157–1163.

Given, R. L., and Enders, A. C. (1989). The implantation reaction. *In* "Biology of the Uterus" (R. M. Wynn and W. P. Jollie, eds.) pp. 175–231. Plenum, New York.

Glasser, S. R., and McCormack, S. A. (1979). Estrogen modulated uterine gene transcription in relation to decidualization. *Endocrinology (Baltimore)* **104**, 1112–1118.

Goldzeiher, J. W. (1981). Polycystic ovarian disease. *Fertil. Steril.* **35**, 371–394.

Gorski, R. A. (1968). Influence of age on the response to paranatal administration of a low dose of androgen. *Endocrinology (Baltimore)* **82**, 1001–1004.

Gorski, R. A. (1971). Gonadal hormones and the perinatal development of neuroendocrine function. *In* "Frontiers in Neuroendocrinolgy" (L. Martini and W. F. Ganong, eds.), pp. 237–290. Oxford Univ. Press, New York.

Gorski, R. A. (1979). The neuroendocrinology of reproduction: An overview. *Biol. Reprod.* **20**, 111–127.

Hahn, D. W., and McGuire, J. L. (1978). The androgen-sterilized rat: Induction of ovulation and implantation by luteinizing hormone-releasing hormone. *Endocrinology (Baltimore)* **102**, 1741–1749.

Ham, E. A., Cirillo, V. J., Zanetti, M. E., and Kuehl, F. A. J. (1975). Estrogen-directed synthesis of specific prostaglandins in uterus. *Proc. Natl. Acad. Sci. U.S.A.* **72**, 1420–1424.

Hamilton, G. S., and Kennedy, T. G. (1994). Uterine vascular changes after unilateral intrauterine infusion of indomethacin and prostaglandin E2 to rats sensitized for the decidual cell reaction. *Biol. Reprod.* **50**, 757–764.

Han, Z., Kokkonen, G. C., and Roth, G. S. (1989). Effect of aging on populations of estrogen receptor-containing cells in the rat uterus. *Exp. Cell Res.* **180**, 234–242.

Harlan, R. E., and Gorski, R. A. (1977). Steroid regulation of luteinizing hormone secretion in normal and androgenized rats at different ages. *Endocrinology (Baltimore)* **101,** 741–749.

Hayashi, S. (1968). Hyperplasia and metaplasia of the uterine epithelium following estrone injections in ovariectomized adult rats given neonatal injections of sex steroids. *Endocrinol. Jpn.* **15,** 229–234.

Heffner, L. J., and van Tienhoven, A. (1979). Effects of neonatal ovariectomy upon ³H-estradiol uptake by target-tissues of androgen sterilized female rats. *Neuroendocrinology* **29,** 237–246.

Hewitt, K., Beer, A. E., and Grinnell, F. (1979). Disappearance of anionic sites from the surface of the rat endometrial epithelium at the time of blastocyst implantation. *Biol. Reprod.* **21,** 691–707.

Hoffman, L. H., Strong, G. B., Davenport, G. R., and Frölich, J. C. (1977). Deciduogenic effect of prostaglandins in the pseudopregnant rabbit. *J. Reprod. Fertil.* **50,** 231–237.

Hoffman, L. H., Davenport, G. R., and Brash, A. R. (1984). Endometrial prostaglandins and phospholipase activity related to implantation in rabbits: Effects of dexamethasone. *Biol. Reprod.* **30,** 544–555.

Holinka, C. F., and Finch, C. E. (1977). Age-related changes in the decidual response of the C57BL/6J mouse uterus. *Biol. Reprod.* **16,** 385–393.

Holinka, C. F., Nelson, J. F., and Finch, C. E. (1975). Effect of estrogen treatment on estradiol binding capacity in uteri of aging rats. *Gerontologist* **15,** 30.

Hsueh, A. J. W., Erickson, G. F., and Lu, K. H. (1979). Changes in uterine estrogen receptor and morphology in aging female rats. *Biol. Reprod.* **21,** 793–800.

Iguchi, T. (1992). Cellular effects of early exposure to sex hormones and antihormones. *Int. Rev. Cytol.* **139,** 1–57.

Iguchi, T., and Takasugi, N. (1976). Occurrence of permanent changes in vaginal and uterine epithelia in mice treated neonatally with progestin, estrogen and aromatizable or non-aromatizable androgens. *Endocrinol. Jpn.* **23,** 327–332.

Janszen, F. H. A., and Nugteren, D. H. (1971). *Histochemie* **27,** 159–164.

Johnson, D. C. (1979). Maintenance of functional corpora lutea in androgenized female rats treated with PMSG. *J. Reprod. Fertil.* **56,** 263–269.

Johnson, D. C., and Witschi, E. (1963). Effects of follicle-stimulating hormone, chronic gonadotropin and estradiol in androgenized female rats. *Endocrinology (Baltimore)* **73,** 467–474.

Jonston, M. E. A., and Kennedy, T. G. (1985). Temporal desensitization of rat uteri for the decidual cell reaction is abolished by cholera toxin acting by a mechanism apparently not involving adenosine 3′:5′-cyclic monophosphate. *Can. J. Physiol. Pharmacol.* **63,** 1052–1056.

Kennedy, T. G. (1979). Prostaglandins and increased endometrial vascular permeability resulting from the application of an artificial stimulus to the uterus of the rat sensitizes for the decidual cell reaction. *Biol. Reprod.* **20,** 560–566.

Kennedy, T. G. (1980). Estrogen and uterine sensitization from the decidual cell reaction: Role of prostaglandins. *Biol. Reprod.* **23,** 955–962.

Kennedy, T. G. (1983a). Embryonic signals and the initiation of blastocyst implantation. *Aust. J. Biol. Sci. Reprod.* **36,** 531–543.

Kennedy, T. G. (1983b). Prostaglandin E2, adenosine 3′: 5′-cyclic monophosphate and changes in endometrial vascular permeability in rat uteri sensitized for the decidual cell reaction. *Biol. Reprod.* **29,** 1069–1076.

Kennedy, T. G. (1985). Evidence for the involvement of prostaglandins throughout the decidual cell reaction in the rat. *Biol. Reprod.* **33,** 140–146.

Kennedy, T. G. (1986a). Intrauterine infusion of prostaglandins and decidualization in rats with uteri differentially sensitized for the decidual cell reaction. *Biol. Reprod.* **34,** 327–335.

Kennedy, T. G. (1986b). Prostaglandins and uterine sensitization for the decidual cell reaction. *Ann. N. Y. Acad. Sci.* **476**, 43–48.

Kennedy, T. G. (1990). Eicosanoids and blastocyst implantation. *In* "Eicosanoids in Reproduction" (M. D. Mitchell, ed.), pp. 123–139. CRC Press, Boca Raton, FL.

Kennedy, T. G., Martel, D., and Psychoyos, A. (1983). Endometrial prostaglandin E2 binding: Characterization in rats sensitized for the decidual cell reaction and changes during pseudopregnancy. *Biol. Reprod.* **29**, 556–564.

Kincl, F. A. (1990). Effect of steroid hormones in the neonate. *In* "Hormone Toxicity in the Newborn," pp. 168–265. Springer-Verlag, Berlin.

Kolena, J., Hácik, T., and Scböková, E. (1977). Postnatal development of gonadotropin binding sites and cAMP synthesis in ovaries and estradiol plasma levels in estrogenized and androgenized female rats. *Endocrinol. Exp.* **11**, 219–225.

Korenbrot, C. C., Paup, D. C., and Gorski, R. A. (1975). Effects of testosterone propionate or dihydrotestosterone propionate on plasma FSH and LH levels in neonatal rats and on sexual differentiation of the brain. *Endocrinology (Baltimore)* **97**, 709–717.

Kramen, M. A. (1974). Implantation in rats treated neonatally with testosterone propionate. *J. Reprod. Fertil.* **38**, 461–463.

Kramen, M. A., and Johnson, D. C. (1971). Mating, fertilization, and ovum viability in the anovulatory, persistent-estrus rat. *Fertil. Steril.* **22**, 745–754.

Kramen, M. A., and Johnson, D. C. (1975). Uterine decidualization in rats given testosterone propionate neonatally. *J. Reprod. Fertil.* **42**, 559–562.

Kurl, R. N., and Borthwick, N. M. (1979). Progesterone receptors and RNA polymerase activity in the rat uterus during the oestrous cycle. *J. Endocrinol.* **83**, 41–51.

Larson, L. I., and Foote, R. H. (1972). Uterine blood flow rates in young and aged rabbits. *Proc. Soc. Exp. Biol. Med.* 141, 67-69.

Larson, L. L., Spilman, C. H., and Foote, R. H. (1972). Uterine uptake of progesterone and estradiol in young and aged rabbits. *Proc. Soci. Exp. Biol. Med.* **141**, 463–466.

Leavitt, W. W. (1989). Cell biology of the endometrium. *In* "Biology of the Uterus" (R. M. Wynn and W. P. Jollie, eds.), pp. 131–173. Plenum, New York.

Le Goascogne, C., and Baulieu, E. E. (1977). Hormonally controlled "nuclear bodies" during the development of the prepuberal rat uterus. *Biol. Cell.* **30**, 195–206.

Lejeune, B., van Hoeck, J., and Leroy, F. (1981). Transmitter role of the luminal uterine epithelium in the induction of decidualization in rats. *J. Reprod. Fertil.* **61**, 235–240.

Lejeune, B., Lamy, F., Lecocq, R., Deschacht, J., and Leroy, F. (1985). Patterns of protein synthesis in endometrial tissues from ovariectomized rats treated with oestradiol and progesterone. *J. Reprod. Fertil.* **73**, 223–228.

Leroy, F., Vansande, J., Shetgen, G., and Brasseur, D. (1974). Cyclic AMP and the triggering of the decidual reaction. *J. Reprod. Fertil.* **39**, 207–211.

Leroy, F., Schetgen, G., and Camus, M. (1980). Initiation of implantation at the subcellular level. *Prog. Reprod. Biol.* **7**, 200–215.

Lobl, R. T. (1975a). Androgenization: Alterations in the mechanism of oestrogen action. *J. Endocrinol.* **66**, 79–84.

Lobl, R. T. (1975b). Estradiol: Alterations in the mechanism of intracellular transport. *Psychoneuroendocrinology* **1**, 131–140.

Lobl, R. T., and Maenza, R. M. (1975). Androgenization: Alterations in uterine growth and morphology. *Biol. Reprod.* **13**, 255–268.

Lobl, R. T., and Mohammed, H. (1975). The influence of androgenization on the tensile properties of the rat uterus. *Biol. Reprod.* **13**, 269–273.

Lobl, R. T., Trotta, P., and Brumberger, B. (1974). Oestrogen-induced proteins and uterine growth in the androgenized female rat. *J. Endocrinol.* **60**, 371–372.

MacLusky, N. J., and Naftolin, F. (1981). Sexual differentiation of the central nervous system. *Science* **211**, 1294–1303.

Maibenco, H. C., and Krehbiel, R. H. (1973). Reproductive decline in aged female rats. *J. Reprod. Fertil.* **32**, 121–123.

Mallampati, R. S., and Johnson, D. C. (1974). Gonadotropins in female rats androgenized by barous treatment: Prolactin as an index to hypothalamic damage. *Neuroendocrinology* **15**, 255–266.

Martel, D., and Psychoyos, A. (1982). Different responses of rat endometrial epithelium and stroma to induction of oestradiol binding sites by progesterone. *J. Reprod. Fertil.* **64**, 387–389.

Martin, L., and Finn, C. A. (1971). Oestrogen-gestagen interactions on mitosis in target tissues. *In* "Basic Action of Sex Steroids on Target Organs," pp. 172–188. Karger, Basel.

Matt, D. W., Lee, J., Sarver, P. L., Judd, H. L., and Lu, J. K. L. (1986). Chronological changes in fertility, fecundity, and steroid hormone secretion during consecutive pregnancies in aging rats. *Biol. Reprod.* **34**, 478–487.

Matt, D. W., Dahl, K. D., Sarkissian, A., and Sayles, T. E. (1993). Apparent absence of negative feedback in middle-aged persistent-estrous rats following luteinizing hormone-releasing hormone agonist treatment: Relation to plasma inhibin and 17β-estradiol. *Biol. Reprod.* **48**, 333–339.

Matton, P., and Maurasse, C. (1985). Deciduomal response in androgenized rats: Effect of age and steroid replacement therapy. *Biol. Reprod.* **32**, 1095–1100.

Maurer, R. A., and Wooley, D. E. (1975). ^{3}H-estradiol distribution in female, androgenized female, and male rats at 100 and 200 days of age. *Endocrinology (Baltimore)* **96**, 755–765.

McCormack, J. T., and Greenwald, G. S. (1974a). Evidence for a preimplantation rise in oestradiol-17β levels on day 4 of pregnancy in the mouse. *J. Reprod. Fertil.* **41**, 297–301.

McCormack, J. T., and Greenwald, G. S. (1974b). Progesterone and oestradiol-17β concentrations in the peripheral plasma during pregnancy in the mouse. *J. Endocrinol.* **62**, 101–107.

McDonald, P. G., and Doughty, C. (1972). Comparison of the effect of neonatal administration of testosterone and dihydrotestosterone in the female rat. *J. Reprod. Fertil.* **30**, 55–62.

McGuire, J. L., and DeDella, C. (1971). In vitro evidence for a progestogen receptor in the rat and rabbit uterus. *Endocrinology (Baltimore)* **88**, 1099–1103.

McGuire, W. L., and Lisk, R. D. (1969). Oestrogen receptors in androgen or oestrogen sterilized female rats. *Nature (London)* **221**, 1068–1069.

McLaren, A. (1969). Stimulus and response during early pregnancy. *Nature (London)* **22**, 793–741.

McLaren, A. (1973). Blastocyst activation. *In* "The Regulation of Mammalian Reproduction" (S. J. Segal, R. Crozier, P. A. Corffman, and P. G. Condliffe, eds.), pp. 321–328. Thomas, Springfield, IL.

Medlock, K. L., Forrester, T. M., and Sheehan, D. M. (1994). Progesterone and estradiol interaction in the regulation of rat uterine weight and estrogen receptor concentration. *Proc. Soc. Exp. Biol. Med.* **205**, 146–153.

Meites, J. (1982). Changes in neuroendocrine control of anterior pituitary function during aging. *Neuroendocrinology* **34**, 151–156.

Milgrom, E., and Baulieu, E.-E. (1970). Progesterone in uterus and plasma. I. Binding in rat uterus 105,000 g supernatant. *Endocrinology (Baltimore)* **87**, 276–278.

Mori, T., Nagasawa, H., and Ohta, Y. (1987). Prolactin and uterine adenomyosis in mice. *In* "Prolactin and Lesions in Breast, Uterus and Prostate" (H. Nagasawa, ed.), pp. 123–139. CRC Press, Boca Raton, FL.

Morikawa, S., Naito, M., Sekiya, S., Takeda, B., and Takamizawa, H. (1982). Fundamental studies on high risk factor of endometrial carcinoma: Functional and histological abnormalities of ovaries and uteri in experimental anovulatory rats. *Acta Obst. Gynaecol. Jpn.* **34**, 165–172.

Moulton, B. C. (1984). Epithelial cell function during blastocyst implantation. *J. Biosci.* **6** Suppl. 2, 11–21.

Moulton, B. C., and Koenig, B. B. (1984). Uterine deoxyribonucleic acid synthesis during preimplantation in precursors of stromal cell differentiation during decidualization. *Endocrinology (Baltimore)* **115**, 1302–1307.

Moulton, B. C., and Koenig, B. B. (1986). Hormonal control of phospholipid methylation in uterine luminal epithelial cells during uterine sensitivity to deciduogenic stimuli. *Endocrinology (Baltimore)* **118**, 244–249.

Moulton, B. C., and Russell, P. T. (1989). Arachidonic acid in uterine phospholipid during early pregnancy and following hormone treatment. *Biol. Reprod.* **41**, 821–826.

Moulton, B. C., Schuler, J. A., and Leftwich, J. B. (1987). Effect of a deciduogenic stimulus on arachidonic acid turnover in uterine phospholipids. *Biol. Reprod.* **36** Suppl. 1, 66.

Mulholland, J., and Jones, C. J. P. (1993). Characteristics of uterine aging. *Microsc. Res. Tech.* **25**, 148–168.

Mulholland, J., and Villee, C. A. J. (1984). Proteins synthesized by the rat endometrium during early pregnancy. *J. Reprod. Fertil.* **72**, 395–400.

Murphy, C. R. (1992). Structure of the plasma membrane of uterine epithelial cells in blastocyst attachment: A review. *Reprod. Fertil. Dev.* **4**, 633–643.

Murphy, C. R., and Rogers, A. W. (1981). Effects of ovarian hormones on cell membranes in the rat uterus. III. The surface carbohydrates at the apex of the luminal epithelium. *Cell Biophys.* **3**, 305–320.

Naftolin, F., and Brawer, J. R. (1978). The effect of estrogens on hypothalamic structure and function. *Am. J. Obstet. Gynecol.* **132**, 758–765.

Naylor, B., and Poyser, N. L. (1975). Effects of oestradiol and progesterone on the *in vitro* production of prostaglandin F2α by the guinea-pig uterus. *Br. J. Pharmacol.* **55**, 229–232.

Neil, J. D. (1972). Sexual differences in the hypothalamic regulation of prolactin secretion. *Endocrinology (Baltimore)* **90**, 1154–1159.

Nilsson, O. (1966). Estrogen-induced increase of adhesiveness in uterine epithelium of mouse and rat. *Exp. Cell Res.* **43**, 239–241.

Nilsson, O. (1974). Changes of the luminal surface of the rat uterus at blastocyst implantation. *Z. Anat. Entwicklungsgesch.* **144**, 337–342.

Ogasawara, Y., Okamoto, S., Kitamura, Y., and Matsumoto, K. (1983). Proliferative pattern of uterine cells from birth to adulthood in intact, neonatally castrated, and/or adrenalectomized mice, assayed by incorporation of [^{125}I] iododeoxyuridine. *Endocrinology (Baltimore)* **113**, 582–587.

Ohta, Y. (1981). Development of uterine ability of deciduoma formation in response to trauma in rats during neonatal life. *Annot. Zool. Jpn.* **54**, 1–9.

Ohta, Y. (1982). Deciduoma formation in rats ovariectomized at different ages. *Biol. Reprod.* **27**, 308–311.

Ohta, Y. (1983). Deciduoma formation in persistent estrous rats produced by neonatal androgenization. *Biol. Reprod.* **29**, 93–98.

Ohta, Y. (1984). Decidual cell reaction in ovariectomized-adrenalectomized rats. *Experientia* **40**, 505–506.

Ohta, Y. (1985a). Deciduomal response in prepuberal rats adrenalectomized-ovariectomized at different ages of early postnatal life. *Zool. Sci.* **2**, 89–93.

Ohta, Y. (1985b). Histochemical localization of prostaglandin synthetase in the rat endometrium with reference to decidual cell reaction. *Proc. Jpn. Acad.* **61**, 467–470.

Ohta, Y. (1987). Age-related decline in deciduogenic ability of the rat uterus. *Biol. Reprod.* **37**, 779–785.

Ohta, Y., and Iguchi, T. (1976). Development of the vaginal epithelium showing estrogen-independent proliferation and cornification in neonatally androgenized mice. *Endocrinol. Jpn.* **23**, 333–340.

Ohta, Y., and Iguchi, T. (1994). Effect of a single injection of estrogen in combination with progesterone treatment on rat uterine epithelial cells. *Zool. Sci.* Suppl., **18**.

Ohta, Y., and Takewaki, K. (1974). Difference in response to uterine trauma between androgen- and estrogen-sterilized rats ovariectomized and given injections of progesterone plus estradiol. *Proc. Jpn. Acad.* **50**, 648–652.

Ohta, Y. and Takewaki, K. (1987). Histochemical localization of adenylate cyclase in hormonally sensitized rat endometrium: Effect of application of deciduogenic stimulus. *Proc. Jpn. Acad.* **63**, 357–360.

Ohta, Y., Iguchi, T., and Takasugi, N. (1989). Deciduoma formation in rats treated neonatally with the anti-estrogens, Tamoxifen and MER-25. *Reprod. Toxicol.* **3**, 207–212.

Ohta, Y., Sato, T., and Iguchi, T. (1993a). Immunocytochemical localization of progesterone receptor in the reproductive tract of adult female rats. *Biol. Reprod.* **48**, 205–213.

Ohta, Y., Suzuki, A., and Iguchi, T. (1993b). The developmental pattern of progesterone receptor expression in rat uterus. *Zool. Sci., Suppl.*, p. 128.

Parr, M. B. (1983). Relationship of uterine closure to ovarian hormones and endocytosis in the rat. *J. Reprod. Fertil.* **68**, 185–188.

Parr, M. B., and Parr, E. L. (1974). Uterine luminal epithelium: Protrusions mediate endocytosis, not apocrine secretion, in the rat. *Biol. Reprod.* **11**, 220–233.

Parr, M. B., and Parr, E. L. (1989). The implantation reaction. *In* "Biology of the Uterus" (R. M. Wynn and W. P. Jollie, eds.), pp. 233–277. Plenum, New York.

Parr, M. B., Tung, H. M., and Parr, E. L. (1986). The ultrastructure of the rat primary decidual zone. *Am. J. Anat.* **176**, 423–436.

Parr, M. B., Parr, E. L., Munaretto, K., Clark, M. R., and Key, S. K. (1988). Immunohistochemical localization of prostaglandin synthase in the rat uterus and embryo during the peri-implantation period. *Biol. Reprod.* **38**, 333–343.

Peng, M.-T., and Peng, Y.-M. (1973). Changed in the uptake of tritiated estradiol in the hypothalamus and adenohypophysis of old female rats. *Fertil. Steril.* **24**, 534–539.

Phillips, C. A., and Poyser, N. L. (1981). Studies on the involvement of prostaglandins in implantation in the rat. *J. Reprod. Fertil.* **62**, 73–81.

Pollard, R. M., and Finn, C. A. (1972). Ultrastructure of the uterine epithelium during the hormonal induction of sensitivity and insensitivity to a decidual stimulus in the mouse. *J. Endocrinol.* **55**, 293–298.

Pollard, R. M., and Finn, C. N. (1974). The effect of ovariectomy at puberty on cell proliferation and differentiation in the endometrium of the aged mouse. *Biol. Reprod.* **10**, 74–77.

Price, D., and Ortiz, E. (1965). The role of fetal androgen in sex differentiation in mammals. *In* "Organogenesis" (R. L. DeHaan and H. Ursprung, eds.), pp. 629–652. Holt, Rinehart & Winston, New York.

Psychoyos, A. (1973a). Endocrine control of egg implantation. *In* "Handbook of Physiology" (R. O. Greep, E. B. Astwood, and S. R. Geiger, eds.), Sect. 7, Vol. 2, Part 2, pp. 187–215. Williams & Wilkins, Baltimore.

Psychoyos, A. (1973b). Hormonal control of ovoimplantation. *Vitam. Horm. (N.Y.)* **31**, 201–256.

Psychoyos, A. (1986). Uterine receptivity for nidation. *Ann. N. Y. Acad. Sci.* **476**, 36–42.

Rabii, J., and Ganong, W. F. (1976). Responses of plasma estradiol and plasma LH to ovariectomy, ovariectomy plus adrenalectomy, and estrogen injection at various ages. *Neuroendocrinology* **20**, 270–281.

Rahima, A., and Soderwall, A. L. (1977). Uterine collagen content in young and senescent pregnant golden hamsters. *J. Reprod. Fertil.* **49**, 161–162.

Rankin, J. C., Ledford, B. E., and Baggent, B. (1977). Early involvement of cyclic nucleotides in the artificially stimulated decidual cell reaction in the mouse uterus. *Biol. Reprod.* **17,** 549–554.

Rankin, J. C., Ledford, B. E., Jonsson, H. T. J., and Baggett, B. (1979). Prostaglandins, indomethacin and the decidual cell reaction in the mouse uterus. *Biol. Reprod.* **20,** 399–404.

Rankin, J. C., Ledford, B. E., and Baggett, B. (1981). The role of prostaglandins and cyclic nucleotides in artificially stimulated decidual cell reaction in the mouse uterus. *In* "Cellular and Molecular Aspects of Implantation" (S. R. Glasser and D. W. Bullock, eds.), pp. 428–430. Plenum, New York.

Reiter, R. J. (1969). Stratified squamous metaplasia of the uterine epithelium in early androgen-treated rats and its inhibition by light deprivation. *Anat. Rec.* **164,** 479–488.

Rodriguez, P., Fernandezgalaz, C., and Tejero, A. (1993). Controlled neonatal exposure to estrogen: A suitable tool for reproductive aging studies in the female rat. *Biol. Reprod.* **49,** 387–392.

Rogers, P. A. W. (1992). Early endometrial microvasular response during implantation in the rat. *Reprod. Fertil. Dev.* **4,** 261–264.

Rudel, H. W., and Kincl, F. A. (1966). The biology of anti-fertility steroids. *Acta Endocrinol. (Copenhagen), Suppl.* **105,** 1–45.

Saiduddin, S., and Zassenhaus, H. P. (1979). Estrous cycles, decidual cell response and uterine estrogen and progesterone receptor in fischer 344 virgin aging rats (40503). *Proc. Soc. Exp. Biol. Med.* **161,** 119–122.

Salazar-Rubio, M., Gil-Recasens, M. E., Hicks, J. J., and Gonzalez-Angulo, Y. A. (1980). High resolution cytochemical study of uterine epithelial cell surface of the rat at identified sites previous to blastocyst-endometrial contact. *Arch. Invest. Méd.* **11,** 117–127.

Sananès, N., and Le Goascogne, C. (1976). Decidualisation in the prepuberal rat uterus. *Differentiation (Berlin)* **5,** 133–144.

Sananès, N., and Psychoyos, A. (1974). Cytochemical localization of adenyl cyclase in the rat uterus. *J. Reprod. Fertil.* **38,** 181–183.

Sananès, N., Baulieu, E. E., and Le Goascogne, C. (1976). Prostaglandin(s) as inductive factor of decidualization in the rat uterus. *Mol. Cell. Endocrinol.* **6,** 153–158.

Sananès, N., Baulieu, E. E., and Le Goascogne, C. (1980). Treatment of neonatal rats with progesterone alters the capacity of the uterus to from deciduomata. *J. Reprod. Fertil.* **58,** 271–273.

Sananès, N., Bauliew, E.-E., and Le Goascogne, C. (1981). A role for prostaglandins in decidualization of the rat uterus. *J. Endocrinol.* **89,** 25–33.

Sanders, R. B., Bekairi, A. M., Abulaban, F. S., and Yochim, J. M. (1986). Uterine adenylate cyclase in the rat: Response to a decidual-inducing stimulus. *Biol. Reprod.* **35,** 100–105.

Sato, T., Okamura, H., Ohta, Y., Hayashi, S., Takamatsu, Y., Takasugi, N., and Iguchi, T. (1992). Estrogen receptor expression in the genital tract of female mice treated Neonatally with diethylstilbestrol. *In Vivo* **6,** 151–156.

Shanbhag, A. B., and Maqueo, M. (1986). Reproduction during initial stages of puberty in neonatally androgenized rats. *Indian J. Exp. Biol.* **24,** 217–223.

Shanbhag, A. B., Nadkarni, V. B., and Nevagi, S. A. (1987). Effect of neonatal androgenization on parturition, lactation and on subsequent fertility in rats. *Indian J. Exp. Biol.* **25,** 5–10.

Shapiro, M., and Talbert, G. B. (1974). The effect of maternal age on decidualization in the mouse. *J. Gerontol.* **29,** 145–148.

Shinoda, K. (1994). Brain aromatization and its associated structures. *Endocrinol. Jpn.* **41,** 115–138.

Shyamala, G., Mori, T., and Bern, H. E. (1974). Nuclear and cytoplasmic oestrogen receptors in vaginal and uterine tissue of mice treated neonatally with steroids an prolactin. *J. Endocrinol.* **63,** 275–284.

Soriero, A. A. (1978). The aging uterus and fallopian tubes. In "The Aging Reproduction System" (E. L. Schneider, ed.), pp. 85–127. Raven Press, New York.

Spicer, S. S. (1960). A correlative study of the histochemical properties of rodent acid mucopulysaccharides. J. Histochem. Cytochem. 8, 18–35.

Suzuki, A., Fukazawa, Y., Nishimura, N., Iguchi, T., and Ohta, Y. (1994). Uterine response to ovarian steroids in neonatally androgenized rats. Zool. Sci. Suppl., 18.

Swanson, H. E., and van der Werff ten Bosch, J. J. (1964). The early-androgen syndrome; differences in response to pre-natal and post-natal administration of various doses of testosterone propionate in female and male rats. Acta Endocrinol. (Copenhagen) 47, 37–50.

Tachi, C., and Tachi, S. (1974). Cellular aspects of ovum implantation and decidualization in the rat. In "Physiology and Genetics of Reproduction, Part B" (E. Coutinho and F. Fuchs, eds.), pp. 263–286. Plenum, New York.

Tachi, C., and Tachi, S. (1976). Cellular aspects of ovum implantation in the rat. Gunma Symp. Endocrinol. 13, 159–178.

Tachi, C., Tachi, S., and Lindner, H. R. (1972). Modification by progesterone of oestradiol-induced cell proliferation, RNA synthesis and oestradiol distribution in the rat uterus. J. Reprod. Fertil. 31, 59–76.

Tachi, S., Tachi, C., and Lindner, H. R. (1970). Ultrastructural features of blastocyst attachment and trophoblastic invasion in the rat. J. Reprod. Fertil. 21, 37–56.

Takasugi, N. (1954). Einflüsse von Androgen und Progestogen auf die Ovarien der Ratten denen sofort nach der Geburt Oestrogeninjektion durchgeführt wurde. J. Fac. Sci., Imp. Univ. Tokyo Sect. 4 7, 153–159.

Takasugi, N. (1966). Persistent changes in vaginal epithelium in mice induced by short-term treatment with estrogen beginning at different early postnatal ages. Proc. Jpn. Acad. 42, 151–155.

Takasugi, N. (1970). Testicular damages in neonatally estrogenized adult mice. Endocrinol. Jpn. 17, 277–281.

Takasugi, N. (1976). Cytological basis for permanent vaginal changes in mice treated neonatally with steroid hormones. Int. Rev. Cytol. 44, 193–224.

Takasugi, N. (1979). Development of permanently proliferated and cornified vaginal epithelium in mice treated neonatally with steroid hormones and the implication in tumorigenesis. Natl. Cancer Inst. Monogr. 51, 57–66.

Takasugi, N., and Furukawa, M. (1972). Inhibitory effect of androgen on induction of permanent changes in the testis by neonatal injections of estrogen in mice. Endocrinol. Jpn. 19, 417–422.

Takewaki, K. (1962). Some aspects of hormonal mechanism involved in persistent estrus in the rat. Experientia 18, 1–6.

Takewaki, K. (1965). Hormone-independent persistent changes in reproductive organs in female rats induced by early postnatal treatment with androgen. Proc. Jpn. Acad. 41, 310–315.

Takewaki, K. (1966). Uteri of adult rats given androgen injections for thirty neonatal days. Annot. Zool. Jpn. 39, 173–178.

Takewaki, K. (1968). Reproductive organs and anterior hypohysis of neonatally androgenized female rats. Sci. Rep. Tokyo Woman's Christ. Coll., 31–47.

Takewaki, K., and Ohta, Y. (1974). Altered sensitivity of uterus to progesterone-estradiol in rats treated neonatally with androgen and ovariectomized as adults. Endocrinol. Jpn. 21, 343–347.

Takewaki, K., and Ohta, Y. (1975a). Changes in uterine sensitivity in rats treated with testosterone propionate at different ages of early postnatal life. Proc. Jpn. Acad. 51, 664–668.

Takewaki, K., and Ohta, Y. (1975b). Deciduoma formation in rats ovariectomized and androgenized during neonatal life. *Endocrinol. Jpn.* **22**, 79–82.

Takewaki, K., and Ohta, Y. (1975c). Decidualization in rats given 5α-dihydrotestosterone or its propionate neonatally. *Endocrinol. Jpn.* **22**, 516–565.

Takewaki, K., and Ohta, Y. (1976). Deciduoma formation in rats with cornified vagina. *Experientia* **32**, 224–225.

Terenius, L., Meyerson, B. J., and Palis, A. (1969). The effect of neonatal treatment with 17β-estradiol or testosterone on the binding of 17β-estradiol by mouse uterus and vagina. *Acta Endocrinol. (Copenhagen)* **62**, 671–678.

Thorpe, L. W., Connors, T. J., and Soderwall, A. L. (1974). Closure of the uterine lumen at implantation in senescent golden hamster. *J. Reprod. Fertil.* **39**, 29–32.

Thrower, S., White, J. O., and Lim, L. (1978). The effects of neonatal administration of testosterone ('androgenization') on sex-hormone receptors in the hypothalamus and uterus of the adult female rat. *Biochem. Soc. Trans.* **6**, 1312–1314.

Tobert, J. A. (1976). A study of the possible role of prostaglandins in decidualization using a nonsurgical method for the instillation of fluids into the rat uterine lumen. *J. Reprod. Fertil.* **47**, 391–393.

Tuohimaa, P., and Johansson, R. (1971). Decreased estradiol binding in the uterus and anterior hypothalamus of androgenized female rats. *Endocrinology (Baltimore)* **88**, 1159–1164.

Uilenbroek, J. J., and van der Werff ten Bosch, J. J. (1972). Ovulation induced by pregnant mare serum gonadotrophin in the immature rat treated neonatally with a low or a high dose of androgen. *J. Endocrinol.* **55**, 533–541.

Uilenbroek, J. T. J., Arendsen de Walff-Exatlo, E., and Blankenstein, M. A. (1976). Serum gonadotrophins and follicular development in immature rats after early androgen administration. *J. Endocrinol.* **68**, 461–468.

van der Schoot, P., and de Greef, W. J. (1983). Regulation of luteal activity in adult female rats treated neonatally with testosterone propionate. *J. Endocrinol.* **96**, 417–425.

van der Schoot, P., van der Vaart, P. D. M., and Vreeburg, J. T. M. (1976). Masculinization in male rats is inhibited by neonatal injections of dihydrotestosterone. *J. Reprod. Fertil.* **48**, 385–387.

van der Werff ten Bosch, J. J., Tuinebreijer, W. E., and Vreeburg, J. M. (1971). The incomplete or delayed early-androgen syndrome. *In* "Hormones in Development" (M. Hamburg, M. X. Zarrow, and D. L. Quinn, eds.), pp. 669–675. Appleton-Century-Crofts, New York.

Vaticón, M. D., Fernández-Galaz, M. C., Tejero, A., and Aguilar, E. (1985). Alteration of prolactin control in adult rats treated neonatally with sex steroids. *J. Endocrinol.* **105**, 429–433.

Velardo, J. T., Dawson, A. B., Olsen, A. G., and Hisaw, F. L. (1953). Sequence of histological changes in the uterus and vagina of the rat during prolongation of pseudopregnancy associated with the presence of deciduomata. *Am. J. Anat.* **93**, 273–305.

Velardo, J. T., Hisaw, F. L., and Bever, A. T. (1956). Inhibitory action of deoxycorticosterone acetate, cortisone acetate, and testosterone on uterine growth induced by 17β-estradiol. *Endocrinology (Baltimore)* **59**, 165–169.

Watanobe, H., Sasaki, S., and Takebe, K. (1991a). A comparative study of the effects of neonatal androgenization and estrogenization on prolactin secretion in adult female rats. *Regul. Pept.* **34**, 149–158.

Watanobe, H., Sasaki, S., and Takebe, K. (1991b). Neonatal androgenization increases vasoactive intestinal peptide levels in rat anterior pituitary: Possible involvement of vasoactive intestinal peptide in the neonatal androgenized-induced hyperprolactinemia. *J. Endocrinol. Invest.* **14**, 875–879.

Watson, J., Anderson, F. B., Alam, M., O'Grady, J. E., and Heald, P. J. (1975). Plasma hormones and pituitary luteinizing hormone in the rat during the early stages of pregnancy and after post-coital treatment with tamoxifen. *J. Endocrinol.* **65,** 7–17.

Weisz, J., and Gunsalus, P. (1973). Estrogen levels in immature female rats true or spurious-ovarian or adrenal? *Endocrinology (Baltimore)* **93,** 1057–1065.

White, J. O., Moore, P. A., Elder, M. G., and Lim, L. (1981). The relationships of the oestrogen and progestin receptors in the abnormal uterus of the adult anovulatory rat: Effects of neonatal treatment with testosterone propionate or clomiphene citrate. *Biochem. J.* **196,** 557–565.

Wrenn, J., Wood, J., and Bitman, J. (1969). Oestrogen responses of rats neonatally sterilized with steroids. *J. Endocrinol.* **45,** 415–420.

Yamada, A. T., and Nagata, T. (1993). Light and electron microscopic radioautographic studies on the RNA synthesis of peri-implanting pregnant mouse uterus during a activation of receptivity for blastocyst implantation. *Cell. Mol. Biol.* **39,** 221–233.

Yamashita, S., Newbold, P. R., McLachlan, J. A., and Korach, K. S. (1989). Developmental pattern of estrogen receptor expression in female mouse genital tracts. *Endocrinology (Baltimore)* **125,** 2888–2896.

Yee, G. M., and Kennedy, T. G. (1993). Prostaglandin E2, cAMP and cAMP-dependent protein kinase isozymes during decidualization of rat endometrial stromal cells in vitro. *Prostaglandins* **46,** 117–138.

Yee, P. M., Squires, P. M., Cejic, S. S., and Kennedy, T. G. (1993). Lipid mediators of implantation and decidualization. *J. Lipid Mediators* **6,** 525–534.

Ying, S.-Y. (1973). Induction of ovulation in rats treated neonatally with androgen. *Proc. Soc. Exp. Biol. Med.* **144,** 822–825.

Yochim, J. M., and DeFeo, V. J. (1963). Hormonal control of the onset, magnitude and duration of uterine sensitivity in the rat by steroid hormones of the ovary. *Endocrinology (Baltimore)* **72,** 317–326.

Yoshinaga, K. (1989). Uterine receptivity for blastocyst implantation. *Ann. N. Y. Acad. Sci.* **541,** 424–431.

Zeilmaker, G. H. (1964). Aspects of the regulation of corpus luteum function in androgen-sterilized female rats. *Acta Endocrinol. (Copenhagen)* **46,** 571–579.

Mechanistic and Developmental Aspects of Genetic Imprinting in Mammals

Keith E. Latham*, James McGrath†, and Davor Solter§
* Fels Institute for Cancer Research and Molecular Biology and the
Department of Biochemistry, Temple University School of Medicine,
Philadelphia, Pennsylvania 19140, † Department of Genetics and Pediatrics,
Yale University Medical Center, New Haven, Connecticut 06511, and § Max
Planck Institute for Immunobiology, D-79108 Freiburg, Germany

Genetic imprinting in mammals allows the recognition and differential expression of maternal and paternal alleles of certain genes. Recent results from a number of laboratories indicate that, at least for some genes, gametic imprints, which must exist in order to mark chromosomes or genes as having been transmitted via sperm or ovum, are not by themselves sufficient to determine allele expression. Other postfertilization events are required, and these events are subject to both tissue-specific and developmental stage-specific regulation. Changes in imprinted gene methylation during preimplantation and fetal life indicate that the establishment of additional allele-specific modifications is likely to contribute to imprinted gene regulation. Disruptions in imprinting processes, loss of imprints, and loss of nonimprinted alleles through uniparental disomy are likely to contribute to a variety of developmental abnormalities and pathological conditions in both mice and humans.

KEY WORDS: Genome imprinting, Gene regulation, Allele inactivation, Mouse Embryo, Cancer.

I. Introduction

The phenotypes of eukaryotic cells from organisms as diverse as yeasts and humans are determined by a combination of two types of information passed through the nucleus: genetic information in the form of DNA base sequence and epigenetic information in the form of DNA modifications or chromosome structure that influence gene expression or, in some cases, gene content. An understanding of the mechanisms by which epigenetic information, which has come to be known as genetic imprinting, affects gene function remains elusive.

Genetic imprinting was first proposed to explain the regulated expulsion of paternal X chromosomes during development of *Sciara* embryos (Crouse, 1960). In this organism, newly fertilized embryos possess three pairs of autosomes and three X chromosomes. Two of the X chromosomes are donated by the sperm. During development, somatic cells eliminate one or two paternal X chromosomes if the embryo is to develop as female or male, respectively. Cells of the germ lineage eliminate the paternal X chromosome after reaching the definitive gonad. Thus, cells of developing *Sciara* are able to distinguish between X chromosomes and autosomes and between maternal and paternal X chromosomes. In monogenic lines of *Sciara,* females produce either female or male progeny, depending upon the genetic composition of their maternal X. Consequently, sexual differentiation, and the fate of paternal X chromosomes, is influenced by maternal genome expression following fertilization. Thereafter, the paternal X chromosomes are treated differentially, depending upon maternal genome expression.

The significance of genetic imprinting in mammals was made clear through two independent lines of evidence. Chief among these was the series of nuclear transplantation experiments by McGrath and Solter (1983, 1984a), and later Surani and co-workers (1984, 1986; Barton *et al.,* 1984), in which embryos bearing two maternal or two paternal genomes were shown to be nonviable. Other experiments (McGrath and Solter, 1984b), in which maternal lethal-specific T^{hp} pronuclei were transplanted to wild-type cytoplasm, also indicated the existence of genetic imprinting. Nuclear transplantation experiments with parthenogenetic embryos further supported this conclusion (Mann and Lovell-Badge, 1984).

Independent evidence for imprinting was provided by Cattanach and co-workers (Cattanach and Kirk, 1985; Cattanach, 1986, 1989; Cattanach and Beechey, 1990), who analyzed the progeny of matings among mice bearing Robertsonian or balanced translocations and demonstrated that both a maternal and a paternal copy of certain chromosomes or chromosome regions was required for embryo viability. These genetic studies complemented the nuclear transplantation experiments by eliminating the possibility that embryo lethality resulted from removal or transfer of nongenetic cytoplasmic or perinuclear components of the early embryo.

Regulation of the X chromosome in *Sciara* exemplifies the major features of genetic imprinting. First, homologous maternal and paternal chromosoms are imprinted or "marked," presumably during gametogenesis, so that the parent of origin can be distinguished. Second, the imprint is stable and heritable through subsequent cell divisions. Third, the imprint results in differential gene function, and in the case of *Sciara,* gene content. Last, the imprint is reversible so that during gametogenesis of each subsequent generation the appropriate imprint is reestablished in accordance with the sex of the developing individual. This chapter discusses the possible molecular mechanisms for marking and inactivating parental alleles.

II. Temporal Aspects of Imprinting

A. Imprinting during Gametogenesis

If maternal and paternal homologs are marked so that a cell can distinguish between the two, this is most likely to occur during gametogenesis. Alternatively, it is possible that imprinting could occur shortly after fertilization when the haploid genomes are physically separate. Another possibility is that maternal genes may be imprinted during oogenesis by factors that are not present after fertilization while naive paternal genes might be modified postfertilization by factors that are unable to displace regulatory factors bound to the maternal gene. The available evidence indicates that imprinted alleles are marked during gametogenesis to distinguish parental origin but that postfertilization events also contribute to the imprinted phenotype.

Forejt and Gregorova (1992) demonstrated that the lethality associated with maternal inheritance of the T^{hp} deletion is not seen when T^{hp} *Mus m. domesticus* females are crossed with *Mus m. musculus* males. This indicated that the two subspecies differ in the activity of an imprinting gene, termed *Imprintor-1*. To test whether the T^{me} locus is imprinted during gametogenesis or postfertilization, matings were performed in which the maternal T^{me} deletion was alternately inherited with a maternal or paternal *Mus m. musculus Imp-1* allele. If *Imp-1* acts postfertilization, inheritance of the paternal *Imp-1* should produce viable progeny. No such progeny were observed (Forejt and Gregorova, 1992), indicating that *Imp-1* acts during gametogenesis. An alternative explanation, however, acknowledged by Forejt and Gregorova (1992), is that *Imp-1* is itself imprinted. Thus, reduced expression in *Mus m. musculus* of another imprinted gene product that interacts with the T^{me} gene product (i.e., IGF2R) could result in viability because the presumed absence of IGF2R could be compensated for by the reduced activity of the imprinted *Igf2r* interactive gene. Forejt and Gregorova also concluded that the T^{me} locus is distinct from the *Igf2r* locus since mice with maternal inheritance of the T^{hp} *M. m. domesticus* deletion are viable but maternal deletion of T*hp* *M. m. musculus* deletion leads to fetal lethality while *Igf2r* is not expressed in either instance. Recent results indicating that a lack of *Igf2r* expression can be compensated for by the absence of *Igf2* expression (Filson *et al.*, 1993), however, raise the possibilities that in *M. m. domesticus* x *M. m. musculus* hybrids the *Igf2* gene is underexpressed, the *Igf2r* gene is more highly expressed, or hybrid mice are less sensitive to the imbalance between these two products. Such possibilities could result in viable mice regardless of the T*hp* maternal deletion.

The *Igf2r* gene has been disrupted by gene knockout with the result that IGF2R-deficient mice survive to birth and therefore do not exhibit the fetal lethality typical for T*me* (Lau *et al.*, 1994; Wang *et al.*, 1994). Imprinting of the Igf2r locus is confirmed by the fact that heterozygotes inheriting the disrupted *Igf2r* gene maternally show essentially the same phenotype as homozygous null progeny. The *Igf2r* knockout mice do, however, exhibit a 30% larger size at birth, as well as organ and skeletal abnormalities and mis-sorting of mannose-6-phosphate tagged proteins (Wang *et al.*, 1994). Most mutant mice die shortly after birth. In some mice, the paternal *Igf2r* allele is re-activated in at least some tissues, allowing survival to adulthood. A higher level of serum *Igf2r* expression has been observed in the IGF2R-deficient mice (Lau *et al.*, 1994). These results indicate that *Igf2r* regulates growth and differentiation during embryogenesis. Although the *Igf2r* knockout mice exhibit some additional phenotypic abnormalities consistent with the T*me* phenotype (Wang *et al.*, 1994), the viability of such mice to birth leaves open the possibility that the T*me* phenotype is due in part to a second unidentified gene. Alternatively, it is possible that interstrain differences or intergenic interactions are responsible for the greater viability of the *Igf2r* knockout mice. Explanations that would invoke a compensatory decrease in IGF2 expression to offset the reduced IGF2R expression, however, would seem at odds with the apparent increase in IGF2 expression in the IGF2R-deficient mice. Resolution of the question of whether the *Igf2r* and T*me* genes are identical awaits the determination of whether an additional gene exists in the T*me* region and is essential for mid-gestational viability. Resolution of this question will also clarify whether the *M. m. musculus* and *M. m. domesticus* strains imprint one or both of these genes differently.

Differences in parental gene imprinting may also contribute to the defective development of DDK strain mouse eggs fertilized by non-DDK sperm. In this case, differences between DDK and other strains in the direction of imprinting for some genes may produce embryos with two identically imprinted and inactivated alleles for these genes, creating a lethal lack of gene function (Sapienza *et al.*, 1992).

Other evidence that imprinting occurs during gametogenesis has been obtained from the analysis of methylation patterns of imprinted genes in mice. For example, one of the methylation sites in the *Igf2r* gene is maternally methylated in oocytes of the adult ovary, in embryos, and in adult tissues. This site is unmethylated, however, in primordial germ cells and oocytes of newborn mice, indicating that this site acquires its maternal-specific methylation status during oogenesis at some time after birth (Stöger *et al.*, 1993; Brandeis *et al.*, 1993).

Studies of parent-specific methylation patterns of transgene DNA further support the occurrence of imprinting during gametogenesis. Chaillet *et al.* (1991) demonstrated that maternal and paternal methylation patterns of two Rsv-Ig-myc transgenic constructs are erased in primordial germ cells and then reestablished during gametogenesis. Interestingly, while the maternal pattern is completely reestablished during oogenesis, establishment of the paternal methylation pattern continues until beyond the blastocyst stage, indicating that while differential methylation of parental transgenes indeed occurs during gametogenesis, further parental allele-specific modification occurs postfertilization. Analyses of another transgene, MPA434 (Ueda *et al.*, 1992), also revealed differences in methylation that are established during gametogenesis, although the relatively insensitive polymerase chain reaction-based assay that was used could not distinguish between partial or complete hypomethylation in sperm DNA. Overall, these studies of transgene methylation also indicate that imprinting occurs during gametogenesis.

B. Genome Modifications by the Egg Cytoplasm

Although these data support genome imprinting during gametogenesis, it is also possible that maternal and paternal genomes can be differentially modified immediately postfertilization and prior to syngamy. Several lines of experimentation in mice reveal the potential importance of interactions between pronuclei and egg cytoplasm.

One notable instance where such interactions occur is in breeding experiments between the DDK mouse strain and other strains. DDK females bred to non-DDK males produce nonviable progeny while the reciprocal cross produces viable offspring (Wakasugi *et al.*, 1967; Wakasugi, 1974; Sapienza *et al.*, 1992; Babinet *et al.*, 1990). As described earlier, some of the inviability of embryos produced by DDK x non-DDK matings may relate to differences in the direction of imprinting for some genes (Sapienza *et al.*, 1992). Nuclear transplantation experiments, however, also indicate that the paternally derived non-DDK genome is negatively modified by the DDK egg cytoplasm. Transplantation of both pronuclei from DDK x BALB/c or DDK x C57BL/6 1-cell embryos to BALB/c or C57BL/6 cytoplasm, respectively, results in normal development (Mann, 1986; Babinet *et al.*, 1990). Transplantation of both pronuclei from BALB/c x DDK embryos to enucleated DDK eggs produces viable progeny, which is consistent with the F_1 matings and indicates that the DDK egg effect is specific for non-DDK male pronuclei (Babinet *et al.*, 1990). Transfer of DDK cytoplasm into non-DDK eggs also produced some reductions in embryo viability (Babinet *et al.*, 1990). These data are consistent with a negative effect of the DDK egg cytoplasm on non-DDK male pronuclei. The putative target for this modification has been mapped to chromosome 11 (Sapienza *et al.*, 1992; Baldacci *et al.*, 1992).

In a more recent study to investigate the molecular basis for this effect of the DDK egg cytoplasm on male pronuclei, it was reported that purified RNA from DDK eggs could negatively affect embryonic development when it was transferred to C57BL/6 homozygous embryos, but not to C57BL/6 x DDK embryos (Renard *et al.*, 1994). This effect could be achieved even using recipients as advanced as the 4-cell stage. It should be noted, however, that the manipulations executed in this study were complex and no control for enhanced viability of F_1 embryos due to hybrid vigor were available. The authors of this study concluded that the DDK egg effect is mediated by an RNA that is present in the egg and that this effect can be achieved at least through the 4-cell stage. They also concluded that these experiments exclude a possible role for imprinting effects. The latter conclusion, however, should not be taken as definitive, since the relevant egg factor may participate in postfertilization mechanisms that affect the expression of genetic imprints (see later discussion).

One aspect of the available DDK data that has not been addressed is the apparent discrepancy between the results of certain microsurgical experiments and the genetic data. The microsurgical experiments mentioned earlier point to a cytoplasmic component of the DDK egg that modifies male pronuclear function, and the recent study by Renard *et al.* (1994) further indicates that this is the result of a maternal RNA deposited in the egg at some point during oogenesis. The genetic data, however, indicate that only half of the offspring from DDKxC57BL/6 F_1 females mated to C57BL/6 males die (Wakasugi, 1974; Sapienza *et al.*, 1992). If the DDK genome encodes an RNA that is deposited in the egg during oogenesis and later affects male pronuclear function, then all of the eggs from the F_1 females should contain this RNA. The viability of 50% of the offspring, however, indicates that this probably is not the case. This raises the possibility that the results obtained in the microsurgical studies reflect embryo lethality related to some other cause. Alternatively, the RNA encoded by the DDK genome may be produced after first polar body extrusion and before the end of the first mitotic cycle.

In addition to interactions between DDK egg cytoplasm and non-DDK paternal genomes, the expression of several transgenes can be influenced by strain-specific modifiers (Sapienza et al., 1989; McGowan et al., 1989; Allen et al., 1990; Surani et al., 1990). One transgene, for example, is differentially expressed at the 2-cell stage when transgene-bearing sperm fertilize eggs of different strains (Allen et al., 1990; Surani et al., 1990), indicating that paternal transgene expression is affected by the egg cytoplasm. A paternally transmitted CMZ 12-lacZ fusion transgene is highly expressed following fertilization of DBA/2 eggs but repressed after fertilization of BALB/c eggs. Expression is also repressed in transgenic females mated to DBA/2 males, indicating that enhanced expression requires DBA/2 ooplasm. Another transgene, TKZ 751, becomes methylated and repressed when transgenic males fertilize BALB/c eggs but not DBA/2 eggs. Repression is not observed when transgenic females are mated to BALB/c males; that is, repression is not due to genetic incompatibilities between genomes but rather to the expression of a maternally transmitted BALB/c modifier. Analyses using reciprocal F_1 hybrid mice indicate that the BALB/c effect is attributable to a single gene or locus which can act zygotically (Allen et al., 1990). Repression is also observed when TKZ 751 transgenic C57BL/6 eggs are fertilized by BALB/c males (Surani et al., 1990). This most likely reflects interactions between maternal and paternal genomes during early embryogenesis. Since the BALB/c modifier affects TKZ751 transgene expression only when maternally transmitted, however, it is possible that the BALB/c modifier interacts with factors in the egg cytoplasm as well as postzygotically.

Striking differences in interactions between egg cytoplasm and pronuclei are also observed for the C57BL/6 and DBA/2 strains. The first demonstration of maternal strain-dependent differences in gene expression for these two strains came from analyses of the expression of the pHRD transgene, which encodes a target sequence for the VDJ recombinase (Engler et al., 1991). For this transgene, the degree of methylation, and hence recombination, is determined by the genotype of the transmitting mother. Passage through the C57BL/6 strain results in a greater degree of methylation while passage through the DBA/2 strain produces undermethylated DNA. It has not been determined whether this effect of maternal genotype on pHRD methylation reflects interactions between the egg cytoplasm and the paternal pronucleus or an interaction between the two genomes during later development.

Nuclear transfer experiments indicate that this strain difference between the C57BL/6 and DBA/2 strains may result from nucleocytoplasmic interactions in the zygote. The developmental potential of androgenetic embryos, for example, is much greater when eggs from C57BL/6 mothers are used than when DBA/2 eggs are used (Latham and Solter, 1991). In addition, transient exposure of C57BL/6 male pronuclei to DBA/2 egg cytoplasm following fertilization compromises the ability of these nuclei to support development to the blastocyst stage, even when these nuclei are transplanted into C57BL/6 recipient eggs (Latham and Solter, 1991). Control embryos prepared by transplanting both maternal and paternal pronuclei to enucleated eggs of the opposite strain develop to the blastocyst stage at a high frequency. Thus, the DBA/2 cytoplasm appears able to permanently modify the paternal genome soon after fertilization (Latham and Solter, 1991). A study, in which the developmental capacities of androgenetic embryos produced with eggs of individual female offspring produced from a backcross of (B6D2)F1 females to DBA/2 males were assessed, indicated that the strain effect on androgenone development involves two independently segregating loci (Latham, 1994).

A more recent study compared the expression of major urinary proteins (MUPs) in the livers of C57BL/6 and DBA/2 adults and in nucleocytoplasmic hybrids prepared by exchanging maternal pronuclei between C57BL/6 and DBA/2 homozygous embryos. Homozygous C57BL/6 and DBA/2 mice expressed a subset of the four MUPs resolved. (C57BL/6xDBA/2)F_1 and (DBA/2 x C57BL/6)F_1 mice expressed all four MUPs. Nucleocytoplasmic hybrids, however, expressed very little of any of the MUPs (Reik *et al.,* 1993). This indicated that the egg cytoplasm can modify maternal pronuclei, resulting in altered gene expression in the adult. It should be noted, however, that *in vitro* culture alone had a similar effect in approximately 5% of the cases analyzed (Reik *et al.,* 1993).

The identity of egg modifiers remains to be determined. Candidate genes, however, may be sought on chromosomes 4 and 17. The gene mapped to chromosome 4 by Engler *et al.* (1991) has not been isolated. Other data indicate that a locus on chromosome 17 may be involved in the imprinting differences between the C57BL/6 and BALB/c strains (J. Walter, N. Allen, and W. Reik, personal communication). We can only speculate about the molecular nature of the factors encoded by these loci. One possibility is that these factors regulate changes in methylation. In this respect, the rapid loss of methyl groups from the paternal genome after fertilization, which is indicative of an active demethylase (Kafri *et al.,* 1992; see later discussion), is intriguing. It is also possible that these factors function analogously to the modifiers of position effect variegation in *Drosophila* (see later discussion).

Several murine genes have been identified that possess a *Drosophila* heterochromatin protein-1-like "chromobox" domain (Singh *et al.,* 1991;

Pearce *et al.*, 1992). At least one of these, the M31 gene, encodes an mRNA that is expressed in the egg and early embryo (K. Peterson, C. Sapienza, and K. Latham, unpublished). This gene maps to chromosome 11 (P. Singh, personal communication) and a homolog encoding a pseudogene is found on the X chromosome (Hamvas *et al.*, 1992). The *bmi*-1 proto-oncogene also shares homology with the *Drosophila Posterior sex combs* and *Su(z)2* genes (van Lohuizen *et al.*, 1991). Analyses of progeny produced by hybrids between C57BL/6 and DDK mice also point to a possible X-linked imprintor locus (C. Sapienza, personal communication).

Additional mouse genes encoding proteins that bind to methylated DNA have been cloned (Lewis *et al.*, 1992). Although a role for egg modifiers in modulating the expression of imprinted genes has not been demonstrated directly, the striking effects of DBA/2 egg cytoplasm on paternal pronuclei uncovered by the analysis of androgenetic embryos (Latham and Solter, 1991) is consistent with such an effect. Other instances where egg modifiers may contribute to phenotype include the effects of maternal strain on the malignancy of teratocarcinomas (Solter *et al.*, 1979, 1981) and the development of parthenogenetic cells in chimeras (Fundele *et al.*, 1991).

III. Mechanisms of Imprinted Gene Regulation

A. A Role for Sequential Events in Imprinted Gene Regulation

Two aspects must be accounted for in considering the molecular basis of genetic imprinting. First, a mechanism must exist for marking the DNA so that a cell can distinguish one parental allele from the other. Second, this "mark" must lead to allelic inactivation. Allelic inactivation might be either the immediate consequence of the mark or require additional processes. Available data from several experimental systems, including mammals, indicate that the latter is most often the case.

In yeast (*Saccharomyces pombe*), for example, mating-type switching, which occurs nearly every other generation, results from a chromosomal imprint that is not by itself sufficient to lead to mating-type switching but merely initiates a process that will be completed two generations later. The mating-type switches in *S. pombe* occur through a gene conversion event that is dependent upon and initiated by a double-stranded DNA break in the *mat-1* gene (Beach and Klar, 1984; Egel *et al.*, 1984; Klar, 1990a,b). The majority of haploid cells dividing mitotically produce two nonequivalent daughter cells. One daughter divides to produce two cells of the same parental cell mating type while the other produces one daughter of the same type and another of the opposite type (Klar, 1990a). This "one-in-four" pattern of inheritance is explicable by a DNA strand-segregation model as opposed to differential inheritance of cytoplasmic or nuclear factors. Since the newly switched cell can produce granddaughter cells that have reverted to the original mating type, the pattern of switching must result from epigenetic differences between the two grandparental DNA strands.

Klar's model (1990a) predicts that in the grandparent of the switched cell, one of the two DNA strands is marked or "imprinted" so that the daughter cell inheriting that strand executes a double-stranded cleavage at that site. At or immediately after DNA replication, gene conversion occurs so that the double-stranded break creates one switched, healed allele and one cleaved, unswitched allele. Thus, the parent of the switched cell inherited an imprinted DNA strand that had to be cleaved before it was a suitable substrate for gene conversion.

This requirement for two sequential epigenetic modifications, that is, imprinting and cleavage, may provide a useful mechanism for regulating the mating-type switching process in response to environmental conditions, since both steps probably require specific enzyme expression (Klar, 1990b). Similarly, the expulsion of one or both paternal X chromosomes in *Sciara* is dependent upon maternal factors, possibly in the egg cytoplasm, and is not an obligate chromosomal feature. In addition, the endogenously imprinted *Igf2* gene in mice escapes inactivation in selected tissues (DeChiara *et al.*, 1991) and both alleles of the *Igf2* and *Igf2r* genes can be expressed in the preimplantation embryo (Latham *et al.*, 1994). Thus, in three different organisms the initial imprint does not by itself determine gene expression, and other controlling mechanisms must operate in the correct temporal sequence. Moreover, the overwhelming pattern for gene regulation by imprinting is one in which cell-specific factors dictate the fate of imprinted genes. Any mechanism that attempts to explain genetic imprinting must therefore include an active role for the cell in recognizing and acting upon the imprint.

B. Methylation and Imprinting

By far, the most often suggested mechanistic explanation for genetic imprinting is DNA methylation. The attractiveness of methylation as the primary mechanism of imprinting relates to the ease with which methylation can accommodate all of the major requirements for the imprinting process, namely, stability, heritability, reversibility, and an ability to affect gene function. Evidence that supports an integral role for methylation in regulating imprinted gene expression has come from studies of X chromosome inactivation, the differential methylation of transgenes, and the differential methylation of endogenous imprinted genes.

To achieve dosage compensation of genes on the X chromosome, female mammalian concepti must inactivate one of their two X chromosomes. This contrasts with the strategy adopted by other organisms, such as *Drosophila,* in which dosage compensation is achieved by modulation of the transcription rates of both X chromosomes (Gorman *et al.,* 1993, and references therein). Imprinting of the X chromosome is seen in marsupials, for which the paternal X is preferentially inactivated in both extraembryonic and somatic cells, and in eutherian mammals, for which the paternal X is inactivated preferentially in extraembryonic lineages (e.g., trophectoderm, yolk sac, parietal endoderm) (Vanderberg *et al.,* 1987; Wake *et al.,* 1976; Harper *et al.,* 1982; West *et al.,* 1977; Takagi and Sasaki, 1982; Riggs and Pfeifer, 1992). Differential expression of maternal and paternal X chromosomes is evident as early as the 8-cell stage (Moore and Whittingham, 1992; Singer-Sam *et al.,* 1992), although heterochromatization does not occur until just after implantation (West *et al.,* 1977; Harper *et al.,* 1982; Takagi and Sasaki, 1982). This suggests that even before heterochromatization the paternal X chromosomes are marked in such a way as to influence gene expression.

X chromosome inactivation is associated with DNA hypermethylation (Mohandes *et al.,* 1981). For some genes, X-inactivation occurs prior to methylation while for other genes methylation has been observed before or coincident with inactivation (Gautsch and Wilson, 1983; Monk, 1986; Singer-Sam *et al.,* 1990; Grant *et al.,* 1992; Kaslow and Migeon, 1987; Lock *et al.,* 1987; Riggs and Pfeifer, 1992). Given the limitations in techniques for analyzing allele-specific patterns of DNA methylation, these observations raise the question of whether allele inactivation requires extensive methylation of the X chromosome or whether a few key methylation sites are sufficient to inactivate X-linked genes, with additional methyl groups being added later to facilitate the establishment and maintenance of the heterochromatic state that is typical for the inactive chromosome. Given the expression of the paternal X chromosome at a reduced level relative to the maternal X chromosome at the 8-cell stage, it is likely that some differences in methylation are established during gametogenesis and confer upon the maternal X chromosome the ability to be preferentially expressed. Differential methylation of the *Xist* gene promoter has been reported (Norris *et al.,* 1994; Kay *et al.,* 1994).

1. Continuity of Methylation between Gametes and Embryonic Cells

Because of the probable role for methylation in contributing to genetic imprinting at some level, it is worthwhile to review what is known of the continuity of methylation between germ cells, gametes, and embryos. In order for methylation to serve as an imprinting signal, parental-specific methylation patterns must exist within the gamete, be propagated through embryonic somatic tissues, and be erased and replaced with the appropriate imprint within the germ lineage.

An early study that evaluated the overall degree of DNA methylation during development was performed by Monk and co-workers (1987). This study revealed that total DNA from primordial germ cells, oocytes, and early embryos is relatively undermethylated while sperm DNA is more highly methylated. A more recent study found that every non-CpG island site was fully methylated in mature sperm DNA but a portion of these sites remained unmodified in oocytes (Kafri *et al.*, 1992). As with the earlier study, specific CpG sites in primordial germ cells of the early fetus were unmethylated, but became methylated during late fetal life. Some underwent demethylation during spermatogenesis and oogenesis. The fully methylated sites detected in sperm DNA underwent complete demethylation during early embryogenesis and were unmethylated in both the trophectoderm and inner cell mass cells of the blastocyst (Kafri *et al.*, 1992).

A recent study demonstrated that methylated transgene constructs also became demethylated upon injection into zygotes, indicating that this demethylation was an active process rather than a passive failure to methylate DNA after replication (Kafri *et al.*, 1993). Interestingly, sites that are methylated only in sperm DNA, but not oocyte DNA, become demethylated sooner following fertilization than sites that are methylated in both sperm and oocyte, and it was suggested that this may reflect an active demethylase that is expressed in the oocyte (Kafri *et al.*, 1992). During later development, many of the CpG sites become remethylated in somatic tissues, although some sites within CpG islands fail to do so. Non-CpG island sites that become methylated during gastrulation may later become demethylated in a tissue-specific manner coincident with gene activation (Shemer *et al.*, 1991).

The extensive demethylation that occurs during early embryogenesis and involves both paternal and maternal alleles poses a problem for a role for methylation in imprinting. A recent study, however, revealed that specific sites within the endogenously imprinted *Igf2* and *Igf2r* genes are methylated in gametes and remain so during early embryogenesis (Brandeis *et al.*, 1993) (see later discussion). This result suggests that some methylated sites are protected from demethylation while others are not. A recent study attempted to test this site-specific protection by injecting an *Igf2* gene construct, in which sites 1 and 2 were methylated, into mouse zygotes (Kafri *et al.*, 1993). Site 1, which is methylated in both gametes and remains so in early embryos, escaped demethylation following injection. Site 2, however, which is only methylated in sperm, became demethylated after injection. This result is reminiscent of the earlier demethylation in zygotes of sites that are methylated only in sperm compared with sites that are methylated in both gametes (Kafri *et al.*, 1992). Although the experiment with injected *Igf2* gene constructs indicates that some CpG sites can escape demethylation, they do not provide experimental proof for specific maintenance of sites that are differentially methylated between sperm and egg. Thus, while abundant evidence exists to indicate that such parental-specific sites can be retained between gamete and early embryo (Brandeis *et al.*, 1993; Stöger *et al.*, 1993), experimental evidence relating to the mechanisms that allow this has not been obtained.

2. Methylation of Transgenes

The parental-specific methylation patterns of mouse transgenes have received a great deal of attention in recent years. The reason for this is that the study of transgenes might reveal endogenously imprinted genes at the insertion site or provide a model system for studying the imprinting process. Although these studies have provided some insight, their applicability to understanding endogenous gene imprinting has been called into question recently, owing to the perceived differences in the behavior between transgenes and endogenous imprinted genes and the striking preponderance of maternally imprinted transgenes. Several detailed reviews of transgene imprinting and methylation have been published (Solter, 1988; Reik *et al.*, 1990; Chaillet, 1992; Allen and Mooslehner, 1992; Pourcel, 1993; Efstratiadis, 1994).

One relevant feature of many transgenes that display some kind of imprinted behavior is the strong influence of the genetic background, so that imprinting is seen in some strains of mice but not others. If we consider imprinting as a mechanism of utmost importance for mammalian development, such strain dependence would argue that imprinting of transgenes is fortuitous and irrelevant for understanding imprinting of endogenous genes. This view is no longer tenable since a few endogenous imprinted genes, at least in humans, display imprinting as a polymorphic trait (Xu *et al.*, 1993; Jinno *et al.*, 1994). Imprinting of transgenes is observed primarily as differences in the degree of methylation, with maternal transmission resulting in hypermethylation and paternal transmission leading to hypomethylation. In the cases where transgenes are expressed, the active allele is always hypomethylated and paternally transmitted. There are several reasons why transgenes might display imprinting behavior and why the analysis of this behavior may be informative. It has been assumed that transgene imprinting reflects the imprinting status of the genome at the site of integration, with the transgenes providing suitable tags for endogenous imprinted regions. Though this is possible for some transgenes, this is not universally applicable. In one case where the integration site was cloned and examined, imprinting behavior was not observed (Sasaki *et al.,* 1991), although allele-specific variation in methylation was present.

An alternative explanation suggested for transgene methylation has been that the transgenes contain (accidentally or deliberately) sequences that can act as a target for the imprinting process, i.e., an "imprinting box." Several examples would fit this hypothesis, most notably the RSV-Ig-myc transgenic lines (Chaillet *et al.,* 1991; Chaillet, 1992; Chaillet, as quoted in Efstratiadis, 1994). In this case, all of the independent transgenic lines produced express the transgene when it is hypomethylated and paternally transmitted, but not when it is hypermethylated and maternally transmitted. Using overlapping deletions, a potential imprinting box has been identified (Chaillet, as quoted in Efstratiadis, 1994). Without the imprinting box, the transgene becomes hypermethylated and silent irrespective of the transmitting parent. These results are of great potential interest and it would be especially informative to determine whether the candidate imprinting box can function in the context of other, unrelated transgenes.

If the analysis of imprinted transgenes can lead to the identification of an "accidental" imprinting box, then it stands to reason that the production of transgenic mice using DNA elements from endogenous imprinted genes could reveal "real" imprinting boxes. Two such attempts have been reported with mixed, though somewhat hopeful results. Transgenic lines containing a complete or modified coding region of the rat *Igf2* gene were produced (Lee *et al.,* 1993) and one out of six lines showed modified expression of the paternally transmitted *Igf2* transgene. This indicates that transgene expression is predominantly affected by the integration site, that the transgene did not contain all of the necessary imprinting signals, and that a partial imprinting signal may suffice in some cases. Among transgenic lines containing modified *H19* transgenes (Bartolomei *et al.,* 1993), several lines maintained almost correct imprinting status of the transgene, although leakiness and genetic background effects were evident. These data are consistent with the possibility that the *H19* transgene contains its own imprinting signal but that this signal is either incomplete or, in some cases, the effects of chromosome position are overpowering. In summary, although the understanding of imprinting from these studies is far from complete, further analysis of some transgenic models might provide valuable information complementary to that emerging from the investigation of endogenous imprinted genes.

3. Methylation of Endogenous Imprinted Genes

The endogenously imprinted *Igf2, Igf2r, H19, Snrpn,* and *Xist* genes exhibit allele-specific methylation patterns in fetal and adult tissues (Sasaki *et al.,* 1992; Stöger *et al.,* 1993; Bartolomei *et al.,* 1993; Brandeis *et al.,* 1993; Ferguson-Smith *et al.,* 1993; Glenn *et al.,* 1993; Norris *et al.,* 1994). Several recent studies (Brandeis *et al.,* 1993; Sasaki *et al.,* 1992) have provided detailed information regarding the sites of parental-specific methylation in these genes as well as the time during development at which these sites become methylated. In addition, results obtained with 5-methylcytosine DNA methyltransferase-deficient mice (Li *et al.,* 1992, 1993), discussed in a later section, have established a close relationship between DNA methylation and the regulation of these genes.

For the maternally imprinted *Igf2* gene, little or no allele-specific methylation is observed within the gene itself or its promoter, and both alleles are sensitive to DNAse digestion (Sasaki *et al.,* 1992). However, analysis of a 600-bp region located 3 kb upstream of the first *Igf2* gene promoter revealed four HpaII sites that were specifically methylated on the paternal allele (Brandeis *et al.,* 1993). Additional methylation of the paternal allele may exist farther upstream (Sasaki *et al.,* 1992). In sperm, only site 3 was fully methylated, while the other sites were partially methylated (Brandeis *et al.,* 1993). Interestingly, all four sites were also methylated in oocytes (Brandeis *et al.,* 1993). During early embryogenesis, sites 4–6 became unmethylated on both alleles while site 3 was methylated in both morulae and blastocysts (Brandeis *et al.,* 1993). Site 3, therefore, satisfies the requirements of an imprinting signal to mark the paternally derived *Igf2* gene.

For the *Igf2r* gene, which is expressed exclusively from the maternal genome in fetal and adult mice (Barlow *et al.,* 1991), two regions exhibiting allele-specific methylation in fetal and adult tissues have been identified (Stöger *et al.,* 1993). Region 1, which encompasses 34 methylation sites within the gene promoter and transcription start site, exhibits paternal-specific methylation. Region 2, located within an intron, encompasses 30 methylation sites and exhibits maternal-specific methylation (Stöger *et al.,* 1993). Paternal-specific methylation of region 1 is not detected in sperm, morulae, or blastocysts, but begins to appear around embryonic day 15, despite the fact that the paternal allele is transcriptionally silent prior to embryonic day 15 (Stöger *et al.,* 1993). Region 2, by contrast, is fully methylated in embryonic stem cells, in day 15 fetuses, and in adult tissues. Moreover, two of the sites examined, HpaII sites 2 and 3, are fully methylated in the oocyte (Stöger *et al.,* 1993; Brandeis *et al.,* 1993). Interestingly, HpaII sites 2 and 3 are not methylated in primordial germ cells or oocytes of neonatal mice, but are methylated in oocytes of adult mice (Brandeis et al., 1993). Two other sites, HpaII sites 1 and 4, are unmethylated in female germ cells but become methylated by the morula stage (Brandeis et al., 1993). Thus, the only sites satisfying the requirements of an imprinting signal of gametic origin that have been observed for the Igf2r gene are sites 2 and 3, which become methylated in adult oocytes. These sites, and two others (sites 1 and 4), are methylated specifically on the maternal allele as early as the blastocyst stage (Brandeis et al., 1993).

Recent results (A. Razin, personal communication), however, demonstrate the additional complexity of the methylation process and the need to examine each of the rapidly changing stages during preimplantation development before a complete understanding of the relationship between methylation and imprinting can be achieved. In region 2 of the Igf2r gene, site 3 is fully methylated in adult oocytes and in the maternal allele in morula and blastocyst. During early embryogenesis, however, this site is demethylated at the 4-cell stage and remethylated at the 8-cell stage. Thus the methylation of that site in oocytes cannot represent the primary imprint unless prior methylation facilitates remethylation, a possiblity for which there is no experimental evidence. An essentially identical example deals with another potential candidate for an imprinting signal. In sperm, the methylated site 3 of the Igf2 gene (see earlier discussion) also undergoes demethylation in zygotes and de novo methylation in 8-cell stage embryo. In contradistinction, site 4 in region 2 of the Igf2r gene, which is unmethylated in the oocyte and methylated in the blastocyst (Brandeis et al., 1993), and which was, therefore, considered an unlikely candidate for the imprinting mark, becomes methylated in the maternal allele in the early zygote while the pronuclei are still separated. All these examples suggest that, although methylation can play an important role in imprinting, there must be some essential prior signal that directs the de novo methylation to the correct (maternal or paternal) allele. These observations also highlight a potential role for egg modifier factors in controlling the establishment of methylation patterns of imprinted genes during early embryogenesis.

For the H19 gene, paternal-specific methylation is observed within the promoter and 5′ portion of the gene in midgestation fetuses, and is associated with DNAse insensitivity and the absence of paternal allele expression (Bartolomei et al., 1993; Ferguson-Smith et al., 1993). A detailed analysis of H19 gene methylation, however, revealed that although the majority of sites examined are fully methylated in sperm, one site is not. In addition, methylation of all sites examined is lost during early embryogenesis and is no longer present in morulae or blastocysts (Brandeis et al., 1993).

Taken together, data for the *Igf2, Igf2r,* and *H19* genes indicate that the complete methylation patterns manifested in fetal and adult tissues are not inherited directly from the gametes. Rather, these patterns are established in a stepwise manner. It has been suggested that during gametogenesis key methylation sites, such as the HpaII site 3 of the *Igf2* gene and site 3 in the *Igf2r* gene are imposed on genes destined to become imprinted and that these key sites in some way direct further methylation and allelic inactivation (Brandeis *et al.,* 1993). Clearly, a great deal remains to be discovered regarding the identity and temporal expression patterns of specific factors that recognize and impose cytosine methylation at appropriate target sequences, as well as factors that later recognize the key methylated sites to elaborate the mature methylation pattern.

Differences in methylation of imprinted genes have been used recently to identify additional imprinted loci (Hatada *et al.,* 1993; Hayashizaki *et al.,* 1994). This involved the use of a new, sophisticated approach for identifying differentially methylated loci—the restriction landmark genomic scanning method. With this method, genomic DNA is initially cleaved with a methylation-sensitive endonuclease, such as NotI, and the cleaved end is isotopically labeled to serve as a landmark. After cleavage with additional nucleases to reduce the size of fragments being analyzed, the DNA is separated by a two-dimensional gel electrophoretic method. If a given landmark site is imprinted and differentially methylated, the corresponding spot will be reduced in intensity by one half relative to other spots. If genetically variant imprinted loci that differ between two strains exist, the result is a spot that is one half the intensity with one hybrid cross and completely absent with the reciprocal cross (Hatada *et al.,* 1993). This latter approach makes it possible to identify and map imprinted loci. This approach was recently applied to the analysis of 3500 landmarks. Eight loci were found that were transmitted in a reciprocal fashion (Hatada *et al.,* 1993). One of these was a locus linked to the glutamine synthetase gene in an imprinted region on chromosome 11 with significant homology to the human *U2af* binding protein (Hatada *et al.,* 1993; Hayashizaki *et al.,* 1994). The discovery of at least one imprinted locus by this approach is further evidence of the link between DNA methylation and genetic imprinting.

C. Possible Mechanisms of Allelic Inactivation

The ability of methyl groups on DNA to affect the binding of regulatory factors both *in vitro* (Watt and Molloy, 1988; Iguchi-Ariga and Schaffner, 1989) and *in vivo* (Ben-Hattar and Jiricny, 1988; Li *et al.,* 1992, 1993) suggests that parental allele-specific methylation groups may regulate gene

expression (Doerfler, 1983; Cedar, 1988). Several issues must be accounted for in considering possible models intended to explain how methylation might regulate imprinted gene function. These include the dependence of allelic inactivation on other processes such as DNA replication,
the role of chromatin structure, the involvement of positive or negative
transcriptional regulatory factors, and the requirement for stage-specific
and tissue-specific effects of imprinting.

1. Replication Timing

It has been postulated that replication timing of imprinted genes may differ
and that this can lead to differences in allelic expression. This notion
derives from several lines of evidence, including the observation that the
level of expression of certain genes varies with their time of replication
during the S phase. Genes that are expressed abundantly tend to be replicated earlier than genes that are silent or expressed at reduced levels
(Goldman *et al.*, 1984; Hatton *et al.*, 1988; Brown *et al.*, 1987; Holmquist,
1987). In addition, the inactive X chromosome replicates later during the
S phase than the active chromosome (Takagi and Oshimura, 1973) and
this has been suggested to play a regulatory role (Grant and Chapman,
1988; Holmquist, 1987; Migeon, 1990; Riggs and Pfeifer, 1992). In this
context, the results of a recent analysis of the timing of replication of the
Igf2r, *Igf2*, and *H19* genes were unexpected. Paternal alleles of these
genes in cells derived from fetal stages were uniformly replicated early,
irrespective of the direction of their imprint (Kitsberg *et al.*, 1993). A
similar result was obtained for loci located in the 15q11-q13 region of the
human genome (Knoll *et al.*, 1994).

The paternal alleles of the *D15S63, D15S10,* and *GABRB3* (γ-aminobu-
tyric acid receptor β3 subunit) loci were all replicated early despite the
fact that the *D15S63* locus lies in a region that is reciprocally (maternally)
imprinted from the region encompassing the *D15S10* and *GABRB3* loci
(Knoll *et al.*, 1994). The former region also includes the maternally imprinted *Snrpn* gene (Leff *et al.*, 1992; Cattanach *et al.*, 1992; Ozcelik *et
al.*, 1992; Reed and Leff, 1994). Interestingly, the nearby nonimprinted
GABRA5 locus exhibited early replication of the maternal allele. These
results indicate that, although clearly parental origin indeed has an effect
on the time of allelic replication, this effect may not be as straightforward
as originally suspected. Since both of the maternally imprinted loci examined (*Igf2* and *D15S63*) lie in close proximity to reciprocally imprinted
loci, a situation that may also contribute to allelic regulation (see later
discussion), it remains to be determined whether alleles for maternally
imprinted genes that are not situated in proximity to paternally imprinted
loci will exhibit differences in replication timing.

2. Chromatin Effects

Mouse homologs of the *Drosophila* modifiers of position effect variegation may also contribute to allelic inactivation by regulating chromatin structure. Position-effect variegation in *Drosophila* is observed following translocation of a wild-type gene to within close proximity to α-heterochromatic domains, which are recognizable in mitotic cells as densely staining material near the centromere and which undergo fewer rounds of replication during the formation of polytene chromosomes (Tartof and Bremer, 1990). For example, expression of the *white* gene, which controls eye color, is altered when that gene is translocated to a region near heterochromatin. Variegation in eye color results from the stochastic inactivation of the gene owing to the spread of heterochromatin in some cells but not others. This inactivation occurs at a discrete stage of development and is then inherited in a clonal fashion, giving rise to variegation.

These effects on gene expression have been observed for a number of *Drosophila* genes (Tartof and Bremer, 1990). The degree of variegation is affected by genetic factors. These include two types of *trans* activating factors, Suppressors of variegation [*Su(var)*] in which mutations inhibit the spread of heterochromatin into the translocated gene, and Enhancers of variegation [*E(var)*], in which mutations promote the spread of heterochromatin. The relevant loci have been further classified into Class I genes, which enhance variegation when duplicated but suppress variegation when deleted, and the reciprocal Class II genes, which enhance variegation when mutated but suppress variegation when deleted. While the exact number of genes that comprise these two classes remains unknown, it is estimated that 20–30 Class I and 2 Class II genes exist (Tartof and Bremer, 1990). It is suggested that differences in the amount or array of these modifiers of variegation act through a combinatorial mechanism to influence the spread of heterochromatin (Locke *et al.*, 1988; Tartof *et al.*, 1989; Tartof and Bremer, 1990). Changes in histone acetylation may also regulate heterochromatization since a decrease in acetylation is associated with suppression of variegation (Mottus *et al.*, 1980). A particularly intriguing possibility is the "dosage assembly" model, which can account for heterochromatic inactivation of the X chromosome (discussed later). In addition to these modifiers of variegation, other loci in *Drosophila* affect the expression of developmental regulatory genes. These include chromobox-containing genes such as *Hetrochromatin protein-1* (*HP-1*), *Polycomb*, and *Hairless* (Tartof and Bremer, 1990).

Four murine genes containing a chromobox domain, a domain of homology shared among the *Drosophila Su(var)*, *HP-1*, and *Polycomb* genes, have been identified (Singh *et al.*, 1991; Pearce *et al.*, 1992; Hamvas *et al.*, 1992). By analogy with their *Drosophila* homologs, these genes may help to regulate heterochromatin formation. The presence of such homologous sequences in mice is consistent with the possibility that factors analogous to enhancers or suppressors of variegation exist in mice and regulate heterochromatin formation and gene expression. The interactions of such molecules with the chromatin may be affected by allele-specific methylation, thereby producing different heterochromatic states between imprinted and nonimprinted alleles. Although position-effect variegation in the mouse has only been observed for X-autosome translocations (Eicher, 1970; Cattanach, 1974), it is possible that such heterochromatic domains exist in autosomes on a much smaller scale than in *Drosophila*. A recent model has been put forth suggesting how methylation may regulate the establishment of repressed chromatin domains in a manner analogous to how certain *Drosophila* genes are regulated (Singh, 1994).

3. Methylation and Imprinted Gene Function

Two models have been proposed to explain how parent-specific patterns of methylation might regulate *Igf2*, *Igf2r*, and *H19* expression. In the case of the *Igf2r* gene, it has been suggested that the methylation of region 2 prevents binding of a repressor to the maternal allele, thereby conferring allele-specific expression (Stöger *et al.*, 1993; Surani, 1993). Support for this idea has come from the analyses of homozygous mutant mice that are defective in DNA methyltransferase expression (Li *et al.*, 1992, 1993). These mice were produced by gene knockout using homologous recombination in embryonic stem cells. Mutant alleles resulting in either partial or severe reductions of methyltransferase activity and 5-methylcytosine content were created (Li *et al.*, 1992, 1993). One mutation, MTaseN, reduced the overall degree of methylation to about one third that of wild-type cells, and homozygosity leads to retarded development and midgestation embryonic lethality, with embryos developing, at most, to the 20-somite stage (Li *et al.*, 1992). A more severe mutation, MTaseS, leads to lethality at the 5–6 somite stage (Li *et al.*, 1993). Expression of the *Igf2r* gene was unaffected in the MTaseN homozygous mutants, but reduced in MTAseN/MTaseS compound heterozygotes and MTaseS homozygotes (Li *et al.*, 1993). This is consistent with the model that failure to methylate the maternal allele would allow the repressor to bind to both alleles.

A more complex model has been proposed for regulation of the reciprocally imprinted *H19* and *Igf2* genes, which are located within 90 kb of one another on mouse chromosome 7 (Bartolomei *et al.*, 1991, 1993; Zemel *et al.*, 1992). This "shared enhancer" model is based on the observation that an enhancer exists downstream of the *H19* gene. The enhancer is proposed to interact with either the *Igf2* or the *H19* genes, but not both.

Owing to differences in promoter strength, the *Igf2* gene may therefore only be expressed when the *cis H19* gene is silent, and vice versa. Support for this model was again obtained from analyses of methyltransferase mutants. Northern blotting analysis of RNA from embryonic day 10.5 MTaseN homozygous mutant fetuses revealed that *H19* RNA, but not *Igf2* mRNA, was expressed (Li *et al.*, 1993). This result suggested that the loss of methylation on the *H19* gene may lead to activation of both *H19* alleles, and hence the inactivation of both *Igf2* alleles. Additional support for the expression of only one of these genes from a given DNA strand comes from a single informative case in which biallelic expression of the *Igf2* gene was observed in human choroid plexus and leptomeninges, with no *H19* mRNA detectable by RNAse protection (Ohlsson *et al.*, 1994).

Observations not explained by this model, however, include expression of both the *H19* and *Igf2* genes in human androgenetic trophoblast cells derived from complete hydatidiform mole (Mutter *et al.*, 1993), expression of both genes in giant trophoblastic cells of mouse androgenetic conceptuses (Walsh *et al.*, 1994) and biallelic expression of both genes in certain tumors (Rainier *et al.*, 1993). In addition, maternal *Igf2* alleles can be activated in cultured cells by treatment with 5-aza-2'-deoxycytidine (Eversole-Cire *et al.*, 1993), which contrasts with the results obtained with MTase-deficient mice. Interestingly, the *H19* gene has been disrupted by gene knockout with the result that heterozygous mice bearing a disrupted maternal allele of *H19* are viable and express significant amounts of *Igf2* transcripts from the maternal allele (P. Leighton and S. Tilghman, personal communication). This experiment confirms a link between *H19* expression and *Igf2* gene regulation, although it does not distinguish between a transcriptional link and a requirement for the *H19* RNA itself.

The observation that the paternally inherited alleles of the oppositely imprinted *H19* and *Igf2* genes are both replicated during early S phase (Kitsberg *et al.*, 1993), which is contrary to a proposed direct relationship between imprinting direction and replication timing, may also be explicable within the context of this model. The paternal imprint of the *H19* or neighboring genes may determine replication timing and thus contribute to allelic *H19* inactivation. Such an effect of replication timing on *Igf2* gene expression would not be required, since this would be controlled by the need for interactions between the *Igf2* gene promoter and the downstream enhancer. With this in mind, it is reasonable to postulate that, since replication domains are generally quite large, 1–2 Mbp (Drouin *et al.*, 1990; Selig *et al.*, 1992), if replication timing indeed affects gene expression, then whenever oppositely imprinted genes are located in close proximity to one another, some other mechanism, such as enhancer use or promoter function, must control gene expression.

4. Postfertilization Events in Imprinted Gene Regulation

The above models may well account for the differential expression of maternal and paternal alleles once the allele-specific methylation patterns are completely established. The postfertilization development of maternal and paternal methylation patterns for both transgenes and endogenously imprinted genes, however, raises the question of whether the patterns of methylation that are established during gametogenesis are sufficient to regulate gene expression or whether differential expression requires additional postfertilization events, including establishment of the complete maternal and paternal methylation patterns. In addition, if postfertilization events are essential for manifestation of the genetic imprint, it is of interest to consider what factors might be required to elicit the imprinted phenotype and when during development such factors might be expressed and exert their effects.

Nuclear transplantation experiments (Latham *et al.*, 1994) revealed that during the preimplantation period neither the *Igf2* nor the *Igf2r* genes appeared to be regulated by imprinting. The *Igf2* gene was abundantly expressed in parthenogenetic blastocysts, even though such embryos have only maternal chromosomes, which should not express this gene. Similarly, the *Igf2r* gene was expressed at the same level in androgenetic (two paternal genomes) 8-cell and blastocyst embryos as in normal and gynogenetic embryos, even though this gene is normally expressed only from the maternal genome. Similar results were obtained by two other laboratories (Gilligan and Solter, 1994; N. Murcia and J. McGrath, unpublished). One could propose that the imprint that is present at fertilization is simply lost during early development in androgenones and gynogenones/ parthenogenones if both maternal and paternal genomes must be present for essential interactions to occur between the two genomes. If this occurred, however, androgenones and gynogenones would have identically nonimprinted chromosomes, with the exception of the original DNA strands contributed by the egg and sperm, and so should exhibit identical phenotypes. Since androgenones and gynogenones/parthenogenones exhibit distinct and complementary phenotypes (Solter, 1988; Howlett, 1991), a loss of the genetic imprint seems an implausible mechanism to explain the expression of imprinted alleles in these embryos. Moreover, embryonic stem cell lines derived from androgenetic and gynogenetic embryos retain the appropriate imprints for *Igf2* and *H19*. These genes are expressed in both cell types, however, until embryoid body formation, at which point the imprints become manifested (Szabo and Mann, 1994). Similarly, analyses of ES cells bearing targeted disruptions of either the maternal or paternal *Igf2r* allele also indicates biallelic expression in ES cells (Wang *et al.*, 1994). A recent study has revealed additional changes in parental allele-specific methylation in conjunction with androgenetic and parthenogenetic ES cell differentiation (Feil *et al.*, 1994). Thus, the differential methylation patterns that exist at the time of fertilization probably cannot by themselves determine the ability of at least some imprinted genes to be expressed. Similarly, analyses of ES cells bearing targeted disruptions of either the maternal or paternal *Igf2r* allele also indicate biallelic expression in ES cells (Wang *et al.*, 1994). Consequently, differential expression of the maternal and paternal alleles of these genes may not occur until peri- or postimplantation, probably in conjunction with the establishment of the complete maternal and paternal methylation patterns.

Given the allele-specific methylation patterns of the endogenously imprinted *Igf2*, *Igf2r*, and *H19* genes (Sasaki *et al.*, 1992; Bartolomei *et al.*, 1993; Stöger *et al.*, 1993), the stepwise establishment of these patterns, the expression of both imprinted alleles of *Igf2* and *Igf2r* in preimplantation embryos, and the altered patterns of expression of these genes in methylation-deficient mice, the establishment of allele-specific patterns of gene expression in response to genetic imprinting must be considered a multistage process requiring the action of regulatory factors both during gametogenesis *and* during early embryogenesis.

5. A Model for Gene Regulation by Imprinting

A model to account for what is known of the methylation and regulation of the *Igf2*, *Igf2r*, and *H19* genes, which are the most extensively studied imprinted genes, is given in Fig. 1. As suggested by Brandeis *et al.* (1993), this model proposes that a few key methylation groups may exist in or around imprinted genes to serve as markers of parental origin and direct the later establishment of complete methylation patterns and eventual allelic inactivation. The methylation patterns thus far revealed for these genes are acquired progressively. Since these methylation patterns are not fully established prior to implantation, this would account for the biallelic expression of the *Igf2* and *Igf2r* genes during the preimplantation period (Latham *et al.*, 1994). We propose that for most, if not all, imprinted genes, biallelic expression will be observed if the genes are expressed in the early embryo and that allele inactivation will require completion of regulatory events such as additional gene methylation and expression of *trans* regulatory molecules. Consistent with this, it has been suggested that for allele-specific expression of transgenes, the gametic methylation pattern is largely irrelevant and the methylation sites that are established postfertilization are of the greatest importance (Chaillet, 1994). An understanding of *Igf2*, *Igf2r*, and *H19* gene regulation should, therefore, offer

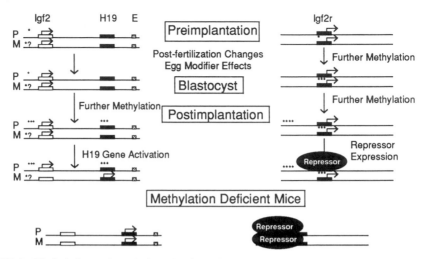

FIG. 1 Methylation and regulation of *Igf2*, *Igf2r*, and *H19* genes. This diagram summarizes the changes in DNA methylation (*) and transcriptional states (arrows) that occur for these imprinted genes during development. The model proposes that the imprints derived from gametes do not directly determine the transcriptional states of these genes, and that a variety of postfertilization events, including those mediated by egg modifiers, zygotic modifiers of imprinting, and changes in DNA methylation, are required for monoallelic expression. These events can be subject to stage-specific and/or tissue-specific control. Expression of the *Igf2r* gene has been proposed to be regulated by binding of a repressor, which is prevented from binding to the maternal allele by DNA methylation within an intron (region 2) (Stöger *et al.*, 1993). Expression of the repressor may be delayed until the complete methylation pattern is established. *Igf2* expression is biallelic initially and may become monoallelic only after induction of the *H19* gene. Methylation of the maternal *Igf2* gene may also contribute to monoallelic expression.

a useful paradigm for understanding the regulation of many imprinted genes. The postfertilization events summarized in Fig. 1 are, therefore, likely to contribute to allele inactivation for other imprinted genes.

Expression of both *Igf2* alleles may also be facilitated by the lack of expression of the *H19* gene, which becomes transcriptionally active around the time of implantation (Bartolomei *et al.*, 1991). Differential regulation of the *Igf2* alleles may not occur until those factors responsible for activating *H19* gene transcription are expressed. The expression of both *Igf2* and *H19* in human and mouse androgenetic trophoblast cells (Mutter *et al.*, 1993; Walsh *et al.*, 1994) and the biallelic expression of both genes in some Wilms' tumors (Rainier *et al.*, 1993), however, indicate that expression of *H19* by itself is not sufficient to *cis*-inactivate the *Igf2* gene and that either a threshold effect or an additional negative regulatory factor is required, as proposed for the *Igf2r* gene (see later discussion). This negative regulatory factor may be unable to bind to the methylated paternal allele. Moreover, such a regulatory factor may cease to be expressed in Wilms' tumor, thereby allowing biallelic *Igf2* gene expression.

The expression in methylation-deficient mice of both *H19* alleles, but neither *Igf2* allele, supports a role for methylation in preventing binding of a repressor to the *Igf2* gene. This scenario does not explain, however, the increased expression of the maternal *Igf2* allele in 5-aza-2'-deoxycytidine-treated cells derived from maternally disomic MatDi7 embryos (Eversole-Cire *et al.*, 1993). Unless this is an artefact of cell culture, this observation raises the possibility that as-yet-undetected methylation of the maternal *Igf2* allele provides an additional level of control, possibly allowing repression of the maternal *Igf2* allele in the absence of *H19* expression. The relative contributions of the maternal *Igf2* gene methylation pattern and *H19* transcription to the imprinted phenotype in cells with a normal chromosomal constitution has not been determined. It would be useful to define precisely the temporal relationship between *H19* gene activation and maternal *Igf2* allele inactivation, and to specifically mutate the *H19* gene promoter or enhancer to prevent *H19* gene expression.

Expression of the *Igf2r* gene is proposed to be wholly determined by binding of a repressor molecule to the paternal allele. Binding of the repressor to the maternal allele is prevented by the maternal methylation pattern in region 2 (Stöger *et al.*, 1993). Methylation-deficient mice fail to methylate their maternal alleles, thereby allowing the repressor to bind to and inactivate both alleles (Li *et al.*, 1992). A key point of this model is that the repressor molecule cannot be expressed until after the complete maternal methylation pattern has been established. This would allow both alleles to be expressed in the early embryo. It is also possible, however, that the entire maternal methylation pattern need not be established in order to prevent repressor binding, and that expression of both alleles in the early embryo results from methylation of both alleles at key sites in region 2. Although data presented by Brandeis *et al.* (1993) indicate that this may occur, it has not been clearly documented whether methylation at these sites is sufficient to prevent repression. The paternal methylation pattern, which is established postimplantation, (Stöger *et al.*, 1993), may also inhibit expression of the paternal allele or merely facilitate the formation or maintenance of an inactive state once repressor binding has occurred. Resolution of this question will most likely require identification of the repressor molecule and examination of paternal *Igf2r* gene expression in its absence.

Tissue-specific expression of factors that recognize autosomal imprinted genes may also be important. In mice, the *Igf2* gene is expressed exclusively from the paternal genome in fetal and adult tissues with the exception of the leptomeninges and the choroid plexus, where both alleles are expressed (DeChiara et al., 1990, 1991). In humans, the IGF2 gene is imprinted, with the paternal allele expressed (Giannoukakis et al., 1993; Ohlsson et al., 1993), but it is biallelically expressed in the liver of children beginning at the end of the first year of life (Davies, 1994). This involves the use of an additional promoter (P1) that appears to be unaffected by imprinting (Vu and Hoffman, 1994). Recently, it has been shown that the mouse insulin 1 and insulin 2 genes are imprinted in the yolk sac, but not pancreas (Giddings et al., 1994). This indicates that the mouse insulin gene is marked in such a way that the yolk sac cells can distinguish maternal and paternal alleles but that this imprint is not recognized or utilized in pancreatic cells. Thus, either the yolk sac expresses a repressor that is not present in the pancreas or the imprint is lost in the pancreas. It has also been reported that major histocompatibility complex class I antigens are imprinted in the placenta of the rat but it is not known if this is tissue specific (Kanbour-Shakir et al., 1993).

Differences among cells or tissues in the expression of factors that recognize genetic imprints may also relate to the apparent relaxation of imprinting effects in certain tumors (Ogawa et al., 1993a,b; Rainier et al., 1993; Feinberg, 1993; Weksberg et al., 1993). Recent reports describe biallelic expression of the IGF2 gene in Wilms' tumor and rhabdomyosarcoma (Ogawa et al., 1993a,b; Rainier et al., 1993; Weksberg et al., 1993). In some Wilms' tumors, biallelic expression of the H19 gene was also observed, sometimes in conjunction with biallelic Igf2 gene expression. Another recent report described a constitutional lack of imprinting regulation of the Igf2 gene in cultured skin fibroblasts of patients with Beckwith-Wiedemann syndrome (BWS) (Ogawa et al., 1993a).

In all of these cases, it was suggested that biallelic expression of these genes in these patients results from a loss of imprinting either during gametogenesis or within the tumor cells themselves. In some cases, translocations with breakpoints in the relevant chromosomal regions may also affect expression (Dao et al., 1987; Mannens et al., 1988). This is an interesting possibility, given the existence of position effect variegation in Drosophila. An alternative explanation for the apparent relaxation of IGF2 and H19 gene imprinting, however, is that the tumor cells may be defective in the expression of factors that recognize and inactivate the imprinted alleles. A defect in the expression of the relevant regulatory factors could be genetically determined (e.g., BWS) or arise during sporadic tumor formation. Such possibilities will be testable only when the relevant factors have been identified. Moreover, even where changes in the methylation status of imprinted genes are observed in tumor cells, it will be essential to determine whether they precede or follow the onset of biallelic expression.

IV. Imprinting and X Chromosome Regulation

A. Tissue-Specific and Developmental Aspects of X Chromosome Regulation

The model explaining *Igf2, Igf2r,* and *H19* gene regulation proposes a central role for regulatory factors that determine monoallelic expression. The existence of such factors might also account for preferential paternal X chromosome inactivation. As mentioned earlier, the temporal relationship between gene inactivation, methylation, and heterochromatin formation varies among different X-linked genes (Monk, 1986; Singer-Sam *et al.,* 1990; Grant *et al.,* 1992; Kaslow and Migeon, 1987; Lock *et al.,* 1987). Thus, it is not clear whether methylation has a causative role in X-linked gene inactivation or stabilizes an inactive state once it is established. It is also noteworthy that the inactive paternal X chromosome in extraembryonic cells is apparently not structurally modified (i.e., hypermethylated) since transfection of this DNA into *Hprt*-deficient cells restores *Hprt* gene function (Kratzer *et al.,* 1983). In keeping with the proposed model, these aspects of imprinted X chromosome regulation are explicable by a tissue-specific regulatory factor that recognizes key methyl groups denoting parental origin. Thus, preferential paternal X chromosome inactivation in extraembryonic lineages (trophectoderm, yolk sac, parietal endoderm) but not in somatic lineages may reflect cell type-specific expression of regulatory factors that specifically recognize the paternal X chromosome. Such factors could directly inactivate the X chromosome or, more likely, bias the inactivation process in the direction of the paternal X by interacting with the same molecular machinery that randomly inactivates X chromosomes in somatic cells (Cattanach and Beechey, 1990). In this regard, it is noteworthy that the paternal X is preferentially inactivated in somatic cells of marsupials, and this appears to be mechanistically similar to the inactivation process observed for eutherian extraembryonic tissues (Vanderberg *et al.,* 1987; Kaslow and Migeon, 1987; Migeon *et al.,* 1989). Moreover, parthenogenetic cells can inactivate an X chromosome in extraembryonic tissues (Rastan *et al.,* 1980) and XO embryos bearing only a paternal X chromosome express that chromosome in extraembryonic tissues (Frels and Chapman, 1979; Papaioannou and West, 1981).

In some cases, XX androgenetic cells inactivate a single X chromosome (Tsukahara and Kajii, 1985), indicating that paternal X inactivation is not an obligate consequence of the parental imprint. Gene products from within or near the X inactivation center (X^{ic}) or the X controlling element (X^{ce}) (e.g., *Xist*) may participate in X inactivation in extraembryonic tissues, but this process may not be as sensitive to X^{ce} alleles as observed with somatic tissues (Cattanach and Beechey, 1990; Bücher *et al.,* 1986).

The factors that mediate X chromosome inactivation in eutherian somatic cells may be regulated in response to the X:autosome ratio, in which case the factors that specifically recognize the paternal X could be constitutively expressed in extraembryonic tissues or also regulated by the X:autosome ratio. As discussed earlier, it appears that some mechanism exists for the maternal X chromosome to be preferentially expressed as early as the 8-cell stage (Moore and Whittingham, 1992; Singer-Sam *et al.*, 1992). This suggests that, not only are the two chromosomes marked in order to distinguish parental origin, but perhaps the factors that recognize and repress the paternal X chromosome in trophectoderm cells are expressed in the preimplantation embryo, possibly in those cells destined to give rise to the trophectoderm. Interestingly, inactivation of one X-linked transgene occurs at different times in different tissues (Krumlauf *et al.*, 1986) and another X-linked transgene on the inactive X chromosome escapes inactivation in a subset of cells (Wu *et al.*, 1992) These results indicate that the expression of at least some of the factors that are required for X inactivation may not be regulated exclusively by the X:autosome ratio, but also by developmental processes.

B. The Role of *Xist* in X Chromosome Regulation

Recent studies with the *Xist* gene provide additional information regarding imprinted regulation of X chromosomes. Norris *et al.* (1994) found that the inactive *Xist* gene promoter present on the active X chromosome is fully methylated while the promoter of the active *Xist* allele on the inactive X chromosome is unmethylated. Thus, unlike the *Igf2r* and *Igf2* genes, for which methylation of the active allele appears to be crucial for expression, it is the inactive *Xist* allele that is methylated. This was found for both embryonic cells, in which X chromosome inactivation is random, and in extraembryonic cells, where the paternal X is preferentially inactivated. Interestingly, methylation is present but not allele specific in undifferentiated ES cells. During differentiation in culture, these cells undergo random X inactivation, as expected for somatic cells. This was observed even in primitive endoderm derivatives of the ES cells. This indicates that continuity of the allelic-methylation pattern from gamete to extraembryonic tissues must be actively maintained and that methylation of the *Xist* promoter itself may determine *Xist* expression. This would further indicate that the early embryo and extraembryonic cells must express one or more factors that recognize the X chromosome imprint and maintain *Xist* methylation. Such a requirement would provide a useful point of regulation that could be exploited in cases where paternal X chromosome inactivation must be circumvented to permit cell survival.

The results of Norris *et al.* (1994) are also of interest with regard to X regulation in the early embryo. The *Xist* gene promoter is apparently unmethylated in sperm, and preliminary data have been cited indicating that the *Xist* gene is fully methylated in the egg (Kay *et al.*, 1994). This indicates that regulation of *Xist* methylation during gametogenesis has evolved to allow the egg to protect expression of its X chromosome. Along with the establishment of a mechanism that directs *Xist* gene methylation during oogenesis would by necessity come the deposition into the egg cytoplasm of factors that recognize and maintain the imprint. This could account for the differences in parental X chromosome expression as early as the 8-cell stage (Moore and Whittingham, 1992), providing that the maternal *Xist* allele is methylated in the early embryo. To ensure maintenance of the *Xist* gene imprint, the expression of these factors in the early embryo and extraembryonic lineages may constitute a normal part of the developmental program regardless of the sex of the embryo. As outlined earlier, however, there is reason to suspect that the expression of these factors is subject to regulation by other parameters.

It has been suggested that a factor encoded by an imprinted gene is expressed exclusively from the maternal genome and that this factor is responsible for promoting *Xist* expression and maintaining preferential paternal X inactivation (Kay *et al.*, 1994). This factor would be present in the egg, thereby promoting *Xist* expression from paternal X chromosomes in the early embryo, and would later be expressed in differentiating cells (Kay *et al.*, 1994). In the model put forth, expression of *Xist* in early gynogenetic embryos, which lack paternally derived X chromosomes, would not be expected prior to loss of the *Xist* imprint around the late morula/early blastocyst stage. Such expression was observed as early as the 4-cell stage, however, but was attributed to contamination of samples with maternal cells. It is difficult to see how such contamination would occur since 4-cell embryos are essentially free of adhering maternal cells. Thus, it appears that the factor that promotes paternal *Xist* expression may also be able to activate maternal *Xist* alleles in the absence of paternal alleles, that is, activation of the paternal *Xist* allele may reflect a bias rather than a predetermined feature of gene regulation. Interestingly, *Xist* expression at the 8-cell stage is less than expected in androgenones (Latham *et al.*, 1994; Kay *et al.*, 1994) and is reduced further in androgenones between the morula and blastocyst stages (Kay *et al.*, 1994). XY androgenones would continue to express their single paternally derived X chromosome, since no maternal genome is present to induce *Xist* gene expression. Failure to express *Xist* in the absence of a maternal genetic contribution would impede the ability to inactivate X chromosomes in XX androgenones, and it appears that these androgenones are selected against during preimplantation development.

It should be noted, however, that XX androgenones can survive, implant, and inactivate an X chromosome, indicating that the block to *Xist* expression and imprinted expression of the factor that is required for *Xist* expression can be overcome in some situations. In addition, it should be noted that gynogenones, which express two maternally derived X chromosomes during the stage at which loss of XX androgenones is first observed, are fully viable during early development. This may reflect *Xist* expression in gynogenomes and/or lack of an effect of two active X chromosomes. Thus, lethality in XX androgenones may involve more than simply failure to express *Xist* and undergo X inactivation. In extraembryonic lineages, the *Xist* imprint would be maintained and expression of this maternally encoded factor would again promote preferential paternal X inactivation, while in the somatic lineages, this factor would lead to random *Xist* expression since the *Xist* imprint is lost. The identification of this factor and discovery of whether it is expressed constitutively or exclusively in female embryos present interesting problems for further study.

It should be noted that the results of Norris *et al.* do not demonstrate conclusively that the methylation status of the *Xist* gene promoter itself constitutes the imprinting signal. It is possible that imprinting at another location actually directs maternal *Xist* promoter methylation. One reason for postulating this is that the random acquisition of *Xist* promoter methylation during ES cell differentiation indicates that this methylation can arise independently of the imprinting mechanism. Additional factors that recognize, maintain, or interpret an imprint, therefore, may bias methylation toward the maternal allele. Such a situation would satisfy the two inherent needs of the system, namely, manifestation of an imprinted phenotype in extraembryonic cells combined with a need to permit random inactivation in somatic cells. Randomization of X inactivation in somatic cells would begin with the suspension of expression of the factors that recognize, maintain, or interpret the imprint. The loss of *Xist* gene promoter methylation observed in ES cells may subsequently facilitate random X inactivation in somatic cells by equalizing the probability that either X chromosome will become methylated and remain active once the expression of such factor(s) has been interrupted.

C. Heterochromatin Effects in X Chromosome Regulation

The mechanism by which differential X chromosome methylation might lead to chromosome inactivation remains unknown. It is possible that heterochromatic regulation of X chromosome function is mediated in part by a mammalian homolog of *Drosophila* modifiers of position-effect variegation. Position-effect variegation has been described only for X-linked genes in the mouse (Cattanach, 1974). In mammals, autosome : X chromosome translocations produce variegated phenotypes consistent with an early clonal determinative event. Unlike the situation in *Drosophila*, however, in which an inactive state for the translocated autosomal gene is stably maintained, X reactivation occurs in mammals so that subclones within each founder clone can be observed (Cattanach, 1974). This reflects a certain degree of plasticity of heterochromatic inactivation in mammalian autosomal DNA.

A "dosage assembly" model has been proposed to explain random X inactivation (Tartof and Bremer, 1990). This model postulates that a large multimeric complex composed of DNA binding proteins encoded by both autosomes and the X chromosome assembles and inactivates one of the X chromosomes. As the X chromosome becomes repressed, the level of expression of the X-encoded components declines precipitously, allowing the remaining X chromosome to remain active. Thereafter, methylation of the inactivated X stabilizes the inactive state. This model is intriguing in that it requires the contribution of both X-linked and autosomal genes whose products act cooperatively to achieve X inactivation. Conceivably, this model could also account for nonrandom X inactivation in extraembryonic tissues if a factor that recognizes the imprint on the paternal X chromosome participates in the complex formation and biases the process toward the paternal X. The X-linked imprinted gene proposed earlier to direct expression from the maternal X chromosome of a factor that is required for *Xist* expression (Kay *et al.,* 1994) could serve this role.

Another interesting possibility is that the X-encoded components of the complex could affect the expression of autosomal imprinted genes. This would be consistent with the proposed existence of an X-linked imprinting gene in humans (Naumova and Sapienza, 1994). The existence of this gene was proposed to account for the observed sex ratio distortions and transmission ratio distortions among the families of patients affected with bilateral retinoblastoma. In these families, males account for a larger percentage of individuals with bilateral retinoblastoma than expected, while no such distortion is observed among patients with unilateral retinoblastoma (Naumova and Sapienza, 1994). The excess of affected males apparently results from a tendency of a portion of the affected males to produce exclusively male offspring, the majority of which are also affected (Naumova and Sapienza, 1994). The proposed explanation for these results predicts that an X-linked imprinting gene that is needed to erase the maternal imprint and/or establish the paternal imprint on a portion of the maternally inherited genome during gametogenesis is defective in males with bilateral retinoblastoma. As a result, female progeny are nonviable because they possess two maternally imprinted (i.e., one maternally and one grandmaternally imprinted) X chromosomes as well as two maternally imprinted copies of certain autosomal genes. Surviving progeny carry the Y chromosome and paternally imprinted (i.e., grandpaternal) autosomal genes, including the mutant retinoblastoma gene, which becomes the expressed allele.

V. Genetic Imprinting in Humans

The intense study of altered human phenotypes provides a powerful system in which to discern imprinted genes. The prototypic example of genetic imprinting in humans is the alternate outcome of deletions on the proximal long arm of human chromosome 15. In this region, three imprinted genes, designated as *SNRPN, PAR-5,* and *PAR-1,* have been identified (Ozcelik *et al.,* 1992; Sutcliffe *et al.,* 1994).

Paternal deletions of 15q11-13 result in the Prader-Willi syndrome (PWS) while maternal deletions result in Angelman syndrome (AS) (Knoll *et al.,* 1989; Clayton-Smith *et al.,* 1992). Detailed analyses have shown that PWS may also result from genetic alterations other than 15q11-13 deletions. First, a significant proportion of PWS individuals result from an inheritance of chromosome 15 from only their mothers (i.e., uniparental disomy, UPD; Nicholls, 1993; Mascari *et al.,* 1992). Presumably, a chromosome 15 trisomic embryo reverts to a euploid state by early embryonic somatic nondisjunction. If this nondisjunction event results in paternal chromosome 15 loss, PWS develops. The combination of paternal chromosome 15 q11-13 deletion and maternal chromosome 15 UPD accounts for approximately 95% of PWS individuals (Mascari *et al.,* 1992). In a rare instance, a child with maternal chromosome 15 UPD possessed both PWS and Bloom syndrome (Woodage *et al.,* 1994). This is explicable if the duplicated maternal chromosome had a mutation for the Bloom syndrome gene, which is located on 15q (telomeric to 15q25).

In contrast to PWS, many AS individuals possess neither maternal deletions nor paternal UPD (Knoll *et al.,* 1991; Malcolm *et al.,* 1991). In a recent study of 93 individuals with AS by Chan *et al.* (1993) , 30 (32%) had neither a maternal deletion of 15q11-13 nor paternal UPD. Significantly, in two of the 60 individuals who possessed maternal deletions, the deletion was outside the AS critical region. This raises the possibility that a long-range regulatory region, perhaps analogous to the globin gene "locus control region" (Epner *et al.,* 1992), may influence the expression of the AS gene. In addition, Chan *et al.* (1993) summarized reports of six instances in which AS recurred in the same family. It is important that in all cases, the affected individuals shared a common maternal chromosome

15 (Wagstaff *et al.*, 1992; Chan *et al.*, 1993). Thus, it is presumed that in instances of familial recurrence, a mutation in the maternally inherited AS gene results in AS. Detailed molecular genetic analyses of affected individuals from such pedigrees should facilitate the identification of the paternally imprinted gene responsible for AS.

In two remarkable families, a balanced translocation involving chromosome 15 exists. Children of balanced translocation parents frequently are deleted for the proximal long arm of chromosome 15. These children alternately exhibit PWS or AS, depending upon whether the transmitting parent is male or female, respectively (Hulten *et al.*, 1991; Smeets *et al.*, 1992).

An additional, very interesting and informative mechanism leading to PWS and AS and not involving the loss of an active allele by deletion or mutation has recently been suggested (Reis *et al.*, 1994). These authors observed several PWS patients who carry a maternal methylation imprint on their paternal chromosome and AS patients who carry a paternal methylation imprint on their maternal chromosome. These findings suggest the existence of the gene(s) which are involved in the imprinting process (analogous to *Imp-1;* Forejt and Gregorova, 1992). Such genes, if mutated, would cause faulty imprinting or absence of germ line resetting of the imprint, thus resulting in pseudo uniparental disomy. Namely, one would observe two (in a functional sense) paternally imprinted alleles although one was inherited from the mother and vice versa.

Other examples of human UPD have been reported. As in the mouse, UPD for some human chromosomal regions is without apparent abnormality, whereas UPD for other chromosomes results in phenotypic consequences. Among the former, paternal UPD for human chromosome 6, detected by homozygous loss of the fourth component of complement, is without phenotypic effect (Welch *et al.*, 1990). Maternal UPD for chromosome 7, detected by homozygosity for either cystic fibrosis (Spence *et al.*, 1988; Voss *et al.*, 1989) or COLIA2 (Spotila *et al.*, 1992) mutations, however, resulted in significant short stature in all three cases reported. Paternal UPD for chromosome 14 is associated with mental retardation and multiple congenital anomalies, including severe kyphoscoliosis (Wang *et al.*, 1991). The latter is perhaps significant because androgenetic mouse chimeras also display vertebral body and thoracic cage skeletal anomalies (Mann *et al.*, 1990). Maternal UPD for human chromosome 16 with trisomy 16 in the placenta is associated with intrauterine growth retardation and, in one instance, an imperforate anus at birth. Whether this is attributable to fetal UPD or to trisomy 16 in the placenta, however, cannot be determined (Kalousek *et al.*, 1993).

An additional example of imprinting in humans is the dramatic pedigree data reported by van der Mey *et al.* (1989), in which relatively rare paraganglioma tumors can be inherited in an autosomal dominant fashion. In these pedigrees, however, susceptibility to carotid body tumors is only paternally inherited, although both males and females can be similarly affected. Analysis of this and related families has established links to human 11q (Heutnik et al., 1992), most likely in the 11q14-q21 interval (Mariman et al., 1993).

One feature in common among a number of these instances in which genetic imprinting contributes to pathological conditions in humans is the alteration in normal growth regulation, leading to malformation or tumor formation. The Beckwith-Wiedemann syndrome, which is also thought to involve genetic imprinting and which has been studied extensively in recent years, provides another excellent example of the importance of genetic imprinting in growth control. Individuals with BWS are large for gestational age, possess congenital malformations, and are predisposed to develop tumors in childhood. Familial BWS maps to human chromosome 11p15.5 (Koufos et al., 1989). Some BWS individuals possess a complete or partial paternal UPD for 11p15.5 (Henry et al., 1991; Ohlsson et al., 1993). Similar to the individual with PWS and Bloom syndrome (Woodage et al., 1994), a child with BWS and thalassemia major has been described (Beldjord et al., 1992). Apparently the paternal chromosome 11 involved in UPD also carried a β-globin gene mutation, leading to thalassemia major.

Human 11p15.5 is syntenic with mouse chromosome 7 and contains the imprinted INS2, IGF2, and H19 genes. The IGF2 gene and H19 genes are also imprinted in humans (Giannoukakis et al., 1993; Ohlsson et al., 1993; Rainier et al., 1993). It is, therefore, reasonable to infer that overexpression of the paternally expressed INS2 and IGF2 alleles during fetal life could result in the fetal overgrowth and the neonatal hypoglycemia that are characteristic of BWS. Overexpression of the IGF2 gene might also explain the childhood tumor predisposition of BWS. The most frequent such tumor is nephroblastoma, or Wilms' tumor. Recent experiments using transgenic mice with an Igf2 transgene under the control of the major urinary protein promoter indicate that overexpression of Igf2 can promote a wide spectrum of tumors through both autocrine and endocrine mechanisms (Rogler et al., 1994).

Another recent study demonstrated reduced tumor growth in RIP-Tag transgenic mice that did not express IGF2 owing to gene disruption (Christofori et al., 1994). In RIP-Tag transgenic mice, the appearance of β-cell tumors in the pancreas is correlated with a high level of local expression of the Igf2 gene, and, interestingly, both maternal and paternal alleles are activated, providing an additional example of the relaxation of imprinting in tumors (see later discussion). The studies mentioned suggest a role for over- and underexpression of the IGF2 gene in tumor formation in mice and also possible local, autocrine effects of IGF2. In this respect it is relevant to note, however, that rodents and humans differ in that humans retain high serum levels of IGF2 during adulthood while rodents do not (Rogler *et al.*, 1994).

Interestingly, when sporadic Wilms' tumors were analyzed for IGF2 and *H19* imprinted gene expression using RNA polymorphisms, biparental gene expression was observed in a significant proportion of tumors (Rainier *et al.*, 1993; Ogawa *et al.*, 1993a,b). This observation was interpreted as a "relaxation of imprinting" (ROI). In another study, DNA analyses indicated that while ROI is associated with variable increases in IGF2 expression, it is consistently associated with loss of H19 expression and methylation of the maternal *H19* allele so that the two parental alleles of the *H19* gene become equally methylated (Steenman *et al.*, 1994; Moulton *et al.*, 1994). The loss of expression of H19, which may function as a tumor suppressor (Hao *et al.*, 1993), may also contribute to tumor formation (Steenman *et al.*, 1994). Fetal overgrowth and childhood tumor predisposition in BWS and sporadic Wilms' tumors might then arise either from paternal 11p15.5 disomy or ROI, both of which could lead to IGF2 overexpression. Given that IGF2 is a fetal growth factor, this would also result in a "large for gestational age" fetus. Continued overexpression of IGF2 postnatally may then be one of several genetic alterations that lead to tumor development.

We propose that the apparent relaxation of imprinting could result from the lack of expression or underexpression of one or more factors that recognize the imprint and promote monoallelic expression. This deficiency could result either from a failure to express the factor at any time during development, or loss of expression of the factor during tumor formation. In the first scenario, there would be a continued proliferation of stem cells that results in fetal overgrowth and an inability to "size regulate," and these stem cells would never establish a fully imprinted phenotype. This explanation could apply to BWS and to cases of Wilms' tumor in which a constitutional loss of the imprinted phenotype is observed for the IGF2 gene. In the second scenario, the actual imprint or "mark" may exist in the tumor cell but not be recognized by the appropriate regulatory factors, which would lead to biallelic expression. Thus, in Wilms' tumors, the kidney stem cells (i.e., nephroblasts) may not have lost their imprint (i.e., chromosomal mark) but rather have ceased to express the factors required for allelic inactivation. A study of other types of stem cells and the methylation status of relevant sites should distinguish between these possibilities.

VI.Conclusions and Perspectives

The results and ideas presented in this chapter lead us to propose a model for the regulation of imprinted gene expression in which the imprint that is imposed on the genomes during gametogenesis serves only to mark each chromosome or chromosome region so that the cell can distinguish maternal and paternal alleles. This mark does not by itself determine the ability of an imprinted gene to be expressed, but rather provides for interactions with specific regulatory factors that affect expression of the imprinted allele. This interaction can involve factors that bind to the undermethylated allele and promote expression or factors that bind to the undermethylated allele and inhibit expression. The expression of these factors can be both stage specific and cell specific, and disrupted in clinical situations so that imprinted genes may not always manifest the appropriate imprints.

It is becoming apparent that several imprinted genes are initially bialleli-cally expressed in the early embryo, and that monoallelic expression of a number of imprinted genes is tissue dependent. The expression of factors that promote allele inactivation may need to be coordinated temporally with the establishment of the complete methylation imprint in order to prevent repression of both alleles. The nature of these factors remains unknown, although egg modifier factors, X-linked imprinting factors, and chromobox-containing homologs of *Drosophila* segmentation genes and modifiers of position-effect variegation are likely candidates. A thorough understanding of how imprinted genes are regulated will require the identi-fication and functional analysis of the genes that encode these factors.

These considerations suggest a possible explanation for the existence of genetic imprinting. Based on the tissue specificity of imprinting effects for such genes as the *Igf2* and *Ins* genes, we suggest that imprinting exists as a reliable method for achieving an incremental increase or reduction in the expression of particular autosomal genes in some tissues but not in others. A prediction inherent in this statement is that for all imprinted genes that encode a protein, there must be some cell type within the embryo or adult in which the imprint is not manifested. This is known to be true for the *Ins* and *Igf2* genes but remains to be tested for other imprinted genes. The need for differential dosage compensation among different cell types, or perhaps different developmental stages, may pre-clude alternative mechanisms of regulating gene expression, for example, by reducing promoter strength, and it may be difficult or impossible to consistently achieve a twofold difference in gene expression by these other mechanisms. Moreover, given the similarities between preferential paternal X inactivation and autosomal imprinted gene regulation, it is possible that the same mechanism that originally provided for dosage compensation of the X chromosome, which is conserved among the different subclasses of mammals, was exploited to achieve conditional dosage compensation for autosomal genes. An analogy between *Xist* and *H19*, another imprinted gene that encodes an RNA that is not translated and may be involved in silencing neighboring genes, follows accordingly. The molecular components of this mechanism may well have been derived in turn from homologous molecules that are expressed in lower organisms and provide for specific determinative events during development.

Acknowledgments

Our research was supported by grants from the National Institutes of Health (GM 49489 to K.E.L. and a National Cancer Institute Cancer Center Support Grant (P30 CA 12227) to the Fels Institute and HD 27533 to J.M.) and a Donaghue Medical Research Foundation Grant in Aid to J.M. We thank our colleagues Drs. Jeff Mann, Aaron Razin, Wolf Reik, Carmen Sapienza, and Prim Singh for sharing their unpublished observations with us

References

Allen, N. D., and Mooslehner, K. A. (1992). Imprinting, transgene methylation and genotype-specific modification. *Semin. Dev. Biol.* **3**, 87–98.

Allen, N. D., Norris, M. L., and Surani, M. A. (1990). Epigenetic control of transgene expression and imprinting by genotype-specific modifiers. *Cell (Cambridge, Mass.)* **61**, 853–861.

Babinet, C., Richoux, V., Guenet, J.-L., and Renard, J.-P. (1990). The DDK inbred strain as a model for the study of interactions between parental genomes and egg cytoplasm in mouse preimplantation development. *Development (Cambridge, UK), Suppl.*, pp. 81–88.

Baldacci, P. A., Richoux, V., Renard, J.-P., Guenet, J. L., and Babinet, C. (1992). The locus Om, responsible for the DDK syndrome, maps close to Sigje on mouse chromosome 11. *Mamm. Genome* **2**, 100–105.

Barlow, D. P., Stöger, R., Herrmann, B. G., Saito, K., and Schweifer, N. (1991). The mouse insulin-like growth factor type-2 receptor is imprinted and closely linked to the T^{me} locus. *Nature (London)* **349**, 84–87.

Bartolomei, M. S., Zemel, S., and Tilghman, S. M. (1991). Parental imprinting of the mouse H19 gene. *Nature (London)* **351**, 153–155.

Bartolomei, M. S., Webber, A. L., Brunkow, M. E., and Tilghman, S. M. (1993). Epigenetic mechanisms underlying the imprinting of the mouse H19 gene. *Genes Dev.* **7**, 1663–1673.

Barton, S. C., Surani, M. A. H., and Norris, M. L. (1984). Role of paternal and maternal genomes in mouse development. *Nature (London)* **311**, 374–376.

Beach, D. H., and Klar, A. J. S. (1984). Rearrangements of the transosable mating-type cassettes of fission yeast. *EMBO J.* **3**, 603–610.

Beldjord, C., Henry, I., Bennani, C., Vanhaeke, D., and Labie, D. (1992). Uniparental disomy: A novel mechanism for thalassemia major. *Blood* **80**, 287–289.

Ben-Hattar, J., and Jiricny, J. (1988). Methylation of single CpG dinucleotides within a promoter element of the Herpes simplex virus tk gene reduces its transcription. *Gene* **65**, 219–227.

Brandeis, M., Kafri, T., Ariel, M., Chaillet, J. R., McCarrey, J., Razin, A., and Cedar, H. (1993). The ontogeny of allele-specific methylation associated with imprinted genes in the mouse. *EMBO J.* **12,** 3669–3677.

Brown, E. H., Iqbal, M. A., Stuart, S., Hatton, K. S., Valinsky, J., and Schildkraut, C. L. (1987). Rate of replication of the murine immunoglobulin heavy-chain locus: Evidence that the region is part of a single replicon. *Mol. Cell. Biol.* **7,** 450–457.

Bücher, T., Linke, I. M., Dunnwald, M., West, J. D., and Cattanach, B. M. (1986). X^{ce} genotype has no impact on the effect of imprinting on X-chromosome expression in the mouse yolk sac endoderm. *Genet. Res.* **47,** 43–48.

Cattanach, B. M. (1974). Position effect variegation in the mouse. *Genet. Res.* **23,** 291–306.

Cattanach, B. M. (1986). Parental origin effects in mice. *J. Embryol. Exp. Morphol.* **97,** Suppl., 137–150.

Cattanach, B. M. (1989). Mammalian chromosome imprinting. *Genome* **31,** 1083–1084.

Cattanach, B. M. and Beechey, C. V. (1990). Autosomal and X-chromosome imprinting. *Development (Cambridge, UK), Suppl.,* pp. 63–72.

Cattanach, B. M., and Kirk, M. (1985). Differential activity of maternally and paternally derived chromosome regions in mice. *Nature (London)* **315,** 496–498.

Cattanach, B. M., Barr, J. A., Evans, E. P., Burtenshaw, M., Beechey, C. V., Leff, S. E., Brannan, C. I., Copeland, N. G., Jenkins, N. A., and Jones, J. (1992). A candidate mouse model for Prader-Willi syndrome which shows an absence of Snrpn expression. *Nat. Genet.* **2,** 270–274.

Cedar, H. (1988). DNA methylation and gene activity. *Cell (Cambridge, Mass.)* **53,** 3–4.

Chaillet, J. R. (1992). DNA methylation and genomic imprinting in the mouse. *Semin. Dev. Biol.* **3,** 99–105.

Chaillet, J. R. (1994). Genomic imprinting: Lessons from mouse transgenes. *Mutat. Res. Fundam. Mol. Mech. Mutagen.* **307,** 441–449.

Chaillet, J. R., Vogt, T. F., Beier, D. R., and Leder, P. (1991). Parental-specific methylation of an imprinted transgene is established during gametogenesis and progressively changes during embryogenesis. *Cell (Cambridge, Mass.)* **66,** 77–83.

Chan, C.-T. J., Clayton-Smith, J., Cheng, X.-J., Buxton, T., Webb, T., Pembrey, M. E., and Malcolm, S. (1993). Molecular mechanisms in Angelman syndrome: A survey of 93 patients. *J. Med. Genet.* **30,** 895–902.

Christofori, B., Naik, P., and Hanahan, D. (1994). A second signal supplied by insuline-like growth factor II in oncogene-induced tumorigenesis. *Nature (London)* **369,** 414–418.

Clayton-Smith, J., Webb, T., Pembrey, M. E., Nichols, M., and Malcolm, S. (1992). Maternal origin of deletion 15q11-13 in 25/25 cases of Angelman syndrome. *Hum. Genet.* **88,** 376–378.

Crouse, H. V. (1960). The controlling element in sex chromosome behavior in Sciara. *Genetics* **45,** 1429–1443.

Dao, D. D., Schroeder, W. T., Chao, L. Y., Kikuchi, H., Strong, L. C., Riccardi, V. M., Pathak, S., Nichols, W. W., and Saunders, G. F. (1987). Genetic mechanisms of tumor-specific loss of 11p sequences in Wilms' tumor. *Am J. Med. Genet.* **41,** 202–217.

Davies, S. M. (1994). Developmental regulation of genomic imprinting of the IGF2 gene in human liver. *Cancer Res.* **54,** 2560–2562.

DeChiara, T. M., Efstratiadis, A., and Robertson, E. J. (1990). A growth-deficiency phenotype in heterozygous mice carrying an insulin-like growth factor II gene disrupted by targeting. *Nature (London)* **345,** 78–80.

DeChiara, T. M., Robertson, E. J., and Efstratiadis, A. (1991). Parental imprinting of the mouse insulin-like growth factor II gene. *Cell (Cambridge, Mass.)* **64,** 849–859.

Doerfler, W. (1983). DNA methylation and gene activity. *Annu. Rev. Biochem.* **52,** 93–124.

Drouin, R., Lemieux, N., and Richer, C. L. (1990). Analysis of DNA replication during S phase by means of dynamic chromosome banding at high resolution. *Chromosoma* **99,** 273–280.

Efstratiadis, A. (1994). Parental imprinting of autosomal mammalian genes. *Curr. Opin. Genet. Dev.* **4,** 265–280.

Egel, R., Beach, D. H., and Klar, A. J. S. (1984). Genes required for initiation and resolution steps of mating-type switching in fission yeast. *Proc. Natl. Acad. Sci. U.S.A.* **81,** 3481–3485.

Eicher, E. M. (1970). X-autosome translocations in the mouse: Total inactivation versus partial inactivation of the X chromosome. *Adv. Genet.* **15,** 175–259.

Engler, P., Haasch, D., Pinkert, C. A., Doglio, L., Glymour, M., Brinster, R., and Storb, U. (1991). A strain-specific modifier on mouse chromosome 4 controls the methylation of independent transgene loci. *Cell (Cambridge, Mass.)* **65,** 939–947.

Epner, E., Kim, C. G., and Groudine, M. (1992). What does the locus control region control? *Curr. Opin. Genet. Dev.* **2,** 262–264.

Eversole-Cire, P., Ferguson-Smith, A. C., Sasaki, H., Brown, K. D., Cattanach, B. M., Gonzales, F. A., Azim Surani, M., and Jones, P. A. (1993). Activation of an imprinted Igf 2 gene in mouse somatic cell cultures. *Mol. Cell. Biol.* **13,** 4928–4938.

Feil, R., Walter, J., Allen, N. D., and Reik, W. (1994). Developmental control of allelic methylation in the imprinted mouse Igf2 and H19 genes. *Development (Cambridge, UK)* **120,** 2933–2943.

Feinberg, A. P. (1993). Genomic imprinting and gene activation in cancer. *Nat. Genet.* **4,** 110–113.

Ferguson-Smith, A. C., Sasaki, H., Cattanach, B. M., and Surani, M. A. (1993). Parental-origin-specific epigenetic modification of the mouse H19 gene. *Nature (London)* **362,** 751–755.

Filson, A. J., Louvi, A., Efstratiadis, A., and Robertson, E. J. (1993). Rescue of the T-associated maternal effect in mice carrying null mutations in Igf2 and Igf2r, two reciprocally imprinted genes. *Development (Cambridge, UK)* **118,** 731–736.

Forejt, J., and Gregorova, S. (1992). Genetic analysis of genomic imprinting: An Imprintor-1 gene controls inactivation of the paternal copy of the mouse T*me* locus. *Cell (Cambridge, Mass.)* **70,** 443–450.

Frels, W. I., and Chapman, V. M. (1979). Paternal X chromosome expression in extraembryonic membranes of XO mice. *J. Exp. Zool.* **210,** 553–560.

Fundele, R., Howlett, S. K., Kothary, R., Norris, M. L., Mills, W. E., and Surani, M. A. (1991). Developmental potential of parthenogenetic cells: Role of genotype-specific modifiers. *Development (Cambridge, UK)* **113,** 941–946.

Gautsch, J. W., and Wilson, M. C. (1983). Delayed de novo methylation in teratocarcinoma cells suggests additional tissue specific mechanisms for controlling gene expression. *Nature (London)* **301,** 32–37.

Giannoukakis, N., Deal, C., Paquette, J., Goodyer, C. G., and Polychronakos, C. (1993). Parental genomic imprinting of the human IGF2 gene. *Nat. Genet.* **4,** 98–101.

Giddings, S. J., King, C. D., Harman, K. W., Flood, J. F., and Carnachi, L. R. (1994). Allele specific inactivation of insulin 1 and 2, in the mouse yolk sac, indicates imprinting. *Nat. Genet.* **6,** 310–313.

Gilligan, A., and Solter, D. (1994). The role of imprinting in early mammalian development. *In* "Parental Imprinting: Causes and Consequences" (R. Ohlsson, M. Ritzen, and K. Hall, eds.). Cambridge Univ. Press, Cambridge, UK (in press).

Glenn, C. C., Porter, K. A., Jong, M. T. C., Nicholls, R. D., and Driscoll, D. J. (1993). Functional imprinting and epigenetic modification of the human SNRPN gene. *Hum. Mol. Genet.* **2,** 2001–2005.

Goldman, M. A., Holmquist, G. P., Gray, M. C., Caston, L. A., and Nag, A. (1984). Replication timing of genes and middle repetitive sequences. *Science* **224,** 686–692.

Gorman, M., Kuroda, M. I., and Baker, B. S. (1993). Regulation of sex-specific binding of the maleless dosage compensation protein to the male X chromosome. *Cell (Cambridge, Mass.)* **72,** 39–49.

Grant, M., Zuccotti, M., and Monk, M. (1992). Methylation of CpG sites of two X-linked genes coincides with X-inactivation in the female mouse embryo but not in the germ line. *Nat. Genet.* **2,** 161–166.

Grant, S. G. and Chapman, V. M. (1988). Mechanisms of X-chromosome regulation. *Annu. Rev. Genet.* **22,** 199–233.

Hamvas, R. M., Reik, W., Gaunt, S. J., Brown, S. D., and Singh, P. B. (1992). Mapping of a mouse homologue of a heterochromatin protein gene to the X chromosome. *Mamm. Genome* **2,** 72–75.

Hao, Y., Crenshaw, T., Moulton, T., Newcomb, E., and Tycko, B. (1993). Tumor-suppressor activity of H19 RNA. *Nature (London)* **365,** 764–765.

Harper, M. I., Fosten, M., and Monk, M. (1982). Preferential paternal X inactivation in extraembryonic tissues of early mouse embryos. *J. Embryol. Exp. Morphol.* **67,** 127–135.

Hatada, I., Sugama, T., and Mukai, T. (1993). A new imprinted gene cloned by a methylation-sensitive genome scanning method. *Nucleic Acids Res.* **21,** 5577–5582.

Hatton, K. S., Dhar, V., Brown, E. H., Iqbal, M. A., Stuart, S., Didamo, V. T., and Schildkraut, C. L. (1988). Replication program of active and inactive multigene families in mammalian cells. *Mol. Cell. Biol.* **8,** 2149–2158.

Hayashizaki, Y., Shibata, H., Hirotsune, S., Sugino, H., Okazaki, Y., Sasaki, N., Hirose, K., Imoto, H., Okuizumi, H., Muramatsu, M., Komatsubara, H., Shiroishi, T., Moriwaki, K., Katsuki, M., Hatano, N., Sasaki, H., Ueda, T., Mise, N., Takagi, N., Plass, C., and Chapman, V. M. (1994). Identification of an imprinted U2af binding protein related sequence on mouse chromosome 11 using the RLGS method. *Nat. Genet.* **6,** 33–39.

Henry, I., Bonaiti-Pellie, C., Chehensse, V., Bekdjord, C., Schwartz, C., Uterman, G., and Junien, C. (1991). Uniparental disomy in genetic cancer predisposing syndrome. *Nature (London)* **351,** 665–667.

Heutink, P., van der Mey, A. G., Sandkuijl, L. A., van Gils, A. P. G., Bardoel, A., Breedveld, G. J., van Vliet, M., van Ommen, G. J., Cornelisse, C. J., Oostra, B. A., Weber, J. L., and Devilee, P. (1992). A gene subject to genomic imprinting and responsible for hereditary paraganglioma maps to chromosome 11q23-qter. *Hum. Mol. Genet.* **1,** 7–10.

Holmquist, G. P. (1987). Role of replication time in the control of tissue-specific gene expression. *Am. J. Hum. Genet.* **40,** 151–173.

Howlett, S. K. (1991). Genomic imprinting and nuclear totipotency during embryonic development. *Int. Rev. Cytol.* **127,** 175–192.

Hulten, M., Armstrong, S., Challinor, P., Gould, C., Hardy, G., Leedham, P., Lee, T., and McKeown, C. (1991). Genomic imprinting in an Angelman and Prader-Willi translocation family. *Lancet* **338,** 638–639.

Iguchi-Ariga, S. M., and Schaffner, W. (1989). CpG methylation of the cAMP-responsive enhancer/promoter sequence TGACGTCA abolishes specific factor binding as well as transcriptional activation. *Genes Dev.* **3,** 612–619.

Jinno, Y., Yun, K., Nishiwaki, K., Kubota, T., Ogawa, O., Reeve, A. E., and Nikawa, N. (1994). Mosaic and polymorphic imprinting of the WT-1 gene in humans. *Nat. Genet.* **6,** 305–309.

Kafri, T., Ariel, M., Brandeia, M., Shemer, R., Urven, L., McCarrey, J., Cedar, H., and Razin, A. (1992). Developmental pattern of gene-specific DNA methylation in the mouse embryo and germ line. *Genes Dev.* **6,** 705–714.

Kafri, T., Gao, X., and Razin, A. (1993). Mechanistic aspects of genome-wide demethylation in the preimplantation mouse embryo. *Proc. Natl. Acad. Sci. U.S.A* **90,** 10558–10562.

Kalousek, D. K., Langlois, S., Barrett, I., Yam, I., Wilson, D. R., Howard-Peebles, P. N., Johnson, M. P., and Giorgiutti, E. (1993). Uniparental disomy for chromosome 16 in humans. *Am. J. Hum. Genet.* **52**, 8–16.

Kanbour-Shakir, A., Kunz, H. W., and Gill, T. J. (1993). Differential genomic imprinting of major histocompatibility complex class I antigens in the placenta of the rat. *Biol. Reprod.* **48**, 977–986.

Kaslow, D. C., and Migeon, B. R. (1987). DNA methylation stabilizes X chromosome inactivation in eutherians, but not in marsupials: Evidence for multi-step maintenance of mammalian X dosage compensation. *Proc. Natl. Acad. Sci. U.S.A.* **84**, 6210–6214.

Kay, G. F., Barton, S. C., Surani, M. A., and Rastan, S. (1994). Imprinting and X chromosome counting mechanisms determine Xist expression in early mouse development. *Cell (Cambridge, Mass.)* **77**, 639–650.

Kitsberg, D., Sellig, S., Brandeis, M., Simon, I., Keshet, I., Driscoll, D. J., Nicholls, R. D., and Cedar, H. (1993). Allele-specific replication timing of imprinted gene regions. *Nature (London)* **364**, 459–463.

Klar, A. J. S. (1990a). The developmental fate of fission yeast cells is determined by the pattern of inheritance of parental and grandparental DNA strands. *EMBO J.* **9**, 1407–1415.

Klar, A. J. S. (1990b). Regulation of fission yeast mating-type interconversion by chromosome imprinting. *Development (Cambridge, UK), Suppl.,* pp. 3–8.

Knoll, J. H. M., Nicholls, R. D., Magenis, R. E., Graham, J. M., Lalande, M., and Latt, S. A. (1989). Angelman and Prader-Willi syndromes share a common chromosome 15 deletion but differ in parental origin of the deletion. *Am. J. Med. Genet.* **32**, 285–290.

Knoll, J. H. M., Glatt, K. A., Nicholls, R. D., Malcolm, S., and Lalande, M. (1991). Chromosome 15 uniparental disomy is not frequent in Angelman syndrome. *Am. J. Hum. Genet.* **48**, 16–21.

Knoll, J. H. M., Cheng, S.-D., and Lalande, M. (1994). Allele-specificity of DNA replication timing in the Prader-Willi syndrome imprinted chromosomal region. *Nat. Genet.* **6**, 41–45.

Koufos, A., Grundy, P., Morgan, K., Aleck, K., Hadro, T., Lampkin, B., Kalbakji, A., and Cavenee, W.K. (1989). Familial Weidemann-Beckwith syndrome and a second Wilms' tumor locus both map to 11p15.5. *Am. J. Hum. Genet.* **44**, 711–719.

Kratzer, P. G., Chapman, V. M., Lambert, H., Evans, R. E., and Liskay, R. M. (1983). Differences in the DNA of the inactive X chromosome of fetal and extraembryonic tissues of mice. *Cell (Cambridge, Mass.)* **33**, 37–42.

Krumlauf, R., Chapman, V. M., Hammer, R. E., Brinster, R. L., and Tilghman, S. M. (1986). Differential expression of alpha-fetoprotein genes on the inactive X chromosome in extraembryonic and somatic tissues of a transgenic line. *Nature (London)* **319**, 224–226.

Latham, K. E. (1994). Strain-specific differences in mouse oocytes and their contributions to epigenetic inheritance. *Development (Cambridge, UK)* **120**, 3419–3426.

Latham, K. E., and Solter, D. (1991). Effect of egg composition on the developmental capacity of androgenetic mouse embryos. *Development (Cambridge, UK)* **113**, 561–568.

Latham, K. E., Doherty, A. S., Scott, C. D., and Schultz, R. M. (1994). Igf2r and Igf2 gene expression in androgenetic, gynogenetic, and parthenogenetic preimplantation mouse embryos: Absence of regulation by genomic imprinting. *Genes Dev.* **8**, 290–299.

Lau, M. M. H., Stewart, C. E. H., Liu, Z., Bhatt, H., Rotwein, P., and Stewart, C. L. (1994). Loss of the imprinted Igf2/cation-independent mannose 6-phosphate receptor results in fetal overgrowth and perinatal lethality. *Genes Dev.* **8**, 2953–2963.

Lee, J. E., Tantravahi, U., Boyle, A. L., and Efstratiadis, A. (1993). Parental imprinting of an Igf2 transgene. *Mol. Reprod. Dev.* **35**, 382–390.

Leff, S. E., Brannan, C. I., Reed, M. L., Ozcelick, T., Francke, U., Copeland, N. G., and Jenkins, N. (1992). Maternal imprinting of the mouse Snrpn gene and conserved linkage homology with the human Prader-Willi syndrome region. *Nat. Genet.* **2**, 259–264.

Lewis, J. D., Meehan, R. R., Henzel, W. J., Maurer-Fogy, I., Jeppesen, P., Klein, F., and

Bird, A. (1992). Purification, sequence, and cellular localization of a novel chromosomal protein that binds to methylated DNA. Cell (Cambridge, Mass.) **69**, 905–914.

Li, E., Bestor, T. H., and Jaenisch, R. (1992). Targeted mutation of the DNA methyltransferase gene results in embryonic lethality. Cell (Cambridge, Mass.) **69**, 915–926.

Li, E., Beard, C., and Jaenisch, R. (1993). Role for DNA methylation in genomic imprinting. Nature (London) **366**, 362–365.

Lock, L. F., Takagi, N., and Martin, G. R. (1987). Methylation of the Hprt gene on the inactive X occurs after chromosome inactivation. Cell (Cambridge, Mass.) **48**, 39–46.

Locke, J., Kotarski, M. A., and Tartof, K. D. (1988). Dosage-dependent modifiers of position effect variegation in Drosophila and a mass action model that explains their effect. Genetics **120**, 181–198.

Malcolm, S., Clayton-Smith, J., and Nicholls, M. (1991). Uniparental-paternal disomy in Angelmann's syndrome. Lancet **337**, 694–697.

Mann, J. R. (1986). DDK egg-foreign sperm incompatibility in mice is not between the pronuclei. J. Reprod. Fertil. **76**, 779–781.

Mann, J. R., and Lovell-Badge, R. H. (1984). Inviability of parthenogenones is determined by pronuclei, not egg cytoplasm. Nature (London) **310**, 66–67.

Mann, J. R., Gadi, I., Harbison, M. L., Abbondanzu, S. J., and Stewart, C. L. (1990). Androgenetic mouse embryo stem cells are pluripotent and cause skeletal defects in chimeras: Implications for genetic imprinting. Cell **62**, 251–260.

Mannens, M., Slater, R. M., Heyting, C., Bliek, J., de Kraker, J., Coad, N., de Pagter-Holthuizen, P., and Pearson, P. L. (1988). Molecular nature of genetic changes resulting in loss of heterozygosity of chromosome 11 in Wilms' tumours. Hum. Genet. **81**, 41–48.

Mariman, E. C. M., van Beersum, S. E. C., Cremers, C. W. R. J., van Baars, F. M., and Ropers, H. H. (1993). Analysis of a second family with hereditary paragangliomas maps to chromosome 11q23-qter. Hum. Genet. **91**, 357–361.

Mascari, M. J., Gottlieb, W., Rogan, P. K., Butler, M. G., Waller, D. A., Armour, J. A., Jeffreys, A. J., Ladda, R. L., and Nicholls, R. D. (1992). The frequency of uniparental disomy in Prader-Willi syndrome. Implications for molecular diagnosis. N. Engl. J. Med. **326**, 1599–1607.

McGowan, R., Campbell, R., Peterson, A., and Sapienza, C. (1989). Cellular mosaicism in the methylation and expression of hemizygous loci in the mouse. Genes Dev. **3**, 1669–1676.

McGrath, J. and Solter, D. (1983). Nuclear transplantation in the mouse embryo by microsurgery and cell fusion. Science 220, 1300-1302.

McGrath, J., and Solter, D. (1984a). Completion of mouse embryogenesis requires both maternal and paternal genomes. Cell (Cambridge, Mass.) **37**, 179–183.

McGrath, J., and Solter, D. (1984b). Maternal T^{hp} lethality in the mouse is a nuclear, not cytoplasmic, defect. Nature (London) **308**, 550–551.

Migeon, B. R. (1990). Insights into X chromosome inactivation from studies of species variation, DNA methylation and replication, and vice versa. Genet. Res. **56**, 91–98.

Migeon, B. R., de Beur, S. J., and Axelman, J. (1989). Frequent derepression of G6PD and HPRT on the marsupial inactive X chromosome associated with cell proliferation in vitro. Exp. Cell Res. **182**, 597–609.

Mohandes, T., Sparkes, R. S., and Shapiro, L. J. (1981). Reactivation of an inactive X chromosome: Evidence for X inactivation by DNA methylation. Science **211**, 393–396.

Monk, M. (1986). Methylation and the X chromosome. BioEssays **4**, 204–208.

Monk, M., Boubelik, M., and Lehnert, S. (1987). Temporal and regional changes in DNA methylation in the embryonic, extraembryonic and germ cell lineages during mouse embryo development. Development (Cambridge, UK) **99**, 371–382.

Moore, T. F., and Whittingham, D. G. (1992). Imprinting of phosphoribosyltransferases during preimplantation development of the mouse mutant, $Hprt^{b-m3}$. Development (Cambridge, UK) **115**, 1011–1016.

Mottus, R., Reeves, R., and Grigliatti, T. A. (1980). Butyrate suppression of position-effect variegation in *Drosophila melanogaster. Mol. Gen. Genet.* **178**, 465–469.

Moulton, T., Crenshaw, T., Hao, Y., Noosikasuwan, J., Lin, N., Dembitzer, F., Hensle, T., Weiss, L., McMorrow, L., Loew, T., Kraus, W., Gerald, W., and Tycko, B. (1994). Epigenetic lesions at the H19 locus in Wilms' tumour patients. *Nat. Genet.* **7**, 440–447.

Mutter, G. L., Stewart, C. L., Chaponot, M. L., and Pomponio, R. J. (1993). Oppositely imprinted genes *H19* and insulin-like growth factor 2 are coexpressed in human androgenetic trophoblast. *Am. J. Hum. Genet.* **53**, 1096–1102.

Naumova, A., and Sapienza, C. (1994). The genetics of retinoblastoma, revisited. *Am. J. Hum. Genet.* **54**, 264–273.

Nicholls, R. D. (1993). Genomic imprinting and uniparental disomy in Angelman and Prader-Willi syndromes: A review. *Am J. Med. Genet.* **46**, 16–25.

Norris, D. P., Patel, D., Kay, G. F., Penny, G. D., Brockdorff, N., Sheardown, S. A., and Rastan, S. (1994). Evidence that random and imprinted Xist expression is controlled by preemptive methylation. *Cell (Cambridge, Mass.)* **77**, 41–51.

Ogawa, O., Becroft, D. M., Morison, I. M., Eccles, M. R., Skeen, J. E., Mauger, D. C., and Reeve, A. E. (1993a). Constitutional relaxation of insulin-like growth factor II gene imprinting associated with Wilms' tumour and gigantism. *Nat. Genet.* **5**, 408–412.

Ogawa, O., Eccles, M. R., Szeto, J., McNoe, L. A., Yun, K., Maw, M. A., Smith, P. J., and Reeve, A. E. (1993b). Relaxation of insulin-like growth factor II gene imprinting implicated in Wilms' tumour. *Nature (London)* **362**, 749–751.

Ohlsson, R., Nyström, A., Pfeifer-Ohlsson, S., Tohonen, V., Hedborg, F., Schofield, P., Flam, F., and Ekström, T. J. (1993). *Igf2* is parentally imprinted during human embryogenesis and in the Beckwith-Weidemann syndrome. *Nat. Genet.* **4**, 94–97.

Ohlsson, R., Hedborg, F., Holmgren, L., Walsh, C., and Ekström, T. J. (1994). Overlapping patterns of IGF2 and *H19* expression during human development: Biallelic IGF2 expression correlates with a lack of *H19* expression. *Development (Cambridge, UK)* **120**, 361–368.

Ozcelik, T., Leff, S., Robinson, W., Donlon, T., Lalande, M., Sanjines, E., Schinzel, A., and Francke, U. (1992). Small nuclear ribonucleoprotein polypeptide N (SNRPN), an expressed gene in the Prader-Willi syndrome critical region. *Nat. Genet.* **2**, 265–269.

Papaioannou, V. E., and West, J. D. (1981). Relationship between the parental origin of the X chromosomes, cell lineage and X chromosome expression in mice. *Genet. Res.* **37**, 183–197.

Pearce, J. J., Singh, P., and Gaunt, S. J. (1992). The mouse has a Polycomb-like chromobox gene. *Development (Cambridge, UK)* **114**, 921–929.

Pourcel, C. (1993). Imprinting and methylation. *In* "Genes in Mammalian Reproduction" (R. B. L. Gwatkin, ed.), pp. 173–184. Wiley-Liss, New York.

Rainier, S., Johnson, L. A., Dobry, C. J., Ping, A. J., Grundy, P. E., and Feinberg, A. P. (1993). Relaxation of imprinted genes in human cancer. *Nature (London)* **362**, 747–749.

Rastan, S., Kaufman, M. H., Handyside, A. H., and Lyon, M. F. (1980). X-chromosome inactivation in extraembryonic membranes of diploid parthenogenetic mouse embryos demonstrated by differential staining. *Nature (London)* **288**, 172–173.

Reed, M. L., and Leff, S. E. (1994). Maternal imprinting of human SNRPN, a gene deleted in Prader-Willi syndrome. *Nat. Genet.* **6**, 163–167.

Reik, W., Howlett, S. K., and Surani, M. A. (1990). Imprinting by DNA methylation: From transgenes to endogenous sequences. *Development (Cambridge, UK), Suppl.,* pp. 99–106.

Reik, W., Römer, I., Barton, S. C., Surani, M. A., Howlett, S. K., and Klose, J. (1993). Adult phenotype in the mouse can be affected by epigenetic events in the early embryo. *Development (Cambridge, UK)* **119**, 933–942.

Reis, A., Dittrich, B., Greger, V., Buiting, K., Lalande, M., Gillessen-Kaesbach, G., Anvret, M., and Horsthemke, B. (1994). Imprinting mutations suggested by abnormal DNA methylation patterns in familial Angelman and Prader-Willi syndromes. *Am. J. Hum. Genet.* **54,** 741–747.

Renard, J.-P., Baldacci, P., Richoux-Duranthon, V., Pourmin, S., and Babinet, C. (1994). A maternal factor affecting mouse blastocyst formation. *Development (Cambridge, UK)* **120,** 797–802.

Riggs, A. D., and Pfeifer, G. D. (1992). X-chromosome inactivation and cell memory. *Trends Genet.* **8,** 169–174.

Rogler, C. E., Yang, D., Rossetti, L., Donohoe, J., Alt, E., Chang, C. J., Rosenfeld, R., Neely, K., and Hintz, R. (1994). Altered body composition and increased frequency of diverse malignancies in insulin-like growth factor -II transgenic mice. *J. Biol. Chem.* **269,** 13779–13784.

Sapienza, C., Paquette, J., Tran, T. H., and Peterson, A. (1989). Epigenetic and genetic factors affect transgene methylation imprinting. *Development (Cambridge, UK)* **107,** 165–168.

Sapienza, C., Paquette, J., Pannunzio, P., Albrechtson, S., and Morgan, K. (1992). The polar-lethal Ovum mutant gene maps to the distal portion of mouse chromosome 11. *Genetics* **132,** 241–246.

Sasaki, H., Hamada, T., Ueda, T., Seki, R., Hagashinkakgawa, T., and Sasaki, Y. (1991). Inherited type of allelic methylation variations in a mouse chromosome region where an integrated transgene shows methylation imprinting. *Development (Cambridge, UK)* **111,** 573–581.

Sasaki, H., Jones, P. A., Chaillet, J. R., Ferguson-Smith, A. C., Barton, S. C., Reik, W., and Surani, M. A. (1992). Parental imprinting: Potentially active chromatin of the repressed maternal allele of the mouse insulin-like growth factor II (*Igf2*) gene. *Genes Dev.* **6,** 1843–1856.

Selig, S., Okumura, K., Ward, D. C., and Cedar, H. (1992). Delineation of DNA replication time zones by fluorescence in situ hybridization. *EMBO J.* **11,** 1217–1225.

Shemer, R., Kafri, T., O'Connell, A., Eisenberg, S., Breslow, J. L., and Razin, A. (1991). Methylation changes in the apolipoprotein AI gene during embryonic development of the mouse. *Proc. Natl. Acad. Sci. U.S.A.* **88,** 11300–11304.

Singer-Sam, J., Grant, M., LeBon, J. M., Okuyama, K., Chapman, V., Monk, M., and Riggs, A. D. (1990). Use of a HpaII-polymerase chain reaction assay to study DNA methylation in the Pgk-1 CpG island of mouse embryos at the time of X-chromosome inactivation. *Mol. Cell. Biol.* **10,** 4987–4989.

Singer-Sam, J., Chapman, V., LeBon, J. M., and Riggs, A. D. (1992). Parental imprinting studied by allele-specific primer extension after PCR: Paternal X chromosome-linked genes are transcribed prior to preferential paternal X chromosome inactivation. *Proc. Natl. Acad. Sci. U.S.A.* **89,** 10469–10473.

Singh, P. B. (1994). Molecular mechanisms of cellular determination: Their relation to chromatin structure and parental imprinting. *J. Cell Sci.* **107,** 2653.

Singh, P. B., Miller, J. R., Pearce, J., Kothary, R., Burton, R. D., Paro, R., James, T. C., and Gaunt, S. J. (1991). A sequence motif found in a *Drosophila* heterochromatin protein is conserved in animals and plants. *Nucleic Acids Res.* **19,** 89–794.

Smeets, D. F. C. M., Hamel, B. C. J., Nelen, M. R., Smeets, H. J. M., Bollen, J. H. M., Smits, A. P. T., Ropers, H.-H., and van Oost, B. A. (1992). Prader-Willi syndrome and Angelman syndrome in cousins from a family with a translocation between chromosomes 6 and 15. *N. Engl. J. Med.* **326,** 807–811.

Solter, D. (1988). Differential imprinting and expression of maternal and paternal genomes. *Annu. Rev. Genet.* **22,** 127–146.

Solter, D., Dominis, M., and Damjanov, I. (1979). Embryo-derived teratocarcinoma: I. The role of strain and gender in the control of teratocarcinogenesis. *Int. J. Cancer* **24,** 770–772.

Solter, D., Dominis, M., and Damjanov, I. (1981). Embryo-derived teratocarcinoma. III.

Development of tumors from teratocarcinoma-permissive and non-permissive strain embryos transplanted to F1 hybrids. *Int. J. Cancer* **28**, 479–483.

Spence, J. E., Periaccante, R. G., Greig, G. M., Willard, H. F., Ledbetter, D. H., Hejtmancik, J. F., Pollack, M. S., O'Brien, W. E., and Beaudet, A. L. (1988). Uniparental disomy as a mechanism for human genetic disease. *Am. J. Hum. Genet.* **42**, 217–226.

Spotila, L. D., Sereda, L., and Prockop, D. J. (1992). Partial isodisomy for maternal chromosome 7 and short stature in an individual with a mutation at the COLIA2 locus. *Am J. Hum. Genet.* **51**, 1396–1405.

Steenman, M. J. C., Rainier, S., Dobry, C. J., Grundy, P., Horon, I., and Feinberg, A. P. (1994). Loss of imprinting of IGF2 is linked to reduced expression and abnormal methylation of H19 in WILMs' Tumour. *Nat. Genet.* **7**, 433–439.

Stöger, R., Kubicka, P., Liu, C.-G., Kafri, T., Razin, A., Cedar, H., and Barlow, D. P. (1993). Maternal-specific methylation of the imprinted mouse *Igf2r* locus identifies the expressed locus as carrying the imprinting signal. *Cell (Cambridge, Mass.)* **73**, 61–71.

Surani, M. A. H. (1993). Genomic imprinting: Silence of the genes. *Nature (London)* **366**, 302–303.

Surani, M. A. H., Barton, S. C., and Norris, M. L. (1984). Development of reconstituted mouse eggs suggests imprinting of the genome during gametogenesis. *Nature (London)* **308**, 548–550.

Surani, M. A. H., Barton, S. C., and Norris, M. L. (1986). Nuclear transplantation in the mouse: Heritable differences between paternal genomes after activation of the embryonic genome. *Cell (Cambridge, Mass.)* **45**, 127–136.

Surani, M. A. H., Kothary, R., Allen, N. D., Singh, P. B., Fundele, R., Ferguson-Smith, A. C., and Barton, S. C. (1990). Genome imprinting and development in the mouse. *Development (Cambridge, UK) Suppl.*, pp. 89–98.

Sutcliffe, J. S., Nakao, M., Christian, S., Orstavik, K. H., Tommerup, N., Ledbetter, D. H., and Beaudet, A. L. (1994). Deletions of a differentially methylated CpG island at the SNRPN gene define a putative imprinting control region. *Nat. Genet.* **8**, 52–58.

Szabo, P., and Mann, J. (1994). Expression and methylation of imprinted genes during in vitro differentiation of mouse parthenogenetic and androgenetic embryonic stem cell lines. *Development (Cambridge, UK)* **120**, 1651–1660.

Takagi, N., and Oshimura, M. (1973). Fluorescence and giemsa banding studies of the allocyclic X chromosome in embryonic and adult mouse cells. *Exp. Cell Res.* **78**, 127–135.

Takagi, N., and Sasaki, M. (1982). Preferential inactivation of the paternally derived X chromosome in the extraembryonic membranes of the mouse. *Nature (London)* **256**, 640–642.

Tartof, K. D., and Bremer, M. (1990). Mechanisms for the construction and developmental control of heterochromatin formation and imprinted chromosome domains. *Development (Cambridge, UK), Suppl.*, 35–46.

Tartof, K. D., Hobbs, C., and Jones, M. (1989). Towards an understanding of position effect variegation. *Dev. Genet.* **10**, 162–176.

Tsukahara, M., and Kajii, T. (1985). Replication of X chromosomes in complete moles. *Hum. Genet.* **71**, 7–10.

Ueda, T., Yamazaki, K., Suzuki, R., Fujimoto, H., Sasaki, H., Sakaki, Y., and Higashinakagawa, T. (1992). Parental methylation patterns of a transgenic locus in adult somatic tissues are imprinted during gametogenesis. *Development (Cambridge, UK)* **116**, 831–839.

Vanderberg, T. L., Robinson, E. S., Samollow, P. S., and Johnston, P. G. (1987). X-linked gene expression and X-chromosome inactivation: Marsupials, mouse and man compared *In* "Current Topics in Biology and Medical Research" (C. L. Market, ed.), pp. 225–253. Alan R. Liss, New York.

van der Mey, A. G. L., Maaswinkel-Mooy, P. D., Cornelisse, C. J., Schmidt, P. H., and van de Kamp, J. J. P. (1989). Genomic imprinting in hereditary glomus tumours: Evidence for a new genetic theroy. *Lancet* **2**, 1291–1294.

van Lohuizen, M., Frasch, M., Wientjens, E., and Berns, A. (1991). Sequence similarity between the mammalian bmi-1 proto-oncogene and the *Drosophila* regulatory genes Psc and Su(z)2. *Nature (London)* **353**, 353–355.

Voss, R., Ben-Simon, E., Avita, A., Godfrey, S., Zlotogora, J., Dagan, J., Tikochinski, Y., and Hillel, J. (1989). Isodisomy of chromosome 7 in a patient with cystic fibrosis: Could uniparental disomy be common in humans? *Am J. Hum. Genet.* **45**, 373–380.

Vu, T. H., and Hoffman, A. R. (1994). Promoter-specific imprinting of the human insulin-like growth factor-II gene. *Nature (London)* **371**, 714–717.

Wagstaff, J., Knoll, J. H. M., Glatt, K. A., Shugart, Y. Y., Sommer, A., and Lalande, M. (1992). Maternal but not paternal transmission of 15q11-13-linked nondeletion Angelman syndrome leads to phenotypic expression. *Nat. Genet.* **1**, 291–294.

Wakasugi, N. (1974). A genetically determined incompatibility system between spermatozoa and eggs leading to embryonical death in mice. *J. Reprod. Fertil.* **41**, 85–94.

Wakasugi, N., Tomita, T., and Kondo, K. (1967). Differences of fertility in reciprocal crosses between inbred strains of mice DDK, KK and NC. *J. Reprod. Fertil.* **13**, 41–50.

Wake, N., Takagi, N., and Sasaki, M. (1976). Non-random inactivation of X-chromosome in the rat yolk sac. *Nature (London)* **262**, 580–581.

Walsh, C., Glaser, A., Fundele, R., Ferguson-Smith, A., Barton, S., Surani, M. A., and Ohlsson, R. (1994). The non-viability of uniparental mouse conceptuses correlates with the loss of the products of imprinted genes. *Mech. Dev.* **46**, 55–62.

Wang, J.-C. C., Passage, M. B., Yen, P. H., Shapiro, L. J., and Mohanfas, T. K. (1991). Uniparental heterodisomy for chromosome 14 in a phenotypically abnormal familial balanced 13/14 Robertsonian translocation carrier. *Am J. Hum. Genet.* **48**, 1069–1074.

Wang, Z.-Q., Fung, M. R., Barlow, D. P., and Wagner, E. G. (1994). Regulation of embryonic growth and lysosomal targeting by the imprinted Igf2/Mpr gene. *Nature (London)* **372**, 464–467.

Watt, F., and Molloy, P. L. (1988). Cytosine methylation prevents binding to DNA of a HeLa cell transcription factor required for optimal expression of the adenovirus major late promoter. *Genes Dev.* **2**, 1136–1143.

Weksberg, R., Shen, D. R., Fei, Y. L., Song, Q. L., and Squire, J. (1993). Disruption of insulin-like growth factor 2 imprinting in Beckwith-Wiedemann syndrome. *Nat. Genet.* **5**, 143–150.

Welch, T. R., Beischel, L. S., Choi, E., Balakrishnan, K., and Bishoff, N. A. (1990). Uniparental isodisomy 6 associated with deficiencey of the fourth component of complement. *J. Clin. Invest.* **86**, 675–678.

West, J. D., Frels, W. I., Chapman, V. E., and Papaioannou, V. E. (1977). Preferential expression of the maternally derived X chromosome in the mouse yolk sac. *Cell (Cambridge, Mass.)* **12**, 873–882.

Woodage, T., Prasad, M., Dixon, J. W., Selby, R. E., Romain, D. R., Columbano-Green, L. M., Graham, D., Rogan, P. K., Seip, J. R., Smith, A., and Trent, R. J. (1994). Bloom syndrome and maternal uniparental disomy for chromosome 15. *Am. J. Hum. Genet.* **55**, 74–80.

Wu, H., Fassler, R., Schnieke, A., Barker, D., Lee, K.-H., Chapman, V., Franke, U., and Jaenisch, R. (1992). An X-linked human collagen transgene escapes X inactivation in a subset of cells. *Development (Cambridge, UK)* **116**, 687–695.

Xu, Y., Goodyer, C. G., Deal, C., and Polychronakis, C. (1993). Functional polymorphism in the parental imprinting of the human Igf2r gene. *Biochem. Biophys. Res. Commun.* **197**, 747–754.

Zemel, S., Bartolomei, M. S., and Tilghman, S. M. (1992). Physical linkage of two mammalian imprinted genes. *Nat. Genet.* **2**, 61–65.

Molecular Mechanisms for Passive and Active Transport of Water

T. Zeuthen

Department of Medical Physiology, The Panum Institute, University of Copenhagen, DK-2200 Copenhagen N, Denmark

Water crosses cell membranes by passive transport and by secondary active cotransport along with ions. While the first concept is well established, the second is new. The two modes of transport allow cellular H_2O homeostasis to be viewed as a balance between H_2O leaks and H_2O pumps. Consequently, cells can be hyperosmolar relative to their surroundings during steady states. Under physiological conditions, cells from leaky epithelia may be hyperosmolar by roughly 5 mosm liter^{-1}, under dilute conditions, hyperosmolarities up to 40 mosm liter^{-1} have been recorded.

Most intracellular H_2O is free to serve as solvent for small inorganic ions. The mechanism of transport across the membrane depends on how H_2O interacts with the proteinaceous or lipoid pathways. Osmotic transport of H_2O through specific H_2O channels such as CHIP 28 is hydraulic if the pore is impermeable to the solute and diffusive if the pore is permeable.

Cotransport of ions and H_2O can be a result of conformational changes in proteins, which in addition to ion transport also translocate H_2O bound to or occluded in the protein. A cellular model of a leaky epithelium based on H_2O leaks and H_2O pumps quantitatively predicts a number of so-far unexplained observations of H_2O transport.

KEY WORDS: Water, Cotransport, H_2O Homeostasis, Epithelia, CHIP 28, H_2O Pumps.

I. Introduction

Water is the most abundant molecule in cells, yet its distribution and transport mechanisms are not very well understood. Movements of H_2O are associated with relatively small driving forces which are difficult to detect and models have to a large extent been based upon slow or inaccurate measurements by osmometers. Only recently have electrophysiological, protein chemical, and molecular methods been combined to reveal a more detailed picture of cellular H_2O homeostasis (see Sections II,D, III, and V,D).

It is customary to consider H_2O as being passively distributed between the intra- and extracellular compartment. In the late 1950s it became clear that what were previously found to be tissue hyperosmolarities (on the order of 100 mosm liter^{-1}) were in fact a result of artificial tissue autolysis (Section II,C). Unfortunately these findings contributed to the dogma that all cells in all steady states were in osmotic equilibrium with their surroundings. However, this conclusion does not apply to all conditions. There is now evidence that cells of H_2O-transporting epithelia, for example, can be hypertonic to the external solution. Current discussions center around how large these hypertonicities are and how they are maintained (Section VI,A).

This review examines the possibility that cellular hyperosmolarity is a result of two types of membrane-bound transport mechanisms (Fig. 1), one passive and downhill (Section III) and one uphill, coupled in membrane proteins to the transport of ions (Section V). As a result, the chemical potential of H_2O may not be the same inside the cell as on the outside in a steady state (Section II,C). This concept can explain a number of observations in a variety of cells ranging from ameba and plant cells to mammalian cells and even cancer cells (Zeuthen, 1992; Zeuthen and Stein, 1994).

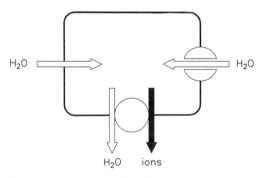

FIG. 1   Passive and active pathways for H_2O. Passive pathways are via the phospholipid bilayer (left) and specific H_2O channels (right). In secondary active transport of H_2O (bottom), the transport is coupled to the transport of ions so that a downhill flux of the ions can energize an uphill flux of H_2O. As a result, the cellular H_2O homeostasis results from a balance between pumps and leaks. In a steady state, the intracellular chemical potential for H_2O can be different from that of the outside solution.

The ability to remain hyperosmolar relative to the surrounding may arise from the cell's absolute requirement for salt; vital intracellular processes require certain concentrations of ions and a constancy of cell volume. Consequently, cells maintain these parameters when exposed to dilute external solutions. In the latter part of this chapter I will use this observation together with findings in regard to passive and secondary active H_2O transport to suggest a molecular model for transport in leaky epithelia, that is, a model which relies more on transport proteins than on the geometry of the tissue (Section VI).

II. The State of H_2O in Cells

In order to describe H_2O transport across the plasma membrane, it is necessary to define a chemical potential for H_2O inside the cell. This is complicated since intracellular H_2O is present in different states. Crudely, most H_2O is either in the free form or bound to surfaces of macromolecules or organelles. In most cell types the free phase dominates, typically 70% (House, 1974) and the chemical potential for H_2O is conveniently equated to this phase. Consequently the driving force is given by the activities of the ions and molecules dissolved here.

A. The Aqueous Environment for Ions and Molecules

Hodgkin and Keynes (1953) observed the movements of radioactive K^+ inside the giant axon of *Loligo* and showed that the intracellular K^+ ions had a diffusion coefficient and a mobility close to those found in free solution. Furthermore, 90% of the intracellular H_2O was free to act as solvent for K^+. Since then, the working hypothesis has been that the intracellular compartment in regard to H_2O and smaller inorganic ions such as K^+, Cl^-, and Na^+ can be considered as a simple free solution. This has been supported in a large number of types of experiments in which the cell membrane is retained while the cytoplasm is removed and replaced with saline: squid axons, resealed red cell ghosts, membrane vesicles, patch clamp experiments and reconstitution of membrane channels into artificial lipid membranes. These preparations provide information that is consistent with the behavior of the intact cell.

The viability of the hypothesis is explained by recent insights into the interactions between biological macromolecules and H_2O. H_2O binds loosely to biological surfaces and has a high rate of exchange with free H_2O in the subanosecond range (Otting *et al.*, 1991). Consequently this

ordered water is available as solvent for ions since it takes little energy
to displace the few H_2O molecules required (Parsegian and Rau, 1984).
The concentration of ions may be different in the free and the ordered
H_2O, but it is wrong to conclude that this gives rise to differences in
activities (Hill, 1994; Section IV,B).

The ordered layers of H_2O do participate in keeping the macromolecules
apart. When, for example, two lipid surfaces or two DNA helixes approach
each other, large entropy forces develop when the ordered H_2O is squeezed
out from between the approaching structures and attains the free form.
The forces have an exponential dependence of distance and a typical
length constant of about 15 Å (Le Neveu et al., 1976; Marčelja, 1976;
Leikin et al., 1993; Rand and Parsegian, 1989; Parsegian and Rau, 1984).
The forces participate in fixing the distance between macromolecules,
since they are balanced by attractive forces such as electrostatic and van
der Waals forces. This is important because the metabolic and reproductive machinery depends on these distances being maintained in a controlled
fashion. The hydration is also important for the dynamics of conformational changes of proteins (see Section II,C).

The distances between larger molecules determine the diffusion rate for
the smaller molecules. For molecules with a radius <15 Å, the cytoplasm
exhibits a viscosity in a range similar to that of free solutions. In Swiss
3T3 fibroblasts Fushimi and Verkman (1991) determined a viscosity 1.2
to 1.4 times that of free solution with 2.7-bis-(2-carboxyethyl)-5-(and-6-)
carboxyfluorescence (BCECF) and 8-hydroxypyrene-1,3,6-trisulfonic
acid (HPTS). The studies suggested that only 20% of intracellular H_2O
was inaccessible to the probes. In cultured cells (CV1 and PtK cells)
Luby-Phelps et al., (1993) found viscosities equal to free solutions by
means of two homologous indocyanine dyes. If the radius of the probe
exceeds 32 Å, the rate of diffusion is only 28% of the rate in free solution.
If the radius exceeds 150 Å, there is no movement at all (Swiss 3T3 cells,
Luby-Phelps et al., 1987). This restriction is probably due to mechanical
barriers. Garner and Burg (1994) have reviewed models for intracellular
crowding and suggest that the cytoplasm behaves as a filamentous network
in which globular proteins are immersed.

Enzymes can function only when intracellular ion activities are within
certain limits. The ions affect the binding and conformational equilibria
of proteins and nucleic acids via mechanisms such as ion association or
release, screening and effects on water activity (Record et al., 1978; Garner
and Burg, 1994; Yancey et al., 1982; Clegg, 1984, 1992; Somero, 1986;
Collins and Washaburg, 1985; Minton, 1983; Fulton, 1982; Teeter, 1991).
Loss of KCl from cells reduces certain enzymatic actions (e.g., DNA
translation in Escherichia coli), an effect which can be offset by increased
levels of Ficoll or albumin (Zimmerman and Harrison, 1987; Cayley et al.,

1991, 1992; Zimmerman and Trach, 1991). Stability can also be obtained by accumulation of so-called nonperturbing ions, e.g., gluconate. This explains why DNA–protein interactions are highly dependent on salt concentrations but tolerate K^+ glutamate better than KCl (Ha et al., 1992; Leirmo et al., 1987). Although many enzyme systems have an absolute requirement for inorganic ions, too high concentrations may have deleterious effects in vitro. Thus Cl^- ions may inhibit protein synthesis (Weber et al., 1977).

It is important to note that while the smaller ions may well have access to the molecular surfaces, many larger solutes are excluded from the ordered H_2O at the surface. This has two consequences. Larger solutes may stabilize the structure of the molecules (Arakawa and Timasheff, 1982, 1985a,b); Timasheff, 1993). Second, the H_2O attached to proteins will be under osmotic stress when, for example, sucrose is present. This is discussed in the following section.

B. Water in Proteins

Proteins change their volume during conformational changes. This can be studied either as a function of the hydrostatic pressure or as a function of the external osmotic pressure. Hydrostatic pressure measurements reflect volume changes of the protein as such, while effects obtained by the osmotic experiments reflect the changes of the protein volume *plus* the volume of the H_2O held by the protein but inaccessible to the osmotic probe.

The hydrostatically induced changes are relatively small. Canfield and Macey (1984) recorded 150 cm^3 mol^{-1} for the Cl^-/HCO_3^- exchanger. Sixty-five cm^3 mol^{-1} has been recorded for the active K^+ transport in red blood cells (Hall et al., 1982) and about 90 cm^3 mol^{-1} for the Na^+/K ATPase (de Smedt et al., 1979). These changes are on the order of 10 $Å^3$ per molecule and most likely comprise several mechanisms: substrate binding, proton binding, and salt bridge formations (Canfield and Macey, 1984). For a review of pressure effects on proteins, see Heremans (1982).

The volume changes associated with changes in osmotic pressure are comparatively large (see Table I). If the osmolarity of the external solutions is increased by osmotic effectors such as sugars or polyethylenglycol (PEG), which are excluded from the hydrated parts of the protein, the H_2O molecules held by the protein will come under osmotic stress (Arakawa and Timasheff, 1982, 1985a). In effect, the hydration water and water held in cavities in the protein will be separated from the external solution by a semipermeable membrane and will be included in the volume measurements.

TABLE I

Changes in the Water Content of Proteins during Conformational Changes

Protein	Process	Δ Volume ($\mathring{A}^3$)	Equivalent number of water molecules	Reference
Voltage-gated anion channel	Opening-closing transition	$2-4 \times 10^4$	660–1320	Zimmerberg and Parsegian (1986)
KCl cotransporter	K^+, Cl^-, H_2O cotransport	1.5×10^4	500	Zeuthen (1994)
Hexokinase	Glucose binding	3300	100	Rand and Fuller (1992); Steitz et al. (1981)
Alamethicin	Opening-closing transition	3000	90	Vodyanoy et al. (1993)
Hemoglobin	T state $\rightarrow$ R state (deoxy- to oxy-state)	1800–2270	60–75	Colombo et al. (1992); Bulone et al. (1991)
Delayed rectifier K^+ channel	Opening-closing transition	1350	40–50	Zimmerberg et al. (1990)
Na^+ channel	Activation Deactivation	730	25	Rayner et al. (1992)
Cytochrome c oxidase	Electron transfer	330	10	Kornblatt and Hoa (1990)

Any conformational change which increases the surface of the protein will lead to a change in the amount of H_2O bound to the protein. This is a major factor in stabilizing the hydrated structure of a protein (Section II,A). The binding energies and change in entropy of the surface water may be significant—25 to 30 cal mol^{-1} $Å^{-2}$ or even more (Eriksson et al., 1991; see also Table I). Hemoglobin and hexokinase shift about 60–100 molecules of H_2O during conformational changes. Colombo et al. (1992) estimate for the oxy-deoxyhemoglobin system that the binding/unbinding to the surface of 60 molecules of H_2O will contribute 0.2 kcal mol^{-1} hemoglobin. In the case of hexokinase, about one half of the H_2O comes from surface areas, the other from more vicinal layers. Such vicinal H_2O will not be seen by nuclear magnetic resonance (NMR) or X-ray because of the short residence time (Rand et al., 1993).

The proteins listed in Table I cover a variety of physiological functions. For membrane-spanning proteins, it has been aruged (Zeuthen and Stein, 1994; Zeuthen, 1994) that changes in H_2O contents during conformational changes could lead to transmembrancous transport of H_2O (see Section V).

C. Cells Can Be Hyperosmolar

It is generally assumed that the intracellular activity of H_2O in steady state is equal to that of the outside solution. In the late 1950s it became clear that what was formerly thought of as cell hyperosmolarity (as much as several hundred mosm 1^{-1}) in fact was a result of artefactual tissue autolysis (Conway et al., 1955; Conway and Geoghegan, 1955; Appelboom et al., 1958; Maffly and Leaf, 1959). These studies dealt with whole heart, muscle, brain, liver, and kidney under resting conditions. Maffly and Leaf (1959) showed that these tissues were isosmotic with plasma within 1.4% (4 mosm $liter^{-1}$). Owing to the hyperosmolarity of the kidney medulla, the accuracy of the kidney measurements was uncertain.

The studies were designed to refute the existence of grotesquely large intracellular osmolarities but cannot automatically be extended to mean that all cells under all conditions are isosmotic with the surroundings. Epithelial cells from leaky epithelia (i.e., small intestine, kidney proximal tubule, gall bladder and choroid plexus) are generally accepted to be hyperosmotic by about 6 mosm $liter^{-1}$ (see Section VI,C) relative to the solution from which they absorb (Fischbarg, 1989). In other words, the first step in the transcellular transepithelial transport of H_2O is passive or purely osmotic. Most important, the cells are also hyperosmolar relative to the final secretion; the exit step in the transcellular transepithelial transport of H_2O is against the direction expected from passive transport (see Section VI). This ability of the epithelial cell to remain hyperosmolar relative to the surroundings is most clearly demonstrated in epithelia adapted to low external osmolarities. Consider some experiments in which gall bladders from bullfrogs or necturi (salamanders) were adapted to transport in salines in which the NaCl concentration had been reduced to give osmolarities of about one half, one quarter, or one eighth of normal (Fig. 2) (Zeuthen, 1981, 1982, 1983). Several points from these studies should be emphasized: (1) The cells are bathed on all sides with a low osmolarity solution. This applies both to the lateral spaces where microelectrodes record no evidence for any hyperosmolarity and to the secretion (Table II). (2) The cell interior can be hyperosmolar to the external solution by more than 40 mosm liter^{-1}. This is found both by microelectrodes and by microosmometric methods (Table III); this double determination excludes methodological artefacts. (3) The hyperosmolarity requires an intact metabolism and functional Na$^+$/K$^+$ ATPase; oxygen removal and ouabain abolished the hyperosmolarity. Similar findings were obtained in the choroid plexus (Fig. 2) and in the freshwater *Hydra* and ameba (Zeuthen, 1992).

Intracellular hyperosmolarity has been suggested to be a result of extracellular hyperosmolarities (e.g., in the lateral intercellular spaces). Such

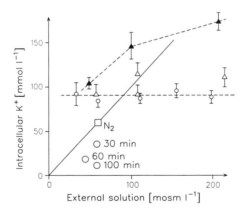

FIG. 2   Apparent intracellular K^1 concentration as a function of external osmolarity in epithelial cells. For external osmolarities below 100 mosm liter^{-1}, the intracellular K$^+$-concentration alone indicates cellular hyperosmolarity (points above the solid line). The contributions from intracellular anions will make the intracellular osmolarity even larger. Ion-selective microelectrodes were applied to gall bladder (open triangles) and choroid plexus (open circles) from necturi kept in tapwater for more than a month. Closed triangles are data from gall bladders taken from animals just caught. The point N$_2$ indicates the effect of bubbling the solutions with nitrogen instead of O$_2$ (gall bladders). The effects of ouabain applied for 30, 60, and 100 min are also shown. The finding and its interpretation were confirmed by direct osmometry (see Table II). (From Zeuthen, 1981, 1982, and unpublished.)

TABLE II

Concentrations, Osmolarities, and Electrical Potentials in Bullfrog Gall Bladder Epithelium Bathed in Low Osmolarity Saline[a]

	E (mV)	Na$^+$ (mmol liter^{-1})	K$^+$ (mmol liter^{-1})	Cl$^-$ (mmol liter^{-1})	Ca^{2+} (mmol liter^{-1})	mosm liter^{-1}
Lumen	0	27	2	27	1	62
Cells	−60.6	6	95	5	~0	87[a]
Spaces	+ 2.7	27	1.66	28.6	0.69	62
Serosa	+ 3.7	28	1.2	29	0.33	62

[a] Calculated by assuming that the activity coefficient and osmotic coefficient are similar to a 95-mM solution of KCl. (From Zeuthen, 1982.)

TABLE III

Osmolarity of *Necturus* Gall Bladder Epithelium

Bathing solution	Animal	Control (mosm liter^{-1})	Ouabain (mosm liter^{-1})	Significance
	1	223 ± 6 (4)	151 ± 5 (3)	$P < .001$
	2	164 ± 5 (4)	144 ± 3 (4)	$P < .001$
116 mosm liter^{-1}	3	181 ± 9 (4)	135 ± 1 (3)	$P < .01$
	4	153 ± 8 (3)	135 ± 1 (3)	$P < .02$
	Mean	180	141	

Determined by Ramsay Osmometer from Zeuthen (1981).

explanations are difficult to maintain. There is no evidence for significant extracellular hyperosmolarities and if there were, such gradients would not give a satisfactory explanation of experimental facts (Zeuthen, 1992; Zeuthen and Stein, 1994).

In conclusion: Some epithelial cells are hyperosmolar to the surrounding solutions in steady state, particularly when bathed in dilute media. Since their membranes are highly water permeable, this poses the conceptual problem of how the imbalance in H_2O is maintained. I suggest that H_2O can be moved by secondarily active transport (Section V) as well as passively (Section III). Thus H_2O homeostasis is a balance between pumps and leaks for H_2O.

D. Recording H_2O Transport

Transport in a steady state can be studied by recording the response of the cell to small abrupt perturbations. The *initial* rate of change in the cellular parameters will then be a property of the cell in the given steady state. The faster the recording is, the more it represents the original steady state and the less secondary responses have occurred. If, for example, cells are exposed to high extracellular K^+ concentrations, a primary response is electrical depolarization and K^+ influx; a secondary response will be the depolarization-induced increase in Ca^{2+}_i which invariably will lead to tertiary responses.

In order to correlate H_2O and ion fluxes across the cell membrane, it is necessary that the perturbations in cell volume and in intracellular ion concentrations be recorded with the same fast time resolution. Ion-selective microelectrodes fulfill both roles, since they can be adapted to record changes in cell volume with a time resolution down to 0.1 sec, which is similar to the rate at which they record ion activities. A small amount of Ch^+ (choline ions) or TMA^+ (tetramethylammonium ions) is added to the superfusion solutions, from which they enter the cells. Since the K^+-ion exchanger is several orders of magnitude more sensitive to Ch^+ and TMA^+ than to K^+, these ions will act as readily recordable markers of extracellular or cellular volume (Nicholson et al., 1979; Hansen and Olsen, 1980; Nicholson, 1993; Ballanyi et al., 1993). Reuss (1985), Cotton et al. (1989), and Cotton and Reuss (1989) introduced the use of TMA^+ as a marker of intracellular volume, and TBA^+ (tetrabutylammonium ion) as a marker of extracellular volume. Ch^+ has been used intracellularly in choroid plexus (Zeuthen, 1991a,b) and in retinal pigment epithelium (laCour and Zeuthen, 1992). The method has also been used in muscle cells and cells from the central nervous system (Walz, 1992).

Optical methods are of several types. In optical slicing introduced by Spring and Hope (1978, 1979) the cell circumference is determined at different levels and the cell volume calculated. It is difficult, however, to achieve time resolutions below 5 sec. Furthermore, the method is applicable only to cells with regular shapes. Among the leaky epithelia, Necturus gall bladder has been preferred (Spring and Ericson, 1982; Davis and Finn, 1985, 1988). Attempts to correlate the initial volume changes with recordings by ion-sensitive microelectrodes seem difficult (Fisher et al., 1981; Fisher and Spring, 1984).

Faster optical methods are based upon the assumption that cell height is a good measure of cell volume. This has been used in frog skin (MacRobbie and Ussing, 1961); in kidney proximal tubule (Carpi-Medina et al., 1984; Welling et al., 1983; Whittembury et al., 1986; Kirk et al., 1987); in vestibular dark cells (Wangemann and Marcus, 1990); in A6 cells (Crowe and Wills, 1991); in the ciliary epithelium (Farahbakhsh and Fain, 1987); in endothelial cells (Mazzoni et al., 1989); in intestine (Siegenbeek van Heukelom et al., 1981); and in Madin-Darby canine kidney (MDCK) cells (Kersting et al., 1991; Schild et al., 1991). Time resolutions as low as 0.5 sec have been employed.

Self-quenching of an intracellular fluorophore has been employed by Chen et al. (1988) and Kim et al. (1988) to measure the volume changes of small cells and vesicles with a time resolution less than 0.5 sec.

Intracellular dyes can be used for simultaneous measurements of cell volume and Ca^{2+} or pH (Tauc et al., 1990; Muallem et al., 1992). The method takes advantage of the fact that the emission of the dyes, obtained by excitation at the isosbestic wavelength, largely reflects the cellular volume changes. This is recorded simultaneously with the excitation obtained at the ion-sensitive wavelength. The time resolution is well within 1 sec.

A good example of the necessity for rapid correlation between H_2O and ion measurements consists of the experiments in which extracellular Na^+ is replaced by K^+ and the effects on cell volume are observed. In epithelia the procedure is used to elucidate the transport mechanisms across the apical and basolateral membranes of gall bladder (Giraldez, 1984; Hermansson and Spring, 1986; Cotton and Reuss, 1989; Davis and Finn, 1988; Parr and Finn, 1989), distal tubule of kidney (Guggino, 1986), descending limb of Henle (Lopez *et al.,* 1988), proximal tubule (Schild *et al.,* 1991), collecting duct (Strange, 1988), choroid plexus (Zeuthen *et al.,* 1987a; Zeuthen, 1994), and vestibular dark cells of the inner ear (Wangemann and Marcus, 1990; Wangeman and Shiga, 1994).

The general picture which emerges is that increases in K^+ concentrations on one side of the epithelium cause an immediate cell swelling (basolateral membrane in gall bladder, kidney, and dark cells, ventricular membrane in choroid plexus) while increases on the other side of the epithelium have only minute effects on cell volume. The usual explanation for the swelling is that K^+ enter with Cl^-, either by electroneutral cotransport or via channels, and that H_2O follows by osmosis. However, the latter conclusion can be drawn only if the water permeability of the membrane is known and if the immediate changes in intracellular osmolarity are recorded as well. In many instances, the change in intracellular osmolarity (estimated from the change in ion concentrations) and the water permeability are far too small to allow an explanation of the cell swelling on the basis of osmosis (see Section V).

III. The Biology of Passive Water Permeation

Most biological membranes have a significant passive water permeability. Take a human red blood cell with an osmotic water permeability of 200 μm sec^{-1}. Given a drop in extracellular osmolarity of 10 mosm liter^{-1}, a difference of 3% from normal, the cell volume increases by 7% sec^{-1}.

The passive transport of H_2O across the plasma membrane takes place via the lipid bilayer and membrane-spanning proteins (Fig. 1). The composition of the lipid phase itself offers some possibilities for varying the water permeability, but the major source of variation, at least in vertebrate cells, results from the presence or absence of membrane proteins. Recently a family of proteins specifically designed for H_2O transport has been discovered (the CHIP 28 proteins). Proteins with other functions, i.e., cotransporters (Section V), have been suggested to possess a passive water permeability. Finally, one may ask if the pathway between membrane proteins and the lipids offers a significant permeability to water. In the case of antibiotic channels, at least, this seems not to be the case (Finkelstein, 1987).

A. Permeation via Phospholipids

In the phospholipid double layer, the hydrophilic phosphate groups face outwards toward the aqueous phase, while the hydrophobic hydrocarbons face inward toward each other. The inner phase has a relatively low concentration of H_2O molecules, the movement of which can be described by diffusion. In this model the rate of diffusion through the membrane is proportional to the product of the partition coefficient and the diffusion coefficient. Other models for permeation include the existence of hydrogen-bonded chains of H_2O located at occasional defects in the lipid double layer (Deamer and Nichols, 1989; Finkelstein, 1987; Forte, 1987).

The permeability of the lipid bilayer can be measured when the proteinaceous water channels are blocked by mercurials. In that case transport is purely diffusive and it should not matter whether the flux is measured by a hydraulic driving force or by diffusion of tritiated H_2O. Table IV lists the water permeability for the lipid membranes of kidney proximal tubule, small intestine, and red blood cells. They are all at or above 10 μm sec^{-1}. This emphasizes that vertebrate cells have significant water permeabilities via their lipid bilayer. Table IV also shows that it is possible to form from biological lipids, bilayers with a wide range of permeabilities (Finkelstein, 1976). For example, fish egg membranes (*Fundulus*) have a permeability which is 1000 times smaller than that of the erythrocyte. Thus *in principle* the values encountered in membranes could be explained by variations in the lipid composition. However, in fact, most lipid membranes, at least in vertebrate cells, have a high water permeability which is in parallel with the additional permeability of the proteinaceous pathway. For the lipid composition of some vertebrate cell membranes, see Tanford (1973). In conclusion, it seems that the lipid bilayer is important when the lower limit of the water permeability is to be defined.

B. Permeation via Proteins, the Aquaporins

It has long been known that cells have membrane proteins designed for passive permeation of H_2O. In human red blood cells, for example, the hydraulic water permeability is larger than the diffusive one and a large fraction of the water permeability is reversibly inhibited by mercurials (Macey, 1984). Within the past few years such proteinaceous pathways

TABLE IV

Water Permeabilities of Lipid Membranes (μm sec^{-1})

	Apical	Basal	Whole cell	References
Rat intestine	40 (0)	50 (0)		Dempster et al. (1991)
Rabbit kidney	49 (60)	62 (63)		van der Goot et al. (1989)
Proximal convoluted	17 (66)			Chen et al. (1988)
tubule	380 (48)	58 (70)		Meyer and Verkman (1987)
			10 (60)	Whittembury et al. (1993)
Rat kidney	453 (25)			Heeswijk and Van Os (1986)
Proximal convoluted				
tubule				
Red blood cell				
Amphiuma			8 (0)	
Chicken			14 (0)	J. Brahm, personal
Human			10 (43)	communication
Duck			12 (81)	
Fundulus				
eggs			0.01	Dunham et al. (1970)
Phosphatidylcholine			22	
Phosphatidylcholine			5.7	Finkelstein (1976)
+ cholesterol				
Sphingomyelin +			0.8	
cholesterol				

Notes: The permeabilities for intestine, kidney, and erythrocytes are measured after inhibition of proteinaceous water channels by mercurials. The numbers in parentheses are the % inhibition obtained by mercurials. Kidney and intestinal values are hydraulic values; red blood cells are diffusional.

have been identified, isolated, and functionally expressed, which has led to defining them as true water channels.

The most prominent group of channels is the aquaporins (Agre et al., 1993a,b), which include CHIP 28, WCH-CD, and λ-TIP. CHIP 28 is the acronym for a channel-forming integral membrane protein with a molecular weight of 28 kDa. CHIP 28 was originally found in red blood cells and kidney but at first had no function attached to it (Denker et al., 1988). It was soon realized, however, that it had similarities to a channel protein (Smith and Agre, 1991; Preston and Agre, 1991) and expression in oocytes or incorporation into liposomes showed that the permeating substance was H_2O (Preston et al., 1992; Van Hoek and Verkman, 1992; Zeidel et al., 1992b; Agre et al., 1993a,b; Dempster et al., 1992; Verkman, 1993; Van Os et al., 1994). WCH-CD is the acronym for a proteinaceous water channel in the collecting duct of the kidney. Its localization suggests that it is a vasopressin-activated water channel that permits renal water conservation (Fushimi *et al.*, 1993; Ma *et al.*, 1994). λ-TIP stands for λ-tonoplast intrinsic protein, which is thought to mediate exchange of H_2O between the vacuole and the cytoplasm in plant cells (Maurel *et al.*, 1993).

Comparison of gene structures and evolutionary distances shows that these three proteins are related to MIP 26, the major intrinsic protein of molecular weight 26 kDa from the mammalian lens (Moon *et al.*, 1993). The MIP family includes an *E. coli* glycerol facilitator, the channel from bovine lens fiber junctions, a plant tonoplast membrane protein, a soybean protein, and a *Drosophila* neurogenic protein (Baker and Saier, 1990; Pao *et al.*, 1991; Gorin *et al.*, 1984; Preston and Agre, 1991; Smith and Agre, 1991). Some of the pore-forming proteins in the MIP family can support an ion flux as well as a water flux; for example, MIP purified from the lens fiber membranes (Ehring *et al.*, 1990) conducts both Cl^- and K^+ as well as H_2O. In the present context I shall deal only with the aquaporins which serve exclusively as water transporters (Zeidel *et al.*, 1992b, 1994; Van Hoek and Verkman, 1992; Maurel *et al.*, 1993).

C. Localization and Function of CHIP 28

CHIP 28 has been found in red blood cells (Denker *et al.*, 1988; Preston *et al.*, 1992) and in the constitutively water-permeable segments of the kidney, such as proximal tubules and thin descending limb of Henle (Nielsen *et al.*, 1993a; Zhang *et al.*, 1993; Hasegawa *et al.*, 1993). It was not found in the ascending thin limbs, thick ascending limbs, collecting ducts, or distal tubules (Nielsen *et al.*, 1993a). One report though (Deen *et al.*, 1992) presents evidence for CHIP 28 in the collecting duct. The choroid plexus apical membrane, the apical and basolateral domains of ciliary epithelium of the eye, the lens epithelium, and the corneal endothelium all contain CHIP 28 (Nielsen *et al.*, 1993b; Echevarria *et al.*, 1993b; Hasegawa *et al.*, 1994a), which suggests a role for this protein in water homeostasis in the accessories to the nervous system; it was not found in whole brain (Nielsen *et al.*, 1993b; Hasegawa *et al.*, 1994a). CHIP 28 was found in some cell types in the male reproductive tract (Brown *et al.*, 1993).

In the digestive tract, the protein was less abundant. It was *not* detected in small intestine or gastric mucosa. It was detected in membranes of the hepatic bile duct and in the neck of the gall bladder where the apical membranes of the epithelial cells exhibited weak immunoreaction and the lateral membranes showed strong immunoreaction (Nielsen *et al.*, 1993b). Hasegawa *et al.* (1994a) report an antibody staining in gall bladder epithelial cells, but do not specify the localization. The protein was not detected in the surface epithelium of the colon (Nielsen *et al.*, 1993b) but was found

in crypts (Hasegawa *et al.*, 1994a). Fenestrated capillaries do not react
to CHIP 28 antibodies while lymphatics do (Nielsen *et al.*, 1993b). CHIP
28 was not present in the secretory cells of pancreas, lacrimal, mammary,
or salivary glands (Nielsen *et al.*, 1993b) which conflicts with the findings
of Hasegawa *et al.* (1994a), who found CHIP 28 in pancreatic acinar cell
and salivary gland epithelia. Some of these discrepancies could originate
from the fact that the continuous capillaries and lymphatic vessels of these
tissues immunoreact with CHIP 28 antibodies while the secretory cells
as such do not (Nielsen *et al.*, 1993a).

Immunolocalization revealed CHIP 28 in endocytic vesicles in
transfected Chinese hamster ovary cells (Ma *et al.*, 1993b), but no intracellular localization was observed in cells from native tissues (Nielsen *et al.*,
1993b). Finally, it should be mentioned that the expression of CHIP 28
increases sharply around birth (Smith *et al.*, 1993; Bondy *et al.*, 1993).

The general picture which emerges for water-transporting tissues is that
barriers which are not exposed to large variations in osmolarity may
contain CHIP 28, while barriers with a water-transporting role as well as
an osmotic barrier role (the gastric ventricle, the small and large intestine)
do not have the pore since this would be potentially lethal. Glands which
can transport against significant hydrostatic gradients may not have the
pore either. For a quantitative discussion, see Section VI.

It should be emphasized, however, that the physiological importance
of CHIP 28 is far from established. Phenotypically normal humans without
functional CHIP 28 suffer no apparent clinical consequence (Preston *et
al.*, 1994a).

D. Molecular Structure of CHIP 28

cDNA cloning suggests that the functional CHIP 28 molecule is a unique
structure with two internal repeats oriented 180° to each other within the
membrane (Preston and Agre, 1991; Preston *et al.*, 1994). The structure
implies six membrane-spanning α-helixes. Similarly, cDNA cloning of
MIP suggests that this protein also has six membrane-spanning segments
(Gorin *et al.*, 1984). Both proteins have their carboxy and amino terminals
in the cytoplasmic solution. These proteins apparently arose by internal
duplication of a three-transmembrane segment (Pao *et al.*, 1991; Wistow
et al., 1991). In contrast to these conclusions, a recent study (Skach *et al.*,
1994) suggests that the CHIP 28 molecule consists of only four membrane-
spanning α-helixes. According to Horwitz and Bok (1987), the structure
of MIP is 50% α-helices and 20% β-sheets, which agrees well with the
findings for CHIP 28 (Van Hoek *et al.*, 1993).

Electron microscopic methods, immunolocalization, and freeze fracture studies show that MIP and CHIP 28 form tetrameric arrays in both natural and artificial membranes (MIP: Zampighi *et al.*, 1989; Ehring *et al.*, 1990; Aerts *et al.*, 1990; CHIP: Verbavatz *et al.*, 1993; Smith and Agre, 1991; Zeidel *et al.*, 1994). In spite of this, the functional unit appears to be the 28-kDa subunit, as shown by radiation inactivation (Van Hoek *et al.*, 1993), and coexpression studies in which mRNA encoded wild-type and certain mutants of CHIP 28 in *Xenopus* oocytes (Preston *et al.*, 1994; Zhang *et al.*, 1993). This view precludes a pathway formed in the center of the four subunits as a candidate for the permeation pathway. Each 28-kDa subunit is about 30 Å wide and the oligomer assembly is 72 Å on the widest diameter. A crude geometrical argument predicts each subunit to be about 50 Å long. There is at present no explanation for this formation of tetramers (Zeidel *et al.*, 1994). Two-dimensional crystals have recently been formed from CHIP 28 (Walz *et al.*, 1993); in this study it was confirmed that CHIP 28 forms tetramers.

C. Other Proteinaceous Pathways

The discovery of CHIP 28 has increased interest in water transport and led to the finding of several other putative water channels. A variety from rat kidney has been expressed in oocytes (Echevarria *et al.*, 1993a). In addition to CHIP 28, a cyclic adenosine monophosphate (cAMP)-sensitive channel was found in medulla and a cAMP-insensitive channel in papilla. Recently a channel insensitive to mercurials has been demonstrated in cells which do not express CHIP 28 (Hasegawa *et al.*, 1994b). Furthermore, a rat kidney cDNA homologous to the ones for CHIP 28 and WCH-CD has been cloned (Ma *et al.*, 1993a). For channels in the collecting duct (see Knepper and Nielsen, 1993; Nielsen and Knepper, 1993).

Not all tissues can afford to have large passive water permeabilities because they may be exposed to large variations in osmotic or hydrostatic pressures. The situation is particularly intriguing if such tissues are able to transport large volumes of H_2O against the prevailing osmotic gradient. The small intestine is a case in point. It is relatively water impermeable (Zeuthen, 1992) yet it reabsorbs up to 10 liters of H_2O daily in humans. It is interesting that the small intestine has *no* CHIP 28, but alternative candidates for water channels are the facilitated and the Na^+-dependent glucose transporters (Fischbarg *et al.*, 1990, 1993; Loike *et al.*, 1993). From some very rough estimates Fischbarg *et al.* (1993) give a permeability of about 1.0×10^{-14} cm^3 sec per molecule for the Na^+/glucose transporter. The permeability of the Na^+/glucose transporter was dependent on the presence of Na^+ but not on glucose (Fischbarg *et al.*, 1989). The finding has been criticized in that known inhibitors such as cytochalasin B and phloretin did not inhibit H_2O transport (Dempster *et al.*, 1991) and that the activation energy for transport was large (Zeidel *et al.*, 1992a); if H_2O crosses the channel passively, the activation energy should be low, on the order of 4 kcal mol^{-1}.

To my mind, these findings and the objection to them suggest another mode of function for the NA^+/glucose transporter: H_2O crosses the protein in cotransport with Na^+ and glucose. The conformational changes which cause the transfer of Na^+ and glucose also cause a transfer of a well-defined volume of H_2O (for further discussion, see Section V).

IV. The Physics of Passive Water Permeation

In aqueous solution, both water and solute molecules perform irregular thermal motions. Each particle exchanges kinetic energy with the surrounding particles in a random fashion. As a result, each particle continuously changes position and velocity. Given physiological salt solutions with concentrations of about 105 mmol liter^{-1}, there are about 400 H_2O molecules per molecule of salt. In such dilute solutions, the individual solute molecules will rarely interact and the properties of the solutions can be approximated by the interaction between the individual solute particles and the H_2O molecules. In order to understand how H_2O permeates a membraneous pore, it is necessary to have a complete description of the molecular-kinetic events in the solution, at the pore mouth and in the pore itself.

Most theories on H_2O permeation, however, are macroscopic theories with parameters which are averages, both temporally and spatially, of molecular kinetic events (Mauro, 1957, 1979; Manning, 1968; Hammel, 1979; Soodak and Iberall, 1979; Hill, 1979, 1982; Finkelstein, 1987). One assumption is that the thermodynamic concepts of activity and pressure can be applied to the microscopic domain of the pore mouth. In fact, however, these concepts have meaning only in volumes so large that spatial and temporal variation are averaged. Another assumption is that certain corrections based on drag factors, Poiseuille flow, and equivalent pore radii can be applied. These are definitely macroscopic parameters and their application to osmosis may not be justified.

A. Molecular-Kinetic Description of Diffusion and Osmosis

If the concentration of solute is nonuniform, diffusion will take place. The movement of each particle is entirely random; more solute particles will move from the concentrated compartment into the diluted one and less from the diluted part into the concentrated one. As a result, there is *no* diffusion force (Einstein, 1956; Hartley and Crank, 1949; Sten-Knudsen, 1978; Hille, 1992). This is stated particularly clearly by Hille (1992, p. 263): The so-called diffusion force "is not a ponderomotive force that can *accelerate* or impart net velocities to molecules. It is a statistical or virtual force describing the increase of 'randomness' due to increasing *entropy* of dilution." The point is worth stressing since one model to explain saltwater coupling in epithelia (Ussing and Eskesen, 1989) begins with the erroneous assumption that the diffusion of salt imparts net velocities to H_2O molecules (Zeuthen, 1992). Any net movement of H_2O is in fact opposite to the direction of solute diffusion owing to the difference in chemical potential for H_2O. This effect is small and is usually compensated for by hydraulic forces.

There is at present no model which combines the microscopic interactions between solutes and solvent during diffusion with the macroscopic or measureable properties. This has forced investigators to apply macroscopic or continuum theories. For example, Einstein (1956) used Stokes' relation to arrive at an expression of the diffusion coefficient D.

In osmosis, a semipermeable membrane separates two solutions of different solute concentrations (Fig. 3). In the present treatment, this membrane is assumed to be infinitely thin and to contain holes with a radius too small for the solute to pass but much larger than the water molecule. In the osmotic case, a solute molecule which has arrived at a position in front of the opening (position 2) can only move away if it returns toward the concentrated solution (position 3). In the case of diffusion, the particle might as well have proceeded toward the dilute solution (position 2 to 3'). Thus, in diffusion, water molecules experience an equal number of solutes moving to one side and to the other, consequently thre is no net transfer of momentum from the solute to the water molecules. In case of osmosis, however, the water molecules in the hole experience more solute molecules moving toward the concentrated side and none to the dilute side. Consequently, there is a net transfer of momentum toward the concentrated side. The osmotic force must be sought in this asymmetry of momentum transfer.

If one had models which could quantitate the microscopic drag between the solute particle and the surrounding H_2O during a jump such as that from 2 to 3 in Fig. 3, it would be possible to arrive at a molecular-kinetic description of osmosis. The osmotic flow would contain the product of the microscopic drag times the rate of reflection of solute molecules. Dainty (1965) attempted a molecular kinetic treatment, but concluded that a satisfactory treatment requires a knowledge of "the statistical mechanics of liquid flow, in which the transfer of momentum to individual molecules arising from the unbalanced jumping process at the pore mouth is considered in details."

In the intermediate case (Fig. 3, right panel), the diameter of the hole

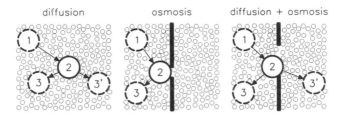

FIG. 3   Molecular-kinetic descriptions of transport: Diffusion (left), osmosis in a perfectly semipermeable membrane (middle) and diffusion plus osmosis in a permeable membrane (right). Small circles indicate the H_2O molecules. Diffusing solute molecules move randomly. Each jump may take place either to the left (from 2 to 3) or to the right (from 2 to 3') and the net transfer of momentum from solute to H_2O will be zero. In case of osmosis, solute cannot cross the membrane. Compared with diffusion, the jumps from 2 to 3 are not balanced by jumps from 2 to 3'. This results in a net transfer of momentum to H_2O in the direction of the concentrated bath. If the membrane is partly permeable, the situation is intermediate. A limited number of jumps into the dilute solution will occur (2-3') (see text).

allows solute to pass. Solute molecules which permeate will behave as they do in diffusion and not give rise to net movements of H_2O. If the solute molecule is reflected from the edge of the hole, this will give rise to osmotic forces. One may conclude that a molecular-kinetic description of the permeable pore must contain both diffusive and osmotic events. This should be reflected in any macroscopic description of the permeable pore; parameters should be continuous functions of the diameter of the hole.

B. Osmosis in Pores

1. Impermeable Solute

Membranes have a thickness which is large compared with the solvent and small solute molecules. Therefore the properties of transport through the pore must be added in series with the processes at the mouth of the pore. For an impermeable pore, all solute molecules are reflected from the orifice. The void left by the solute molecule is filled by neighboring water molecules which come, in part, from the pore itself. This leads to transport through the pore.

In an attempt to describe this essentially statistical phenomena in thermodynamic terms, chemical potentials and hydrostatic pressures have been used (Fig. 4A). This approximation is called the Vegard-Mauro model (Vegard, 1908; Mauro, 1957; Hill, 1982; Finkelstein, 1987). It postulates a linear gradient of hydrostatic pressure through the pore; in one end the

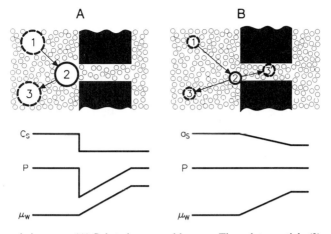

FIG. 4   Osmosis in pores. (A) Solute-impermeable pore. The solute particle (2) cannot enter the pore and is reflected by the membrane, a jump from 2 to 3. This kinetic process can be averaged in time and space to give the profiles for solute concentration (C_s), hydrostatic pressure P, and chemical potential for H_2O (μ_w). Note the sharp jump in pressure at the mouth of the pore, which is a consequence of this averaging process. (B) Permeable pore. The solute particle which has arrived at the mouth of the pore (2) can either be reflected (3) as in the impermeable pore or it can enter the pore (3'). These processes can be averaged in time and space to express conventional thermodynamic parameters. Since there are no sharp transitions in solute activities (a_s), the transport in the permeable pore is described by diffusion (see text).

pressure equals the pressure of the solvent-free bath while it describes a sharp jump at the interphase with the solute-containing bath. This unphysical abrupt change in hydrostatic pressure is the price for using temporal and spatial averages for forces which are stochastic by nature. A more realistic model must take into account the fact that the driving force originates outside the pore in the vicinity of the solute molecule and not at the pore solution interphase (see Section III). What the model emphasizes is that the driving force in its averaged form is hydrodynamic and not diffusive, which was also shown experimentally (Mauro, 1957).

2. Permeable Solute

There are two opposing macroscopic theories for H_2O transport in a solute-permeable pore. One is based upon the postulate of a negative hydrostatic pressure gradient in the pore; the other is based upon diffusion. In the hydraulic description (Garby, 1957; Finkelstein, 1987), the pressures and concentrations are in principle similar to those depicted for the impermeable pore (Fig. 4A). The hydrodynamic model has been expanded to pores in which only the pore wall rejects the solute (Anderson and Malone, 1974). Here the hydrodynamically driven flow along the pore side leads to convective flow in the center.

These hydrodynamic models have been criticized by Hill (1982, 1989a,b), who argues, correctly in my view, that if solute can enter the pore, the activity of the solute at equilibrium will be the same outside and just inside the pore. Consequently there can be no difference in osmotic pressure and therefore no balancing of hydrostatic pressure. Hill emphasizes that activities, not concentrations, appear in the expressions for flows and in the Gibbs-Duhem formalisms. In a pore, the solute concentration may differ significantly from that outside the pore. However, it is the difference in activities which contributes to the driving forces for transport. With no hydrostatic pressure differences in the system, the flows are entirely driven by differences in activities, which means that both solute and H_2O diffuse (Fig. 4B).

The hydrodynamic and diffusional theories are divided on whether a surface field in the pore can reduce the number of solute molecules at the surface in such a way as to diminish the osmotic pressure. A simple argument shows that such surface pressures *cannot* exist (Fig. 5). Consider a slab of porous material, the internal surfaces of which attribute a potential energy to the solute molecules. Let the slab be separated from the outer solution on the left by a semipermeable membrane (reflection coefficient

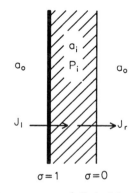

FIG. 5 Do surface hydrostatic pressures exist? A slab of porous material is soaked in a solution of a permeable solute with an activity a_o. The internal surfaces of the porous material exert a force on the solute molecules. Is the hydrostatic pressure P_i and the activity a_i inside the slab different from the pressure and activity on the outside? The answer is no. This can be seen by covering the slab by a semipermeable membrane ($\sigma = 1$) on one side, while σ is low ($=0$) on the other side. If there was a difference in activity and pressure between the inside and outside fluxes, J_i and J_r would be different from zero and indefinite energy could be drawn from the system (see text).

$\sigma = 1$) while the right side is covered with a membrane with $\sigma = 0$. The flow of volume across an interphase is proportional to $\Delta P - \sigma \Delta \pi = \Delta P - \sigma RT \Delta a$, where ΔP and $\Delta \pi$ and Δa are the differences in hydrostatic pressure, osmotic pressure, and solute activity. Now the flow is zero across both the left and right interface. For the left interface, this means that $\Delta P = \sigma RT \Delta a$, while application to the right interface leads to $\Delta P = 0$ since $\sigma = 0$. So Δa must also be zero. The argument can be phrased in different ways: If there existed a hydrostatic pressure inside a porous material which was balanced by an activity gradient, then one could derive energy from the system *ad infinitum* by closing part of the surface by a semipermeable membrane and connect this surface to the open one by an external pathway.

The hydrodynamic model for the impermeable pore combined with the diffusional model for the permeable pore has been put forward by Hill (1982, 1994) as the bimodal theory. In this description, the transition from the impermeable case to the permeable one is characterized by abrupt changes in transport parameters, such as reflection coefficients. This is a consequence of the fact that the two macroscopic theories begin from different physical descriptions, neither of which applies to the transition situation.

The dichotomy between a hydrodynamic and a diffusive description dissolves in a molecular kinetic model. If a solute particle collides with the orifice of the pore and is reflected (Fig. 4B), forces will arise which move water from the pore into the void left by the particle. Such events lead to hydrodynamic movements. If the solute particle enters the pore, water moves from the pore into the concentrated bath by interdiffusion. In short, the permeable pore will be characterized by a mixture of diffusive and hydrodynamic events. For the very wide pore, most of the solute molecules will enter the pore and an entire diffusive description will be adequate. In the other limit, the narrow impermeable pore, all solute molecules will be reflected and the hydrodynamic description can be used.

C. Experiments versus Theories

In pores with diameters much larger than the H_2O molecule, hydrodynamic flow is described by Poiseuille's equation. In the absence of alternatives, modified Poiseuillean descriptions have been used for pores of molecular dimensions (Finkelstein, 1987). The modifications involve empirical corrections for effective pore radius and transport within the pore (Paganelli and Solomon, 1957; Ferry, 1936; Faxén, 1922). As pointed out by Finkelstein (1987), there is no theoretical justification for these corrections. In the bimodal theory (Hill, 1994), water flow generated by impermeable solutes is also treated as Poiseuille flow, but other empirical correction factors are employed (Wang and Skalak, 1969; Haberman and Sayre, 1958). The great schism arises when the solute can penetrate the pore. Here Finkelstein continues to use the Poiseuille equation while Hill uses a diffusive description: Fick's equations modified by drag factors (Wang and Skalak, 1969; Haberman and Sayre, 1958).

Poiseuille's law may not apply for narrow pores. It requires the existence of a velocity profile and flow lines which cannot be defined in pores of molecular dimensions. Therefore, it should be no surprise when the agreement between theory and experiment is poor; likewise, good agreement could be fortuitous (Table V). In the gramicidin A pore ($r = 2.5$ Å), mannitol is impermeable and the Poiseuillean approximation used by Hill (1994) gives an accurate description. Urea ($r \leq 2.4$ Å) is just on the edge of being permeable: Bean (1972) lists urea radii in Å: 2.35 (from density measurements), 1.8 (from Stokes-Einstein considerations) 2.3 (from other model considerations) and 1.8 (from viscosity measurements). Some authors claim urea can penetrate (Cohen, 1975; Finkelstein, 1974; Hill, 1994; others say it cannot (Finkelstein, 1987). If urea is assumed to be permeable, the diffusional approximation gives the best prediction.

With nystatin and amphotericin, only permeable solutes have been tested and Poiseuillean flow gives a poor prediction of the measured water flow. The diffusive approximation is somewhat better but still off by a factor of 5. Na^+ and Cl^- have dehydrated diameters of about 2 Å and it is uncertain whether these permeate in a hydrated or dehydrated state. In the present context they are assumed to permeate dehydrated. Sucrose has an average radius of 4.5 Å (Bean, 1972) but is assumed to be able to enter end on.

The hydraulic permeability of CHIP 28 has been determined by solutes of varying size—NaCl, mannitol, and sucrose. Preston et al. (1992) obtained 1.0 (in units of 10^{-14} cm^3 sec^{-1}) in oocytes with NaCl as solute. The authors state that the value could be an underestimate because it is not known if some copies of CHIP 28 are expressed in intracellular membranes and not exclusively on the outer plasma membrane. The value is much smaller than the one obtained from intact red blood cells where values of 16 to 20 were obtained (Preston et al., 1992; Smith et al., 1993; Finkelstein, 1993). The permeability of single CHIP 28 channels reconstituted in artificial lipid membranes has also been determined. With mannitol, roughly 10 (in units of 10^{-14} cm^3 sec^{-1}) was obtained by Van Hoek and Verkman (1992), later referred to as 6.8 in Zhang et al. (1993); at 10°C 3.6 has been obtained (Verbavatz et al., 1993). Sucrose gave values of 11.7 (Zeidel et al., 1992b); later more accurate estimates by the same group gave 4.6 (Zeidel et al., 1994) and 5.4 (Walz et al., 1993).

Apparently the permeability obtained with NaCl is lower than that

TABLE V

Osmotic Permeabilities of Membrane Pores[a]

	Osmolyte		Experiments	Model calculations	
				Hydrodynamic[b]	Bimodal[c]
Gramicicin A	Mannitol	imp[i]	6[d]	3	6.2
(r = 2.5 Å)	Urea	perm?	1[e]	3	1.6
Nystatin, double length (r = 4Å)	Nacl Glucose Sucrose Glycerol Ethylene Glycol Urea	perm.	1.5[b]	31	7.8
Amphotericin B double length (r = 4)	NaCl Glucose Urea	perm.	2.2 − 4.5[b]	31	7.8
CHIP 28 (r = 2 to 2.5)[f]	Nacl	imp?	1[g]		1.6
	Mannitol Sucrose	imp.	3.8–10[h]		6.2

[a] Units are 10^{-14} cm^3 sec^{-1}

[b] For references see Finkelstein (1987). The theoretical values are obtained from Poiseuille's law. The values are not corrected with drag factors (Wang & Skalak 1969) which would decrease them by 10 to 20%.

[c] For references see Hill (1994). The values calculated for impermeable solutes are treated as hydraulic (Poiseuillian) and are corrected with the drag factors of Wang & Skalak (1969) which contribute a reduction of 10 to 20%. The values calculated for permeable solutes are obtained from diffusive theory. In fact they should be multiplied with a reflection coefficient, smaller than one, which brings them closer to the experimental values.

[d] Dani and Levitt (1981).

[e] Rosenberg and Finkelstein (1978).

[f] Assumed.

[g] From Preston, et al. (1992). Probably an underestimate, see also text.

[h] For references, see text.

[i] imp: impermeable; perm: permeable.

obtained with sucrose and mannitol. This agrees with the bimodal theory, which states that use of impermeable solutes leads to higher osmotic permeabilities than use of permeable solutes. Yet three points must be emphasized: First, the determination with NaCl as solute relies on one report only and it might be an underestimate. Second, it is not known whether Na$^+$ or Cl$^-$ enters CHIP 28 or if it does, how. What is the role of charges and degree of hydration? Third, in red blood cells, NaCl and larger impermeable solutes give the same water permeability; in other words, the reflection coefficient (σ_{NaCl}) equals 1. Similar results have been obtained in vesicle preparations from kidney proximal tubule (Pearce and Verkman, 1989; Shi *et al.*, 1991); in intact kidney proximal tubules (Carpi-Medina *et al.*, 1984; González *et al.*, 1982), and in choroid plexus apical membranes (Zeuthen, 1991a), all membranes which contain CHIP 28 (see Section III). In one study only (Welling *et al.*, 1987), was σ_{NaCl} significantly less than one. The bimodal theory predicts σ_{NaCl} significantly lower than one (Hill, 1994), which contrasts with most experimental findings.

In conclusion, the bimodal theory with its emphasis on diffusion gives a better theoretical treatment of the permeable pore than do hydrodynamic models. Yet there are still too few available data to allow a decision on the physical model for water permeation in pores of molecular dimensions.

V. Secondary Active Transport of Water in Membrane Proteins

Secondary active transport of H_2O requires a transfer of energy from a downhill flux of solutes to the uphill flux of H_2O. The total change in free energy for the fluxes should be negative and stem entirely from the transmembraneous differences in solute and H_2O concentrations; there is no energy requirement from chemical bonds (e.g., from ATP). Secondary active transport may have the additional property of cotransport: chemically coupled transport for which the ratio of the fluxes is constant, independent of the imposed driving forces. Furthermore, a downhill flux of any one of the involved species should be able to drive uphill fluxes of all the others. The proteinaceous character of the cotransporter should be manifest by saturation of the fluxes at increasing driving forces, by specificity for the ions, and by specific drug inhibition.

Three possible examples of cotransport of H_2O will be discussed: K^+, Cl^-, H_2O; Na^+, K^+, Cl^-, H_2O; and Na^+, organic solute, H_2O. The examples are all of the symport type. Cotransporters of the antiporter type, Cl^-/HCO_3^- (Cotton and Reuss, 1991) and the Na^+/H^+ exchange (T. Zeuthen, unpublished) have not been found to translocate H_2O.

K^+/Cl^- cotransport is particularly well studied in red blood cells (Lauf, 1985; Lauf *et al.*, 1992; Parker, 1993). It has also been found in epithelial cells and many other cell types (Lang *et al.*, 1993; Zeuthen and Stein, 1994). Generally K^+/Cl^- cotransport is associated with conditions where cells lose H_2O, such as in regulatory volume decrease. In red blood cells, K^+/Cl^- cotransport is activated with the cell volume increases. K^+/Cl^- cotransport has also been implicated in volume loss in certain plant cells (Hill and Findlay, 1981) and in freshwater ameba and *Hydra* (Zeuthen, 1992). $Na^+/K^+/2Cl^-$ contransport is usually associated with a regulatory volume increase (Lang *et al.*, 1993; O'Grady *et al.*, 1987; Zeuthen and Stein, 1994), but also with many steady-state situations (for references, see Section VI,C). The discussion of cotransport of Na⁺, organic substrate, and H₂O deals with brush border membranes of kidney and small intestine. The glucose transporters here form a separate family (Wright *et al.*, 1992) (see Section V,D).

A. Experimental Evidence for Cotransport of H_2O

1. $K^+/Cl^-/H_2O$ Cotransport in Leaky Epithelia

I consider first the ventricular membrane of the choroid plexus epithelium. This is equivalent to the basolateral membrane of the gall bladder and will be called the exit membrane, that is, the membrane across which H₂O is secreted. Ion and water movements were investigated by means of intracellular microelectrodes sensitive to K^+, Cl^-, Na^+, and cell volume (see Section II,D and Zeuthen, 1991a,b, 1993a,b, 1994). An illustrative example is shown in Fig. 6. The osmolarity of the exit compartment was increased by addition of mannitol or NaCl and the cells shrank as expected. However, if the osmolarity was increased by the addition of KCl, the cell immediately *swelled* despite the fact that the osmotic difference *per se* favored a cell shrinkage; simultaneous recordings of the intracellular ion content showed that the swelling took place before the osmolarity of the intracellular compartment had changed significantly. In consequence, the movement of H₂O took place in the direction opposite to passive movement. In the presence of furosemide (or bumetanide) the addition of KCl

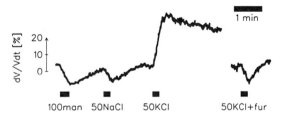

FIG. 6 Osmotic behavior of epithelial cells. Cellular volume changes were recorded by microelectrodes (Section II,D) in choroid plexus of *Necturus* in response to the addition of 100 mmol liter⁻¹ of mannitol (100 man), 50 mmol liter⁻¹ of NaCl, 50 mmol liter⁻¹ of KCl and 50 mmol liter⁻¹ of KCl to which 10⁻⁴ mol liter⁻¹ of furosemide had been added. The recordings are from the same cell and solution changes were performed on the exit side of the epithelium. The last event was recorded 7 min after the KCl event, at which time the cell had returned to its control volume. Please note that KCl causes a cell *swelling*, although the osmotic gradient favors a cell shrinkage (from Zeuthen, 1994, with permission).

induced the swelling expected from a normal osmoticum. The conclusion is that the flux of H_2O is not driven by the osmotic difference but instead by the flux of KCl; energy is transferred from a downhill flux of KCl to an uphill flux of H_2O. That is, the coupling of flows appears direct (i.e., chemical), not indirect (i.e., osmotic).

Several lines of evidence support the concept that H_2O is cotransported with K^+ and Cl^-: (1) *Specificity:* Cotransport of K^+, Cl^- and H_2O requires K^+ and is specific for Cl^-; anion substitutions with SCN^-, acetate$^-$, and NO_3^- inhibit the transport. Na^+ is not cotransported. (2) *Electrical properties:* The coupling is electroneutral; there are no effects of changes in membrane potential. (3) *Transfer of energy:* Increases in either extracellular K^+ or Cl^- result in downhill influxes of K^+ or Cl^- and an uphill influx of H_2O; the latter can proceed against osmotic gradients of up to 100 mosm l^{-1}. When the osmolarity of the extracellular compartment is increased abruptly by mannitol, effluxes of K^+ and of Cl^- are initiated in parallel with an efflux of H_2O. When the mannitol is removed again, the acquired cell hyperosmolarity will cause a passive influx of H_2O accompanied by influxes of K^+ and Cl^-. These ion influxes take place against their respective electrochemical gradients. Now a downhill flow of H_2O drives an uphill flow of ions. In sum, a downhill gradient of any one of the involved species, K^+, Cl^-, or H_2O, is able to drive the uphill flux of the two others. (4) *Coupling ratio:* If the passive water fluxes which proceed through the lipid membrane and water channels which have a total water permeability of 0.6×10^{-4} cm sec^{-1} (osm liter^{-1})$^{-1}$ (Fig. 7) are subtracted, then the ratio of coupling between K^+, Cl^-, and H_2O in all types of experiments is found to be close to 110 mmol l^{-1}. This is equivalent to about 500 molecules of H_2O for each K^+ and Cl^- ion (Zeuthen, 1994). This number will be used in the model considerations that follow. The invariant stoichiometry applied to the wide range of differences in osmolarities (up to 600 mosm liter^{-1} of mannitol) and in KCl or K^+ concentrations (up to 50 mmol liter^{-1}) which were tested. Tissues adapted to transport in salines with a lower or higher osmolarity than physiological also exhibited K^+, Cl^-, and H_2O cotransport with the same coupling ratio—about 110 mmol liter^{-1} (T. Zeuthen, unpublished). (5) *Saturation:* The coupled transport of K^+, Cl^-, and H_2O apparently exhibits saturation. The transport of H_2O and K^+ saturate when the extracellular osmolarities are increased to about 200 mosm liter^{-1} (Fig. 7) and at extracellular K^+ activities of about 50 mmol liter^{-1}. Any additional H_2O and K^+ transport is through passive systems. A final conclusion on saturation will have to take inactivation of the cotransporter at elevated osmolarities into account (Parker, 1993). (6) *Inhibition:* The coupled transport of K^+, Cl^-, and H_2O was inhibited by furosemide (10^{-4} mol liter^{-1}) and bumetanide (10^{-5} mol liter^{-1}). In conclusion, the coupling between the K^+, Cl^-, and the H_2O fluxes fulfill all formal criteria for cotransport by a membrane protein.

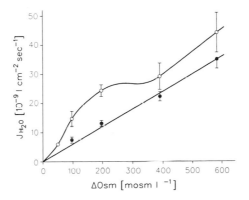

FIG. 7 H_2O fluxes during osmotic shrinkage. The initial rate of water flux (J_{H_2O}) was recorded in response to abrupt increases in osmolarity (Δosm) implemented by mannitol (details as in Fig. 6). Cells were adapted to control solutions (open circles) or to Cl^- free solutions (closed circles). The extra H_2O flux seen in the presence of Cl^- is carried by $K^+/Cl^-/H_2O$ cotransport. (From Zeuthen, 1994, with permission.) At increases of 400 and 600 mosm liter^{-1} in the presence of Cl^-, the effect is affected by a concomitant inactivation of the $K^+/Cl^-/H_2O$ cotransport protein. (From T. Zeuthen, unpublished.)

2. Dark Cell Epithelium

The high K^+ concentration of the endolymph in the vestibular system of the equilibrium organ is maintained by secretions from the dark cell epithelium (Fig. 8). In analogy with transport in the *stria vascularis* of the cochlea of the inner ear, vestibular dark cells maintain a vestibular K^+ concentration of about 140 mmol liter^{-1} by transport from the perilymph.

Vestibular dark cell epithelium from gerbils (small rodents) were bathed in physiological salines which could be changed (90%) in about 15 sec (Wangemann and Marcus, 1990; Wangemann et al., 1992). When K^+ was increased from 3.6 mmol liter^{-1} to 25 mmol liter^{-1} (replacing Na^+), there was a cell swelling of about 1 to 2% sec^{-1} (Fig. 8); when K^+ was reduced again, the cell shrank at roughly the same rate. The effect of K^+ on cell volume results from $Na^+/K^+/Cl^-$ cotransport located to the basolateral membrane (Marcus et al., 1987; Marcus and Marcus, 1987; Wangemann and Marcus, 1990). The K^+-induced cell swelling was inhibited by bumetanide and piretanide, known blockers of cotransport, but not by blockers of K^+ and Cl^- channels.

One may ask which water permeabilities (L_p) and osmotic driving forces are required if these very rapid volume changes are to be explained by simple osmosis. Unfortunately no direct study of these parameters exists, but some directions for further research can be obtained from considerations of data such as those in Fig. 8. Assume that the L_p can be obtained from the rate of volume changes elicited by removal of mannitol or NaCl

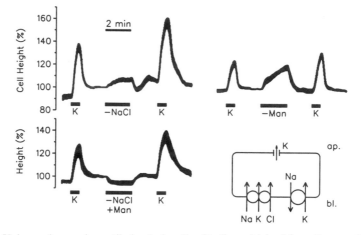

FIG. 8 Volume changes in vestibular dark cell epithelium obtained from the ampulla of the semicircular canal of the inner ear. The events (K) are changes in cell height in response to isosmolar increases in external K^+ concentration from 3.6 to 25 mmol liter^{-1} (replacements of Na$^+$). Hypoosmolarities were obtained (upper left) by removal of 75 mmol liter^{-1} of NaCl or (upper right) removal of mannitol. Lower left: isosmotic replacement of NaCl by mannitol. Lower right: Cell model. Vestibular dark cells contain in their apical (ap) membrane (facing the endolymph) a K^+ channel and in their basolateral membrane (facing the perilymph) a Na^+/K^+ ATPase, a $Na^+/K^+/2Cl^-$ cotransport, and a Cl^- channel (not shown). K^+-induced cell volume changes are thought to take place via the $Na^+/K^+/2Cl^-$ cotransporter (see text). (From Wangemann and Shiga, 1994, with permission.)

from the external solutions. An estimate of 0.16 to 0.3 $\times$ 10^{-4} cm sec^{-1} (osm liter^{-1})$^{-1}$ for the whole cell can be obtained from the data (Fig. 8), disregarding unstirred layer effects (see later discussion). In K^+-free or Cl^--free solutions, this estimate of L_p was larger but no more than twice this value (P. Wangemann, personal communication). Incidently, these estimates are close to those for the ventricular membrane of the chorioid plexus.

In order to explain the rate of swelling by osmosis on the basis of this L_p, the intracellular osmolarity would have to increase by some 600 mosm liter^{-1} in a few seconds when K^+ was increased extracellularly. There are several reasons why such rapid changes cannot take place. First, if the additional intracellular osmolarity was a result of increases in intracellular K^+ and Cl^- concentrations, this in itself would cause the driving forces for the influx of the ions to be abolished. Second, it is unlikely that the transport rate of the responsible $Na^+/K^+/2Cl^-$ cotransporters could be that fast, given known turnover rates and density of these proteins. The explanation based on simple osmosis becomes even more untenable if one considers the efflux of H_2O (cell shrinkage) which results when extracellular K^+ is replaced by Na^+ again. Since the rate of shrinkage is about the same as the swelling, the cell has to become about 600 mosm liter^{-1} *hyposmolar* relative to the external solution to expel the H_2O by simple osmosis. This means that the cell, from being hyperosmolar by 600 mosm liter^{-1} when extracellular K^+ is high, must change its osmolarity by 1200 mosm liter^{-1} in a few seconds when K^+ is returned from 25 to 3.6 mmol liter^{-1}.

Obviously, these hypothetical osmotic differences will be smaller if the L_p is, in fact, larger than the one estimated directly from Fig. 8. It would therefore be important to obtain an estimate of unstirred layers, which might cause the observed cell swellings to be smaller than those given by the *true* membrane water permeability. The effect is likely to be small, however, judging from quantitative studies in other epithelia (Zeuthen, 1992).

Instead of an explanation based on osmosis, I suggest that the K^+-induced fluxes of H_2O take place via $Na^+/K^+/2Cl^-/H_2O$ cotransport. The involvement of Cl^- and Na^+ stems from several experimental findings (Wangemann and Shiga, 1994): There was a rapid cell shrinkage when Cl was replaced isotonically by gluconate and a rapid swelling when Cl^- was returned again. Using arguments analogous to those discussed earlier, these changes are difficult to explain by osmosis. In addition, cell swelling implemented by hyposmolar removal of NaCl was only half as fast as changes obtained by reducing external mannitol with Cl^- maintained constant (Fig. 8). The reduction of external NaCl causes an influx of H_2O via a passive water permeability, but when Cl^- is reduced, there is an opposing efflux of H_2O by the cotransport system. When conformational changes in proteins were slowed by cooling and mild glutaraldehyde fixation, NaCl reduction was as effective in inducing cell swelling as was mannitol reduction.

One way to resolve the issue would be to implement osmotic challenges by additions of KCl, as was done in the experiments shown in Fig. 6. Since intracellular K^+ and Cl^- cannot rise faster than the extracellular concentrations, the cell should shrink initially if simple osmosis took place. If, on the other hand, the cell began to swell initially, this would imply that H_2O moved against the direction given by the osmotic difference and would be strong evidence for secondary active transport of H_2O.

3. Kidney Proximal Tubule

Repeated hyperosmolar application of 30 mmol liter^{-1} of urea to the basolateral surface of the isolated rabbit proximal straight tubules induced instantaneous and rapid increases of cell volume (González *et al.*, 1993). The swelling was caused by Ca^{2+}-mediated activation of an inwardly directed $Na^+/K^+/2Cl^-$ cotransporter in the basolateral membrane. Since urea increased the osmolarity of the external solution, and since one must assume that the urea concentration inside the cell cannot rise within 1 sec above the extracellular urea concentration, it seems that the influx of H_2O across the basolateral membrane during urea application is uphill (Kirk *et al.*, 1987). One explanation of this effect would be that H_2O was transported into the cell together with the ions by $Na^+/K^+/2Cl^-/H_2O$ cotranport.

4. Na^+/L-Alanine/H_2O, Na^+/D-Glucose/H_2O Cotransport in Small Intestine

The small intestine is relatively water impermeable. This prevents changes in body osmolarity in the face of variations in the osmolarity of the intestinal lumen. Pappenheimer (1993) has suggested that the enzymatic breakdown of sugars and peptides at the mucosal brush border membrane might raise the osmolarity near the apical membrane by as much as 200 to 300 mmol liter^{-1}. The question is this: How can the intestine maintain high rates of H_2O reabsorption in the face of adverse osmotic gradients? (Fig. 9A).

Simple osmosis is not sufficient. Given the low water permeability, it would require inward transepithelial differences of around 100 mosm liter^{-1} to explain inward transport by osmosis (Zeuthen, 1992). This contradicts the fact that isolated intestinal epithelium can transport H_2O at physiological rates when bathed on two sides by the same solution. I suggest that the $Na^+/$glucose as well as the $Na^+/$amino acid transporters cotransport H_2O.

This idea was tested in an epithelial cell from *Necturus* small intestine. The mucosal surface is suddenly exposed to 25 mmol liter^{-1} of L-alanine (replacing an equivalent amount of mannitol). This causes an immediate increase in cell volume as recorded with microelectrodes (see Fig. 10). The rate of increase was equivalent to the swelling which would be obtained if the cell was made hyperosmolar relative to the mucosal solution by 25 mosm liter^{-1}. Yet there was no indication of such an increase in cellular osmolarity. In fact, the cellular K^+ concentration decreased by 0.5 mmol liter^{-1} sec^{-1} while the intracellular Na^+ concentration increased at the same rate. So alanine causes an immediate cell swelling despite the apparent absence of an osmotic driving force. The ratio of Na^+ influx to the influx of H_2O was in the range of 50 to 100 mmol of Na^+ per liter of H_2O. Similar volume increases have been observed in mammalian enterocytes (MacLeod and Hamilton, 1991). The suggestion of $Na^+/$organic substrate/H_2O cotransport is compatible with the data of Fischbarg *et al.* (1993), who found that the transport of H_2O by the $Na^+/$glucose cotransporter depended on the presence of Na^+ and had

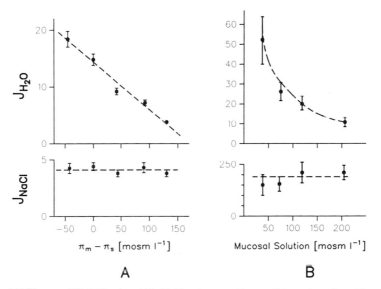

FIG. 9 (A) Fluxes of H_2O (J_{H_2O}) and NaCl (J_{NaCl}) across the small intestine of rat. The values are plotted as a function of the osmotic difference across the epithelial cell layer maintained by adding or removing NaCl from the liminal solution. (Data from Parsons and Wingate, 1961.) Units for J_{H_2O} are $\mu l\ hr^{-1}\ mgdw^{-1}$ and J_{NaCl}, $\mu osm\ hr^{-1}\ mgdw^{-1}$. (B) The rate of transepithelial transport of H_2O and NaCl as a function of external osmolarity. *Necturus* gall bladders were adapted to transport in normal and dilute external solutions. The bladders were mounted unilaterally and have virtually the same solution on both sides. J_{H_2O} was larger the larger the degree of dilution and followed an inverse relationship while the J_{NaCl} remained almost constant. (Data from Zeuthen, 1982). Units J_{H_2O} are $\mu l\ cm^{-2}\ hr^{-1}$ and J_{NaCl}, $mmol\ cm^{-2}\ hr^{-1}$).

a high activation energy, typical of conformational changes in a membrane protein (see Section III,E).

B. Thermodynamics of Cotransport of H_2O

The energetic requirements for uphill transport of H_2O can be assessed using the Gibbs equation. Such an analysis will be independent of the molecular mechanisms of the cotransport. Consider the substrates A and B which are translocated from the inside (i) to the outside (o) of a membrane in proportions a to b together with n molecules of H_2O (concentration C_w). At equilibrium:

$$[A_i]^a \times [B_i]^b \times [C_{w,i}]^n = [A_o]^a \times [B_o]^b \times [C_{w,o}]^n \qquad (1)$$

The brackets indicate concentrations. For the $K^+/Cl^-/H_2O$ cotransporter

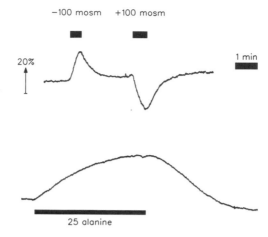

FIG. 10 Volume changes in enterocytes in response to alanine. The cell volume from *Nect-urus* small intestine was recorded by a choline ion-sensitive microelectrode (Section II,D) while bathed in physiological saline. The luminal solution was first changed rapidly to one with half normal osmolarity (-100 mosm), which caused an immediate cell swelling, then to a solution to which 100 mmol liter^{-1} of mannitol had been added ($+100$ mosm). In the lower trace the cell was exposed to 25 mmol liter^{-1} alanine, which isomotically replaced mannitol. This caused an immediate cell swelling. (From T. Zeuthen, unpublished.)

in a typical cell, $A_i = K_i^+ = 80$ mmol liter^{-1}, $a = 1$; $B_i = Cl_i = 40$ mmol liter^{-1}, $b = 1$; $A_o = K_o^+ = 2$ mmol liter^{-1}; $B_o = Cl_o^- = 110$ mmol liter^{-1}, $n = 500$ (Section V,A and Zeuthen, 1994). The ratio $[C_{w,i}]^n/[C_{w,o}]^n$ is to a good approximation equal to $\exp[n(osm_i - osm_o)/n_w]$ where osm is the osmolarity in (osm liter^{-1}) and n_w is the number of moles of H_2O per liter ($= 55$). As can now be calculated, these differences in ion concentrations can balance an osmotic difference of about 300 mosm liter^{-1} (see also Fig. 11, curve 3).

The rate of H_2O transport by the $K^+/Cl^-/H_2O$ cotransporter ($J^*_{H_2O}$) will to a first approximation be proportional to the difference between the product (Eq. 1) taken inside the cell and outside. Since C_w is proportional to $\exp(-osm/n_w)$, $J^*_{H_2O}$ can be expressed as:

$$J^*_{H_2O} = B\left[K_i \times Cl_i \times \exp\left(-\frac{n}{n_w} \times osm_i\right) - K_o \times Cl_o \exp\left(-\frac{n}{n_w} osm_o\right)\right] \quad (2)$$

The factor B can be determined from Fig. 7. Here the experimentally determined $\Delta J^*_{H_2O}/\Delta osm_o$ equals 0.6×10^{-4} cm sec^{-1} (osm liter^{-1})$^{-1}$. This can be compared with the theoretically determined $\Delta J^*_{H_2O}/\Delta osm_o$ from Eq. 2, which equals $-B \times K_o \times Cl_o [-n/n_w \exp(-n/n_w osm_o)]$. With $K_o = 2$ mmol liter^{-1} and $Cl_o = 110$ mmol liter^{-1}, B is calculated to be

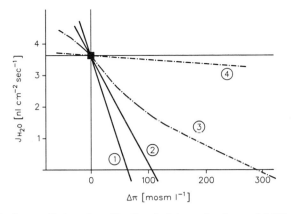

FIG. 11 Quantitative predictions from Eq. 2 and of the molecular model (Fig. 13). Transepithelial water transport (J_{H_2O}) is plotted as a function of the transepithelial osmotic difference implemented by either adding mannitol to the entry side (curve 1) or by removal of NaCl from the exit bath (curve 2). The calculations proceed as follows: The rate of H_2O transport by the $K^+/Cl^-/H_2O$ cotransport itself is calculated from Eq. 2. If mannitol is applied to the entry side, and under the assumption that the cell osmolarity remains close to the elevated osmolarity, curve 3 (stipled) is obtained. This secondary active flux of H_2O will be superimposed by an adverse osmotic flow via the passive pathways given by the measured L_pS. When this passive flux is subtracted from curve 3, curve 1 is obtained. If NaCl is removed from the exit bath, the H_2O transport by the $K^+/Cl^-/H_2O$ cotransporter itself will be described by curve 4. When the passive H_2O fluxes are taken into account and subtracted from the values given by curve 4, the H_2O transport across the epithelium is obtained as given by curve 2 (see text).

1.8 × 10^{-2} cm sec^{-1} (osm liter^{-1})$^{-2}$. With this value of B reintroduced in Eq. 2, it is found that $J^*_{H_2O}$ equals 3.6 nl cm^{-2} sec^{-1} for osm$_c$ = 300 mosm liter^{-1} and K_i = 80 mmol liter^{-1} and Cl_i = 40 mmol liter^{-1}. This value for $J^*_{H_2O}$ is very close to that observed for frog choroid plexus when surface foldings are taken into account (Wright *et al.*, 1977; Zeuthen and Wright, 1981) and for other amphibian epithelia (Zeuthen, 1992). $J^*_{H_2O}$ calculated as a function of osmolaritic differences across the membrane is given in Fig. 11, curves 3 and 4.

The energy which would be dissipated (E_d) by the passive KCl exit (J_{KCl}) is far in excess of that required to move H_2O uphill, out of the cell,

$$E_d = J_{KCl} \times RT \ln [K_i] \times [Cl_i]/[K_o] \times [Cl_o] \tag{3}$$

with J_{KCl} = 0.3 × 10^{-9} mol cm^{-2} sec^{-1}, a typical amphibian value, E_d amounts to 2 × 10^{-6} J cm^{-2} sec^{-1}. This is more than three times the effect needed to promote the equivalent volume flow of 2.7 × 10^{-9} liter^{-1} sec^{-1} cm^{-2} of H_2O against an osmotic difference of 100 mosm liter^{-1} (Zeuthen, 1994).

In the case of Na^+/glucose/H_2O cotransport, assume $A_o = Na_o =$ 115 mmol liter^{-1} and $A_i = Na_i = 10$ mmol liter^{-1}. Experiments suggest a stoichiometry of $2Na^+ : 1$ glucose for the Na^+/glucose cotransporter so $a = 2$ (Loo et al., 1993). Assume that the glucose concentration in the intestinal lumen, $B_o = 10$ mmol liter^{-1} and that glucose in the cell amounts to 25 mmol liter^{-1}. n was determined in the pilot experiments (Fig. 10) to be in the range of 300–500. With these numbers, Eq. 1 shows that the Na^+/glucose/H_2O cotransporter can easily transport H_2O from an intestinal lumen 300 mosm liter^{-1} hyperosmolar relative to the cell.

It is appropriate to ask how much the binding of H_2O to the membrane protein affects the activation energies and therefore the rate of transport. If 500 molecules of H_2O were attached to the surface of the protein, they would contribute 2 kcal mol^{-1} or about 3.4 RT (see Section II,B). This would affect the turnover rate by one to two orders of magnitude. This contribution is significant but not prohibitive. The activation energies for K^+/Cl^- cotransport have been estimated to be 27 kcal mol^{-1} and for Na^+/K^+/$2Cl^-$ cotransport, 17.7 kcal mol^{-1} (Ellory and Hall, 1988).

C. Molecular Model for Cotransport of H_2O

The molecular model for cotransport of H_2O by membrane proteins (Zeuthen, 1993a, 1994; Zeuthen and Stein, 1994) depends on one fact and several assumptions. The fact is that conformational changes in proteins involve movements of complete domains relative to each other and thereby bring about large changes in the amount of bound or enclosed water (Table I). This effect was first demonstrated for hexokinase (Anderson et al., 1979) and for an L-arabinose-binding protein (Mao et al., 1982). The uptake and release of H_2O can explain secondary active transport of H_2O in membrane proteins if the following assumptions hold (Fig. 12):

(1) When the membrane protein is in its inward-facing configuration, the substrate(s) approaches the binding site by diffusing through a hydrated access pathway. Thus the protein exposes more surface to the inside than to the outside of the membrane. (2) When the substrates bind to the protein, conformational changes cause the permeability barrier to move and the substrates and their surrounding water become exposed to the other side of the membrane. (3) When the substrates leave the protein, the access channel closes and the water held on the surfaces or/and in the access channel is transferred to the same bath as the substrates. (4) In this closed conformation, the protein may open to the inside of the membrane again and the transport cycle will repeat.

As a consequence, a well-defined amount of H_2O is transferred across the membrane together with the substrates. The cotransport takes place

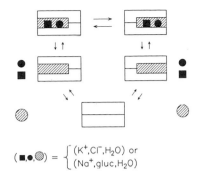

$$(\blacksquare, \bullet, \oslash) = \left\{ \begin{array}{l} (K^+, Cl^-, H_2O) \text{ or} \\ (Na^+, gluc, H_2O) \end{array} \right.$$

FIG. 12 Molecular model for secondary active transport of H_2O. The mobile barrier hypothesis of Mitchell (1957) can be applied to the cotransport of K^+, Cl^-, and H_2O, or alternatively to Na^+, glucose (alanine), and H_2O. The binding of the substrates (left) in the protein induces a conformational change which shifts the permeability barrier from one side of the protein to the other. This allows simultaneous transport of K^+, Cl^-, and the H_2O occluded in the access pathway. The protein also exists in a closed state where access volumes are small (see text).

via quasi-equilibria between allosteric conformations of the membrane protein. In this connection H_2O can be considered as an allosteric ligand (Colombo *et al.*, 1992) and equilibria will be determined by Gibbs's equation.

It is not a new idea that substrates move through a wide hydrated vestibule in order to reach the binding sites. It has been implied for K^+ channels (Latorre and Miller, 1983), for Na^+/K^+ ATPase (David *et al.*, 1992; Gadsby *et al.*, 1993; Goldshleger *et al.*, 1994) and for the F_oF_1-class of proteins (Tanford, 1983). It is relevant to ask how large these access channels should be. If a hydrated K^+ ion (diameter 8 Å) is separated from the surface of the protein by one layer of water molecules, the access channel must have a diameter of at least 12 Å. Membrane proteins protrude from the surface of the lipid bilayer into the external solutions. If the length of the membrane protein is 100 Å, the maximal volume of the access channel would be about 11,000 Å^3. This is small, about 2% of the volume of a protein of mole weight 260 kDa as determined for the K^+/Cl^- cotransporter (Cherksey and Zeuthen, 1987) or 3% of the $Na^+/K^+/2Cl^-$ cotransporter (Lytle *et al.*, 1992). An access volume of 11,000 Å^3 would contain about 300 molecules of H_2O. It would take an access volume of 15,000 Å^3 (or a diameter of 14 Å) to contain the 500 molecules translocated by the $K^+/Cl^-/H_2O$ cotransporter.

The mobile barrier model was suggested by Mitchell (1957) and reviewed recently by Nikaido and Saier (1992). It suggests that the barrier for transport moves from one side of the binding site to the other. Different binding constants in the different conformations could, if present, compensate for different substrate concentrations on the two sides, as discussed by Tanford (1983). For kinetic descriptions, see Stein (1986, 1990).

D. Structure of Cotransport Proteins

Is there any structural evidence for H_2O-filled vestibules in cotransport proteins? Unfortunately such information is indirect, derived mostly from sequence data and hydropathy diagrams. Most data are from bacterial systems, but the information gained here apparently applies to eukaryotic systems as well (Baldwin, 1989; Ames 1992; Kaback, 1992; Harvey and Oxender, 1992; Higgins, 1992; Nikaido and Saier, 1992; Kaplan, 1993). In prokaryotes the α-helix is the fundamental unit and two times six of these is suggested as the general motif. This also applies to the $Na^+/$ glucose transporter, the $Na^+/Cl^-/GABA$ transporter of the eukaryotes (Wright *et al.*, 1992) and the $Na^+/K^+/2Cl^-$ cotransporter (Xu *et al.*, 1994; Palfrey and Cossins, 1994). The α-helix consists of a spiral of about 25 amino acids—long enough to span the membrane (100 Å) and with an approximate diameter of 5 Å. These partly hydrophobic α-helices are joined by hydrophilic regions.

The membrane protein bacteriorhodopsin from *Halobacterium halobium* (Branden and Tooze, 1991) has been investigated by X-ray crystallography (resolution 3 Å) and confirmed to be built by transmembraneous α-helixes. In other membrane proteins, the evidence for membrane-spanning α-helixes is conjectural. The primary sequence data do suggest a series of hydrophobic regions, each sufficiently long to span the membrane as an α-helix, but other configurations have been suggested. α-Helixes of various lengths could be surrounded by membrane-spanning helixes which form the boundaries to the lipid (Lodish, 1988).

Proteins can have similar functions as long as the shape of the constituent subunits are similar and as long as essential chemical properties located on the surface are maintained. It is becoming increasingly clear that a number of gross features such as major groves or folds may be the same among a variety of proteins, while the amino acids which form the bulk of the subunits are completely different (Orengo *et al.*, 1944). L-Arabinose, D-xylose, D-galactose, L-frucose, L-ramnose, and raffinose are transported with H^+ in *E. coli* and several enterobacteri; D-glucose with H^+ in cyanobacteria; and H^+ with lactose in streptococci. These cotransporters are homologous to some types of $Na^+/$glucose transporters in yeast, algae, plants, protozoa, and mammals.

Other cotransporters have primary sequences that are not recognizable at all. The $Na^+/$glucose cotransporters in kidney and intestine of rabbit and humans are similar to the Na^+/proline, Na^+/pantothenate in *E. coli* but do not resemble any of those mentioned above (Griffith *et al.*, 1992; Wright *et al.*, 1992). In conclusion, a protein's ability to have water-containing groves or folds should not be sought in the amino acid sequence but rather in its quarternary structure.

Twelve parallel membrane-spanning α-helices of diameter 5 Å packed central-symmetrically perpendicular to the lipid membrane will form a pore with a diameter of 14 Å. Such an internal diameter suggests that large volumes of H_2O could be held in the protein either by occlusion or by surface binding. It is highly suggestive, but possibly fortuitous, that the calculations based on experimental evidence from the K^+/Cl^-/H_2O cotransporter outlined above (Section V,B) also arrive at a diameter of the access pathway of about 14 Å. This diameter would also explain how larger substrates penetrate (i.e., oligopeptides) (Higgins, 1992; Ames, 1992). Electron microscopic techniques applied to the lac permease, which cotransports β-galactosides and H^+ in a ratio of one to one (Kaback, 1992), suggest that this protein has a large solvent-filled cleft (Costello *et al.*, 1984, 1987; Li and Tooth, 1987).

In conclusion, neither electrophysiological evidence nor evidence from molecular biology rule out significant amounts of H_2O, up to 500 molecules, being held and translocated in cotransport proteins. How these H_2O molecules are "sucked in and squirted out by the crevices" (Mitchell, 1990, p. 288) during conformational changes remains to be elucidated.

VI. A Molecular Model for Epithelial Transport

The intracellular environment is vital to the cell, and membrane transport is orchestrated to maintain intracellular salt concentrations, H_2O contents, and thereby working distances between macromolecules at constant values (Parker, 1993; Garner and Burg, 1994; Häussinger *et al.*, 1994; see also Section II). Epithelial cells have their membrane proteins distributed asymmetrically between their two membranes, so the mechanisms of cellular homeostasis may lead to vectorial transport: uphill H_2O transport by cotransport proteins at one membrane and a passive, downhill transport at the other (see Fig. 13).

The epithelial model to be presented here does not include other anatomical properties. In most conventional models, coupling between salt and H_2O fluxes results from salt accumulation in the lateral intercellular spaces. Significant accumulations have never been observed, however, and are unlikely to occur; the lateral spaces are not long and narrow enough (Hill, 1980; Schafer, 1990) and the basement membrane may be too leaky to

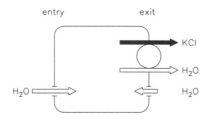

FIG. 13   Molecular model for epithelial water transport. The cell is hyperosmolar (by some 6 mosm liter^{-1} at normal physiological osmolarities) and H_2O enters passively across both the entry and exit membrane. H_2O is pumped (secondarily active) across the exit membrane by cotransport with KCl. The length of arrows is quantitatively correct for the choroid plexus epithelium (see text and Fig. 11 for quantitative predictions).

retard salt diffusion (Welling *et al.*, 1972; Persson and Spring, 1984; Moreno and Diamond, 1972; Wright and Diamond, 1968; Hill and Hill, 1987; Tripathi and Boulpaep, 1989; Whittembury and Reuss, 1992; Zeuthen, 1992; Zeuthen and Stein, 1994). In the following section I discuss the most central properties of H_2O transport in leaky epithelia which any model, cellular or molecular, must address. Finally in Sections VI,B and C I suggest how a molecular model, that is, a model based on the properties of the membrane transporters only, might explain these features.

A. Three Properties of Transport in Leaky Epithelia

Three features characterize transepithelial H_2O transport: (1) Uphill transport. H_2O can be transported across the cell layer from a low chemical potential for H_2O to a high chemical potential. (2) Isotonic transport or near-isotonic transport. The ratio of the amount of transported salt (typically NaCl) to the amount of transported H_2O is close to the ratio of salt and H_2O in the bathing solutions. (3) Inverse relationship between the rate of volume transport and extracellular osmolarity. In tissues adapted to transport in dilute solutions, the rate of H_2O transport is high. If the external solutions are diluted by a factor of, say two, the rate of transport is increased by a factor close to two compared with the one at normal osmolarities. The transported solution remains isotonic or near isotonic to the external solutions.

1. Uphill Transport

Any model must predict the relation between the rate of transepithelial water transport J_{H_2O} and the transepithelial osmotic difference $\Delta\pi$, as shown in Fig. 9A for rat small intestine (Parsons and Wingate, 1961). J_{H_2O}

decreases linearly as the opposing osmotic difference is increased. This is characteristic of most leaky epithelia (Zeuthen, 1992; Zeuthen and Stein, 1994). At a certain osmotic difference $\Delta\pi = \hat{C}$, J_{H_2O} is zero. Here the inert ability of the tissue to transport H_2O is exactly matched by the externally imposed osmotic flow. $\hat{C}$ is called "the strength of transport" (Weinstein and Stephenson, 1981a,b) and has been recorded as 30 mosm liter^{-1} in *Necturus* gall bladder (Persson and Spring, 1982); 40 mosm liter^{-1} in fish gall bladder (Diamond, 1962); 80 mosm liter^{-1} in rabbit gall bladder (Diamond, 1964a); and 120 mosm liter^{-1} in dog gall bladder (Grim and Smith, 1957). In dog small intestine, $\hat{C}$ equals 200 mosm liter^{-1} (Hakim *et al.*, 1963) while 50 (Pappenheimer and Reis, 1987) and 150 mosm liter^{-1} (Parsons and Wingate, 1961) have been recorded in the rat small intestine. In the mammalian kidney proximal tubule, 25 to 60 mosm liter^{-1} are needed to stop transport (Frömter *et al.*, 1973; Alpern *et al.*, 1985; Bomsztyk and Wright, 1986; Green *et al.*, 1991). For the choroid plexus epithelium of rabbit, $\hat{C}$ equals 150 mosm liter^{-1} (Heisey *et al.*, 1962) (Zeuthen, 1992; Zeuthen and Stein, 1994).

2. Isotonic Transport or Near-Isotonic Transport

The ability for isotonic transport has been reviewed extensively (House, 1974; Hill, 1980; Schafer, 1990; Triphathi and Boulpaep, 1989; Whittembury and Reuss, 1992; Zeuthen, 1992; Zeuthen and Stein, 1994). In assessing the underlying coupling mechanism, it is important to take into account that, while some epithelia transport precisely isotonically (e.g., gall bladder: Diamond, 1964b; Hill and Hill, 1978; Zeuthen, 1982), others do not. Thus the choroid plexus epithelium secretes a fluid which is hyperosmolar by 50 mosm liter^{-1} in frog (Wright *et al.*, 1977) and 6–9 mosm liter^{-1} in mammals (Cserr, 1971; Rapoport, 1976). The mammalian kidney proximal tubule produces a solution which is 3 to 4 mosm liter^{-1} hypertonic (Green and Giebisch, 1989).

3. Inverse Relationship between the Rate of Volume Transport and Extracellular Osmolarity

The inverse relationship between transport rate of H_2O and external osmolarity (Fig. 9B) is retained at dilutions of a factor of at least ten in the *Necturus* gall bladder (Hill and Hill, 1978). In mammalian gall bladder, the inverse relationship is retained only to a dilution of two (Diamond, 1964b); for higher dilutions the transport is smaller. The inverse relationship could arise from the cellular regulation of the intracellular osmolarity close to its physiological value. In that case, the rate of influx (and therefore the transepithelial H_2O transport) would be proportional to the dilution of the external solutions. This proposal is supported by the finding that the intracellular K^+ activity and osmolarity remain relatively constant despite adaption to low osmolarities (Fig. 2; Section II,D and Tables II and III). Since the secretion is isosmolar to the bathing solutions, it remains to be explained how H_2O leaves the hyperosmolar cell into the dilute fluid secreted.

Isotonic transport is retained in dilute solutions in gall bladder (Diamond, 1964b; Hill and Hill, 1978; Zeuthen, 1982) and amphibian kidney proximal tubule (Whittembury and Hill, 1982). The H_2O transport goes up in proportion to the dilution and the Na^+ transport remains constant (Fig. 9B). $K_{\frac{1}{2}}$ for Na^+ entry is relatively low for the entry step—about 10 mmol liter^{-1} (Davis and Finn, 1985; Altenberg and Reuss, 1990). Accordingly Na^+ entry is relatively unaffected by dilution. The onus is on the exit membrane, which must maintain the export of Na^+ and H_2O as fast as they enter.

In conclusion, many transport phenomena in leaky epithelia can be explained by a model in which the cell regulates its intracellular osmolarity to a normal physiological value. The influx of H_2O could then be explained by osmosis. The transport mechanisms at the exit membrane should be adjusted in a concerted way to maintain the intracellular osmolarity while keeping the H_2O and Na^+ efflux equal to the H_2O and Na^+ influxes across the entry membrane. Possibly the intracellular osmolarity constitutes a reference point for these transport mechanisms.

B. Uphill Transport of H_2O in the Molecular Model

I suggest that transepithelial transport of H_2O in its simplest form is maintained by osmotic transport at the entry membrane of the cell and secondary active transport with KCl at the exit membrane (Zeuthen, 1993a).

The measurements from choroid plexus (Zeuthen, 1991a,b, 1994) allow a quantitative description. Here the membrane which contains the $K^+/Cl^-/H_2O$ cotransporter may in addition have a passive leak (Fig. 13). The L_p of the entry membrane is about 10^{-4} cm sec^{-1} (osm liter^{-1})$^{-1}$, while the L_p of the ventricular membrane, the exit membrane, is 0.5×10^{-4} cm sec^{-1} (osm liter^{-1})$^{-1}$. The rate of H_2O transport by the cotransporter ($J^*_{H_2O}$) is given by Eq. 2 (Section V,C).

When combined, these values predict *precisely* three tissue properties: (1) the rate of H_2O transport, (2) the ability for uphill transport of H_2O, and (3) the reflection coefficient for NaCl (σ_{NaCl}) of the epithelium.

With no transepithelial osmotic difference, the rate of epithelial H_2O transport is given by the rate of the $K^+/Cl^-/H_2O$ cotransport, which, given values of ion concentrations in the cell and external fluids, transports at a rate of 3.6 nl cm^{-2} sec^{-1}. This is close to experimentally determined values. For further discussion and references, see Section V,B. It is represented in Fig. 11 by the intercepts at a transepithelial osmotic difference ($\Delta\pi$) of zero.

The ability for uphill transport (or determination of the strength of transport $\hat{C}$, see Section VI,A) can be determined in two ways: by adding mannitol to the entry compartment or by diluting the exit compartment. If mannitol is added to the entry compartment and if the cell osmolarity follows this osmolarity, the cotransporter itself will be slowed by the osmotic gradient (curve 3 in Fig. 11). When the H_2O transport via the passive pathways is considered as well and the total transepithelial water flux calculated, it can be seen that the total transepithelial H_2O transport is zero at a transepithelial osmotic difference of 65 mosm liter^{-1} (curve 1), which is within the range of observed values (references in IV,A,1). In experiments where the exit compartment is diluted, the Cl$^-$ concentration is reduced as well. As a result, the K$^+$/Cl$^-$/H_2O cotransporter will tend to be slowed by the osmotic difference but increased by the Cl$^-$ concentration difference; in total $J^*_{H_2O}$ will be only slightly affected (curve 4). The passive flux of H_2O through the cell will be opposite to $J^*_{H_2O}$ and determined by the L_p s. When the opposing osmotic difference equals 110 mosm liter^{-1}, the sum of the passive and the secondary active fluxes of H_2O will be zero (curve 2). This is in excellent agreement with the observation that the plexus can transport against 150 mosm liter^{-1} (Heisy et al., 1962).

It is important to note that the model predicts that the epithelium can transport against larger osmotic differences when these are implemented by NaCl than with mannitol; that is, the reflection coefficient σ_{NaCl} for the whole epithelium is lower than $\sigma_{mannitol}$, which equals one. In the model σ_{NaCl} is 0.6, equal to 65 mosm liter^{-1} divided by 110 mosm liter^{-1} (see Fig. 11). This agrees with experimental observations; in the proximal tubule where this is particularly well studied, σ_{NaCl} is between 0.5 and 0.8 (Zeuthen, 1992).

The molecular model also explains how cells can remain hyperosmolar relative to the outside solutions, as discussed in relation to Fig. 2. Water enters passively by osmosis but because the secretion is isotonic with the bathing solutions, the efflux of H_2O is uphill. The intracellular hyperosmolarity can be assessed from the rate of transport and the water permeability (L_p) of the entry membrane. In amphibians this is about 5×10^{-4} cm sec^{-1} (osm liter^{-1})$^{-1}$ for the apical membrane in proximal tubule, intestine, and gall bladder, and roughly a factor of ten higher in mammalian cells. At normal osmolarities, the rates of transport are about 3 μl cm^{-2} sec^{-1} for amphibians and a factor of ten larger in mammals (Tripathi and Boulpaep, 1989; Zeuthen, 1992). Hence the intracellular hyperosmolarity required to pull in H_2O at the desired rate amounts to some 6 mosm liter^{-1} for both amphibians and mammals at physiological osmolarities. At dilute external solutions, the cell retains its osmolarity, and hyperosmolarities of 40 mosm liter^{-1} have been recorded (Section II, Tables II and III). As discussed in Section V, the $K^+/Cl^-/H_2O$ cotransporter can easily move H_2O from such hyperosmolar cellular compartments into the exit bath.

In conclusion, the numerical predictions of the molecular model constitute substantial support for a central role of the $K^+/Cl^-/H_2O$ cotransporter in epithelial transport. Hence, it is encouraging that it is possible from a method of perturbations, such as those presented in Figs. 6 and 7, to derive properties of membrane transporters which, when taken as a whole, predict the properties of a leaky epithelium in a steady state.

C. Isotonic Transport by the Molecular Model

The simple molecular model (Fig. 13) did explain how H_2O left the cell in an uphill fashion in cotransport with K^+ and Cl^- but it did not explain how the intracellular osmolarity was maintained. Furthermore, the simple model will not by itself explain isotonic transport. Given the coupling ratio of 500 water molecules per molecule of KCl (Section V,A,1), KCl and H_2O leave the cell via cotransport in a ratio of 110 mmol liter^{-1}. Since the Na^+/K^+ ATPase takes up 2 K^+ ions for each 3 Na^+ ions released, the final secretion would have a concentration of 165 mmol liter^{-1} of NaCl. In amphibians this solution would be hyperosmolar to the bathing solutions by about 90 mosm liter^{-1}. Recirculation at the exit membrane would be a way to obtain isotonic transport (Fig. 14); a reuptake of about 50% of the initially secreted Na^+ would be required at normal osmolarities. In this model isotonic transport is equivalent to the question of which transporters are found at the exit membrane and how their rate of transport is controlled.

K^+/Cl^- cotransport has been located at the exit membrane of leaky epithelia in gall bladder (Garzia and Armstrong, 1983; Reuss, 1983; Larson and Spring, 1983; Hill and Hill, 1987), in kidney tubules (Greger and Schlatter, 1983; Gullans et al., 1986; Eveloff and Warnock, 1987; Baum and Berry, 1984; Sasaki et al., 1988), and in choroid plexus (Zeuthen, 1987, 1991a,b, 1994; Johanson et al., 1992, 1993; Johanson and Preston, 1994). The $Na^+/K^+/Cl^-$ cotransporter is located at the entry membrane of leaky epithelia in the intestine (O'Grady et al., 1987), in the gall bladder (Davis and Finn, 1985), in the choroid plexus (Johanson et al., 1993; Johanson and Preston, 1994), and in the vestibular dark cells (Wangemann and Marcus, 1990), or at the entry membrane of secretory systems such as the basolateral membrane of glands (Petersen, 1988; Lauf et al., 1987;

entry   exit

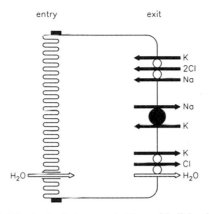

FIG. 14 Molecular model for isotonic transport. The epithelial cell is supplied with Na^+/ K^+/ATPase and K^+/Cl^-/H_2O cotransport at the exit membrane (basolateral in gall bladder, ventricular in choroid plexus). The cell has also a reuptake system for Na^+ and Cl^-, in the present example a Na^+/K^+/$2Cl^-$ cotransporter; Na^+/HCO_3^- in combination with Cl^-/ HCO_3^- was another possibility K$^+$ and Cl^- channels are not drawn. H_2O enters passively across the entry membrane (apical in gall bladder, blood-facing in choroid plexus), which is determined by the osmotic gradient; entry mechanisms for ions are not shown (see text).

Haas, 1989; Lang *et al.*, 1993). There is evidence, however, that Na^+/ K^+/$2Cl^-$ cotransport is present or can be induced at the exit membrane of absorptive epithelia in frog skin (Ussing, 1985), choroid plexus (Johanson *et al.*, 1990; Zeuthen, 1994), intestine (O'Brien *et al.*, 1993; MacLeod and Hamilton, 1990), MDCK cells (Lauf *et al.*, 1987), and kidney proximal tubules (González *et al.*, 1993). Na^+,$2HCO_3^-$/Cl^- exchange and Na^+/ $3HCO_3^-$ exchange have been located at the exit membrane of the renal proximal tubule (Aronson, 1989) and Cl^-/HCO_3^- exchange at the exit membrane of the choroid plexus (Johanson *et al.*, 1985; Zeuthen, 1987) and proximal tubule (Aronson, 1989). Finally, the exit membrane is the exclusive site for Na^+/K ATPase in leaky epithelia (Zeuthen and Wright, 1981).

Accordingly there are several possibilities for reuptake of Na^+, Cl^-, and K^+. A Na^+,$2HCO_3$/Cl^- exchange in cooperation with Cl^-/HCO_3^- exchange would take up NaCl, while the Na^+/K ATPase would take up K^+. Another possibility, illustrated in Fig. 14, is the Na^+/K^+/$2Cl^-$ cotransporter. The driving force for this transporter is for reuptake, particularly so in dilute solutions [for values see Table II and Zeuthen (1981, 1982)].

For isotonic or near-isotonic transport to be explained under dilute conditions, the appropriate membrane transporters would have to be stimulated: The K^+/Cl^-/H_2O cotransporter should work faster to account for the increased rate of H_2O transport. This would require an increased reuptake of K^+ and Cl^- at the exit membrane. In regard to the K^+ uptake, this could be accomplished by increased Na^+/K^+ ATPase activity and by the inward-directed $Na^+/K^+/2Cl^-$ cotransport which has been found to be stimulated by reduced external Cl^- (Haas and McBrayer, 1994; O'Grady *et al.*, 1987). This would also cause the reuptake of Na^+. It is in accord with the model that the Cl^- conductance in the exit membrane increases by as much as a factor of 10 when external Cl^- concentrations are reduced (Zeuthen, 1987; Zeuthen *et al.*, 1987b) or during cell swelling (Davis and Finn, 1988). This would allow for the increased Cl^- recirculation. This scheme agrees with the finding that oxygen consumption goes up when epithelia are adapted to dilute solutions (Frederiksen and Leyssac, 1969).

Channels are not included explicitly in the model. Although the exit membrane contains Cl^- channels (Stoddard and Reuss, 1988, 1989; Christensen *et al.*, 1989; Zeuthen, 1987; Zeuthen *et al.*, 1987b; Davis and Finn, 1988) and K^+ channels (Christensen and Zeuthen, 1987; Brown *et al.*, 1988; Loo *et al.*, 1988; Kawahara *et al.*, 1987; Parent *et al.*, 1986; Sackin and Palmer, 1987; Sheppard *et al.*, 1988a,b, 1991; Valverde *et al.*, 1991; Sepúlveda *et al.*, 1991), quantitative studies show that the conductances mediated by the channels during steady state in themselves are too small to account for transepithelial transport (Christensen and Zeuthen, 1987; Christensen *et al.*, 1989). This does not mean that channels play no role in the control of transport. Furthermore both K^+ and Cl^- channels can be activated during acute volume regulatory decrease.

How a cell senses its own volume or osmolarity and how these signals are transferred to the membrane transporters is currently being investigated in a variety of cell types (Lang *et al.*, 1993; Parker, 1993; Lauf *et al.*, 1992; Cossins 1991; Schultz, 1989; Ellory and Hall, 1988; Häussinger *et al.*, 1994; Grinstein *et al.*, 1992; Grinstein and Foskett, 1990). Two schools of thought seem to exist. One suggests that stretch-activated channels deliver the initial signal while the other suggests that the signal of volume change is a change in concentration of cytoplasmic protein, that is, a change in macromolecular crowding (Parker, 1993). The latter view is similar to the suggestion presented here that cell osmolarity defines a setpoint for transport.

VII. Summary: A Pump-Leak Model for Cellular H_2O Homeostasis

This chapter combines recent electrophysiological, protein chemical, and molecular biological knowledge in a model for cellular H_2O homeostasis. It deals briefly with the state of H_2O in cells and extensively with the passive and secondarily active transport of H_2O across the cell membrane.

Most intracellular H_2O is either in a free state or loosely bound to organic macromolecules. Both phases serve as solvent for small inorganic ions such as K^+. Hydration forces in the bound H_2O help to keep macromolecules apart, which facilitates the movements of smaller molecules. The effects of bound H_2O are central for transport across the membrane. The amount of H_2O held in and released from membrane proteins during conformational changes can be significant.

If large passive H_2O permeabilities are required, membranes are equipped with proteinaceous water channels. Current interest focuses on CHIP 28. This protein is found in leaky epithelia which can utilize small transepithelial osmotic differences (e.g., kidney proximal tubules). Significantly, it is not found in epithelia which may encounter large changes in osmolarities, such as gastric mucosae and small intestine. In such cases the passive water permeability must be kept low in order to avoid excessive deleterious transport of H_2O. It should be emphasized, however, that phenotypically normal humans without functional CHIP 28 suffer no apparent clinical consequences (Preston *et al.*, 1994a).

There is no satisfactory physical model of passive water permeation in membrane pores. A realistic physical description of osmosis must eventually consider molecular-kinetic events. In the absence of such theories, macroscopic descriptions have been employed; if the pore is impermeable to the solute, the flow is driven by hydrostatic pressure gradients; if the pore is permeable, no differences in hydrostatic pressures develop and H_2O flow is diffusive. The intermediate case, where some solute particles are reflected while some enter the pore, cannot be treated by macroscopic theories and must await the development of molecular-kinetic models.

H_2O can be cotransported with ions in membrane proteins. In the case of $K^+/Cl^-/H_2O$ cotransport, one K^+ ion, one Cl^- ion, and about 500 molecules of H_2O act as allosteric ligands: A downhill gradient of each of the species can energize an uphill transport of the others in a constant coupling ratio. The cotransport exhibits specificity, saturation, and inhibition. There is also evidence that other cotransporters may translocate H_2O.

The molecular model suggested for secondary active transport of H_2O uses established concepts for protein dynamics. Conformational changes in the membrane protein shift a mobile barrier from one side of the protein to the other. As a result, occluded H_2O molecules are translocated across the membranes. It seems that cotransporters often contain 12 membrane-spanning α-helices. This allows for spatial models of the protein, which may occlude 500 molecules of H_2O.

Cells in steady states can be hyperosmolar relative to the external solution owing to pumps and leaks for H_2O. The leaks are the passive permeabilities such as CHIP 28; the pumps are cotransporters which translocate ions and H_2O. The dogma that cells in steady states are isosmotic with their surroundings is based upon osmometric observations in a few tissues under a limited range of conditions and does not apply for all tissues under all conditions.

Transport of H_2O by leaky epithelia can be explained by the molecular properties of the membrane transporters. The passive influx of H_2O is balanced by secondary active transport of H_2O at the exit membrane. This molecular model gives a precise quantitative explanation of the rate of transepithelial H_2O transport, of the ability for transepithelial uphill H_2O transport, and of the low reflection coefficient for NaCl.

There is a close connection between the maintenance of intracellular osmolarity and isotonic transport in leaky epithelia. If the tissues are adapted to transport in dilute solutions, the intracellular osmolarity remains high. This results in an increased passive entry of H_2O and accords with the observation that transepithelial H_2O transport goes up proportionally with the extent of dilution. To maintain isotonic transport and high intracellular osmolarity under dilute conditions, there must be an increased reuptake of ions at the exit membrane. The signals for the coordinated response required from the transporters could be parameters linked to the cytoplasm, i.e., osmolarity.

Acknowledgments

The technical expertise of Anni Thomsen and S. Christoffersen has been invaluable. Financial support was received from the Danish Research Council, the NOVO Foundation, and the Bio Membrane Centre. I am very grateful for the criticism and constructive suggestions I have received from Drs. P. B. Dunham, A. E. Hill, W. D. Stein, and P. Wangemann.

References

Aerts, T., Xia, J.-Z., Slegers, H., Block, J., and de Clauwaert, J. (1990). Hydrodynamic characterization of the major intrinsic protein from the bovine lens fiber membranes. *J. Biol. Chem.* **265,** 8675–8680.

Agre, P., Sasaki, S., and Chrispeels, M. J. (1993a). Aquaporins: A family of water channel proteins. *Am. J. Physiol.* **265,** F461.

Agre, P., Preston, G. M., Smith, B. L., *et al.* (1993b). Aquaporin CHIP: The archetypal molecular water channel. *Am. J. Physiol.* **265,** F463–F476.

Alpern, R. J., Howlin, K. J., and Preisig, P. A. (1985). Active and passive components of chloride transport in the rat proximal convoluted tubule. *J. Clin. Invest.* **76,** 1360–1366.

Altenberg, G. A., and Reuss, L. (1990). Apical membrane Na+/H+ exchange in necturus gallbladder epithelium. *J. Gen. Physiol.* **95,** 369–392.

Ames, G. F.-L. (1992). Bacterial periplasmic permeases as model systems for the superfamily of traffic ATPases, including the multidrug resistance protein and the cystic fibrosis transmembrane conductance regulator. *In* "Molecular Biology of Receptors and Transporters Bacterial and Glucose Transporters" (M. Friedlander and M. Mueckler, eds.), pp. 1–32. Academic Press, San Diego, CA.

Anderson, C. M., Zucker, F. H., and Steitz, T. A. (1979). Space-filling models of kinase clefts and conformational changes. *Science* **204**, 375–380.

Anderson, J. L., and Malone, D. M. (1974). Mechanism of osmotic flow in porous membranes. *Biophys. J.* **14**, 957–982.

Appelboom, J. W. T., Brodsky, W. A., Tuttle, W. S., and Diamond, I. (1958). The freezing point depression of mammalian tissues after sudden heating in boiling distilled water. *J. Gen. Physiol.* **41**, 1153–1169.

Arakawa, T., and Timasheff, S. N. (1982). Stabilization of protein structure by sugars. *Biochemistry* **21**, 6536–6544.

Arakawa, T., and Timasheff, S. N. (1985a). Mechanism of poly(ethylene glycol) interaction with proteins. *Biochemistry* **24**, 6756–6762.

Arakawa, T., and Timasheff, S. N. (1985b). The stabilization of proteins by osmolytes. *Biophys. J.* **47**, 411–414.

Aronson, P. S. (1989). The renal proximal tubule: A model for diversity of anion exchangers and stilbene-sensitive anion transporters. *Annu. Rev. Physiol.* **51**, 419–441.

Baker, M. E., and Saier, M. H., Jr. (1990). A common ancestor for bovine lens fiber major intrinsic protein, soybean nodulin-26 protein, and E. coli glycerol facilitator. *Cell (Cambridge, Mass.)* **60**, 185–186.

Baldwin, S. A. (1989). Homologies between sugar transporters from eukaryotes and prokaryotes. *Annu. Rev. Physiol.* **51**, 457–471.

Ballanyi, K., Strupp, M., and Grafe, P. (1993). Electrophysiological measurement of volume changes in neurons, glial cells, and muscle fibers in situ. *In* "Practical Electrophysiological Methods" (H. Kettenmann and R. Grantyn, eds.), pp. 363–366. Wiley, New York.

Baum, M., and Berry, C. A. (1984). Evidence for neutral transcellular NaCl transport and neutral basolateral chloride exit in the rabbit proximal convoluted tubule. *J. Clin. Invest.* **74**, 205–211.

Bean, C. P. (1972). The physics of porous membranes—neutral pores. *In* "Membranes" (G. Eisenman, ed.), pp. 1–54. Dekker, New York.

Bomsztyk, K., and Wright, F. S. (1986). Dependence of ion fluxes on fluid transport by rat proximal tubule. *Am. J. Physiol.* **250**, F680–F689.

Bondy, C., Chin, E., Smith, B. L., Preston, G. M., and Agre, P. (1993). Developmental gene expression and tissue distribution of the CHIP28 water-channel protein. *Proc. Natl. Acad. Sci. U.S.A.* **90**, 4500–4504.

Branden, F., and Tooze, G. W. (1991). "Introduction to Protein Structure." Gerland Publishing, New York and London.

Brown, D., Verbavatz, J.-M., Valenti, G., Lui, B., and Sabolic, I. (1993). Localization of the CHIP28 water channel in reabsorptive segments of the rat male reproductive tract. *Eur. J. Cell Biol.* **61**, 264–273.

Brown, P. D., Loo, D. D. F., and Wright, E. M. (1988). Ca2+-activated K+ channels in the apical membrane of necturus choroid plexus. *J. Membr. Biol.* **105**, 207–219.

Bulone, D., Donato, I. D., Palma-Vittorelli, M. B., and Palma, M. U. (1991). Density, structural lifetime, and entropy of H-bond cages promoted by monohydric alcohols in normal and supercooled water. *J. Chem. Physiol.* **94**, 6816–6826.

Canfield, V. A., and Macey, R. I. (1984). Anion exchange in human erythrocytes has a large activation volume. *Biochim. Biophys. Acta* **778**, 379–384.

Carpi-Medina, P., Lindemann, B., González, E., and Whittembury, G. (1984). The continuous measurement of tubular volume changes in response to step changes in contraluminal osmolarity. *Pfluegers Arch.* **400**, 343–348.

Cayley, S., Lewis, B. A., Guttman, H. J., and Record, M. T., Jr. (1991). Characterization of the cytoplasm of *Escherichia coli* K-12 as a function of external osmolarity. Implications for protein-DNA interactions in vivo. *J. Mol. Biol.* **222,** 281–300.

Cayley, S., Lewis, B. A., and Record, M. T., Jr. (1992). Origins of the osmoprotective properties of betaine and proline in *Escherichia coli* K-12. *J. Bacteriol.* **174,** 1586–1595.

Chen, P.-Y., Pearce, D., and Verkman, A. S. (1988). Membrane water and solute permeability determined quantitatively by self-quenching of an entrapped fluorophore. *Biochemistry* **27,** 5713–5718.

Cherksey, B. D., and Zeuthen, T. (1987). A membrane protein with a K+ and a Cl− channel. *Acta Physiol. Scand.* **129,** 137–138.

Christensen, O., and Zeuthen, T. (1987). Maxi K$^+$ channels in leaky epithelia are regulated by intracellular Ca^{2+}, pH and membrane potential. *Pfluegers Arch.* **408,** 249–259.

Christensen, O., Simon, M., and Randlev, T. (1989). Anion channels in a leaky epithelium. A patch-clamp study of choroid plexus. *Pfluegers Arch.* **415,** 37–46.

Clegg, J. S. (1984). Properties and metabolism of the aqueous cytoplasm and its boundaries. *Am. J. Physiol.* **246,** R133–R151.

Clegg, J. S. (1992). Cellular infrastructure and metabolic organization. *Curr. Top. Cell. Regul.* **33,** 3–14.

Cohen, B. E. (1975). The permeability of liposomes to nonelectrolytes. *J. Membr. Biol.* **20,** 205–234.

Collins, K. D., and Washaburg, M. W. (1985). The Hofmeister effect and the behaviour of water at interfaces. *Q. Rev. Biophys.* **18,** 323–422.

Colombo, M. F., Rau, D. C., and Parsegian, V. A. (1992). Protein solvation in allosteric regulation: A water effect on hemoglobin. *Science* **256,** 655.

Conway, E. J., and Geoghegan, H. (1955). Molecular concentration of kidney cortex slices. *J. Physiol. (London)* **130,** 438–445.

Conway, E. J., Geoghegan, H., and McCormack, J. I. (1955). Autolytic changes at zero centigrade in ground mammalian tissues. *J. Physiol. (London)* **130,** 427–437.

Cossins, A. R. (1991). A sense of cell size. *Nature (London)* **352,** 667–668.

Costello, M. J., Viitanen, P., Carrasco, N., Foster, D. L., and Kaback, H. R. (1984). Morphology of proteoliposomes reconstituted with purified lac carrier protein from Escherichia coli. *J. Biol. Chem.* **259,** 15579–15586.

Costello, M. J., Escaig, J., Matsushita, K., Viitanen, P. V., Menick, D. R., and Kaback, H. R. (1987). Purified lac permease and cytochrome o oxidase are functional as monomers. *J. Biol. Chem.* **262,** 17072–17082.

Cotton, C. U., and Reuss, L. (1989). Measurement of the effective thickness of the mucosal unstirred layer in necturus gallbladder epithelium. *J. Gen. Physiol.* **93,** 631–647.

Cotton, C. U., and Reuss, L. (1991). Effects of changes in mucosal solution Cl⁻ or K$^+$ concentration on cell water volume of necturus gallbladder epithelium. *J. Gen. Physiol.* **97,** 667–686.

Cotton, C. U., Weinstein, A. M., and Reuss, L. (1989). Osmotic water permeability of necturus gallbladder epithelium. *J. Gen. Physiol.* **93,** 649–679.

Crowe, W. E., and Wills, N. K. (1991). A simple method for monitoring changes in cell height using fluorescent microbeads and an Ussing-type chamber for the inverted microscope. *Pflüegers Arch.* **419,** 349–357.

Cserr, H. F. (1971). Physiology of the choroid plexus. *Physiol. Rev.* **51,** 273–311.

Dainty, J. (1965). Osmotic flow. *Symp. Soc. Exp. Biol.* **19,** 75–85.

Dani, J. A., and Levitt, D. G. (1981). Water transport and ion-water interaction in the gramicidin channel. *Biophys. J.* **35,** 501–508.

David, P., Mayan, H., Cohen, H., Tal, D. M., and Karlish, S. J. D. (1992). Guanidinium derivatives act as high affinity antagonists of Na$^+$, K$^+$-ATPase. *J. Biol. Chem.* **267,** 1141–1149.

Davis, C. W., and Finn, A. L. (1985). Effects of mucosal sodium removal on cell volume in necturus gallbladder epthelium. *Am. J. Physiol.* **249**, C304–C312.

Davis, C. W., and Finn, A. L. (1988). Potassium-induced cell swelling in Nectutus gallbladder epithelium. *Am. J. Physiol.* **254**, C643–C650.

Deamer, D. W., and Nichols, J. W. (1989). Proton flux mechanisms in model and biological membranes. *J. Membr. Biol.* **107**, 91–103.

Deen, P. M. T., Dempster, J. A., Wieringa, B., and Van Os, C. H. (1992). Isolation of a cDNA for rat CHIP28 water channel: High mRNA expression in kidney cortex and inner medulla. *Biochem. Biophys. Res. Commun.* **188**, 1267–1273.

Dempster, J. A., Van Hoek, A. N., de Jong, M. D., and Van Os, C. H. (1991). Glucose transporters do not serve as water channels in renal and intestinal epithelia. *Pflüegers Arch.* **419**, 249–255.

Dempster, J. A., Van Hoek, A. N., and Van Os, C. H. (1992). The quest for water channels. *NIPS* **7**, 172–176.

Denker, B. M., Smith, B. L., Kuhajda, F. P., and Agre, P. (1988). Identification, purification, and partial characterization of a novel M_r 28,000 integral membrane protein from erythrocytes and renal tubules. *J. Biol. Chem.* **263**, 15634–15642.

de Smedt, H., Borghgraef, R., Ceuterick, F., and Heremans, K. (1979). Pressure effects on lipid-protein interactions in $(Na^+ + K^+)$-ATPase. *Biochim. Biophys. Acta* **556**, 479–489.

Diamond, J. M. (1962). The mechanism of water transport by the gall-bladder. *J. Physiol. (London)* **161**, 503–527.

Diamond, J. M. (1964a). Transport of salt and water in rabbit and guinea pig gall bladder. *J. Gen. Physiol.* **48**, 1–14.

Diamond, J. M. (1964b). The mechanism of isotonic water transport. *J. Gen. Physiol.* **48**, 15–42.

Dunham, P. B., Cass, A., Trinkaus, J. P., and Bennett, M. V. L. (1970). Water permeability of fundulus eggs. *Biol. Bull. (Woods Hole, Mass.)* **139**, 420–421.

Echevarria, M., Frindt, G., Preston, G. M., *et al.* (1993a). Expression of multiple water channel activities in *Xenopus* oocytes injected with mRNA from rat kidney. *J. Gen. Physiol.* **101**, 827–841.

Echevarria, M., Kuang, K., Iserovich, P., *et al.* (1993b). Cultured bovine corneal endothelial cells express CHIP28 water channels. *Am. J. Physiol.* **265**, C1349–C1355.

Ehring, G. R., Zampighi, G., Horwitz, J., Bok, D., and Hall, J. E. (1990). Properties of channels reconstituted from the major intrinsic protein of lens fiber membranes. *J. Gen. Physiol.* **96**, 631–664.

Einstein, A. (1956). "Investigations on the Theory of the Brownian Movements." Dover, New York.

Ellory, J. C., and Hall, A. C. (1988). Human red cell volume regulation in hypotonic media. *Comp. Biochem. Physiol. A* **90A**, 533–537.

Eriksson, A. E., Baase, W. A., Zhang, X.-J., *et al.* (1991). Response of a protein structure to cavity-creating mutations and its relation to the hydrophobic effect. *Science* **255**, 178–183.

Eveloff, J., and Warnock, D. G. (1987). K-Cl transport systems in rabbit renal basolateral membrane vesicles. *Am. J. Physiol.* **252**, F883–F889.

Farahbakhsh, N. A., and Fain, G. L. (1987). Volume regulation of non-pigmented cells from ciliary epithelium. *Invest. Ophthalmol. Visual Sci.* **28**, 934–944.

Faxén, H. (1922). Der Widerstand gegen die Bewegung einer starren Kugel in einer zahen Flussigkeit, die zwischen zwei parallellen Kugel in einer Wanden eingeschlossen ist. *Ann. Phys. (Leipzig)* [4] **68**, 89–119.

Ferry, J. D. (1936). Statistical evaluation of sieve constants in ultrafiltration. *J. Gen. Physiol.* **20**, 95–104.

Finkelstein, A. (1974). Aqueous pores created in thin lipid membranes by the antibiotics nystatin, ampholerisin B and gramicidin A: Implications for pores in plasma membranes.

In "Drugs and Transport Processes" (B. A. Callingham, ed.), pp. 241–250. Macmillan, London.

Finkelstein, A. (1976). Water and nonelectrolyte permeability of lipid bilayer membranes. *J. Gen. Physiol.* **68**, 127–135.

Finkelstein, A. (1987). "Water Movement Through Lipid Bilayers, Pores, and Plasma Membranes." Wiley, New York.

Finkelstein, A. (1993). The water permeability on narrow pores. *In* "Isotonic Transport in Leaky Epithelia" (H. H. Ussing, J. Fischbarg, O. Sten-Knudsen, E. H. Larsen, and N. J. Willumsen, eds.), pp. 487–497. Munksgaard, Copenhagen.

Fischbarg, J. (1989). On the theory of solute-solvent coupling in epithelia. *In* "Water Transport in Biological Membranes" (G. Benga, ed.), pp. 153–167. CRC Press, Boca Raton, FL.

Fischbarg, J., Kuang, K., Hirsch, J., *et al.* (1989). Evidence that the glucose transporter serves as a water channel in J774 macrophages. *Proc. Natl. Acad. Sci. U.S.A.* **86**, 8397–8401.

Fischbarg, J., Kuang, K., Vera, J. C., *et al.* (1990). Glucose transporters serve as water channels. *Proc. Natl. Acad. Sci. U.S.A.* **87**, 3244–3247.

Fischbarg, J., Kuang, K., Li, J., Arant-Hickman, S., Vera, J. C., Silverstein, S. C., and Loike, J. D. (1993). Facilitative and sodium-dependent glucose transporters behave as water channels. *In* "Isotonic Transport in Leaky Epithelia" (H. H. Ussing, J. Fischbarg, O. Sten-Knudsen, E. H. Larsen, and N. J. Willumsen, eds.), pp. 432–449. Munksgaard, Copenhagen.

Fisher, R. S., and Spring, K. R. (1984). Intracellular activities during volume regulation by necturus gallbladder. *J. Membr. Biol.* **78**, 187–199.

Fisher, R. S., Persson, B.-E., and Spring, K. R. (1981). Epithelial cell volume regulation: Bicarbonate dependence. *Science* **214**, 1357–1359.

Forte, J. G. (1987). Introduction to mini-reviews on H^+/OH^- permeability through membranes. *J. Bioenerg. Biomembr.* **19**, 409–411.

Frederiksen, O., and Leyssac, P. P. (1969). Transcellular transport of isosmotic volumes by the rabbit gall-bladder in vitro. *J. Physiol. (London)* **201**, 201–224.

Frömter, E., Rumrich, G., and Ullrich, K. J. (1973). Phenomenologic description of Na^+, Cl^- and HCO_3^- absorption from proximal tubules of the rat kidney. *Pfluegers Arch.* **343**, 189–220.

Fulton, A. B. (1982). How crowded is the cytoplasm? *Cell (Cambridge, Mass.)* **30**, 345–347.

Fushimi, K., and Verkman, A. S. (1991). Low viscosity in the aqueous domain of cell cytoplasm measured by picosecond polarization microfluorimetry. *J. Cell Biol.* **112**, 719–725.

Fushimi, K., Uchida, S., Hara, Y., Hirata, Y., Marumo, F., and Sasaki, S. (1993). Cloning and expression of apical membrane water channel of rat kidney collecting tubule. *Nature (London)* **361**, 549–552.

Gadsby, D. C., Rakowski, R. F., and De Weer, P. (1993). Extracellular access to the Na,K pump: Pathway similar to ion channel. *Science* **260**, 100–103.

Garby, L. (1957). Studies on transfer of matter across membranes with special reference to the isolated human amniotic membrane and the exchange of amniotic fluid. *Acta Physiol. Scand.* **40**, 1–84.

Garner, M. M., and Burg, M. B. (1994). Macromolecular crowding and confinement in cells exposed to hypertonicity. *Am. J. Physiol.* **266**, C877–C892.

Garzia, A., and Armstrong, McD. W. (1983). KCl cotransport: A mechanism for basolateral chloride exit in necturus gallbladder. *J. Membr. Biol.* **76**, 173–182.

Giraldez, F. (1984). Active sodium transport and fluid secretion in gallbladder epithelium of *Necturus*. *J. Physiol. (London)* **348**, 431–455.

Goldshleger, R., Tal, D. M., Shainskaya, A., Or, E., Hoving, S., and Karlish, S. J. D. (1994). Organization of the cation binding domain of the Na$^+$/K$^+$-pump. *In* "The Sodium Pump Structure Mechanism, Hormonal Control and its Role in Disease" (E. Bamberg and W. Schoner, eds.), pp. 309–320. Steinkopff, Darmstadt.

González, E., Carpi-Medina, P., and Whittembury, G. (1982). Cell osmotic water permeability of isolated rabbit proximal straight tubules. *Am. J. Physiol.* **242,** F321–F330.

González, E., Gutierrez, A. M., Echevarria, M., and Whittembury, G. (1993). Urea triggers (Ca^{2+})$_i$ related Na-K-2Cl co-transport in kidney proximal tubules (PST). *Proc. Congr. Int. Union Physiol. Sci.* **32,** 91.17 (abstr.).

Gorin, M. B., Yancey, S. B., Cline, J., Revel, J.-P., and Horwitz, J. (1984). The major intrinsic protein (MIP) of the bovine lens fiber membrane: Characterization and structure based on cDNA cloning. *Cell (Cambridge, Mass.)* **39,** 49–59.

Green, R., and Giebisch, G. (1989). Osmotic forces driving water reabsorption in the proximal tubule of the rat kidney. *Am. J. Physiol.* **257,** F669–F675.

Green, R., Giebisch, G., Unwin, R., and Weinstein, A. M. (1991). Coupled water transport by rat proximal tubule. *Am. J. Physiol.* **261,** F1046–F1054.

Greger, R., and Schlatter, E. (1983). Properties of the lumen membrane of the cortical thick ascending limb of Henle's loop of rabbit kidney. *Pflüegers Arch.* **396,** 315–324.

Griffith, J. K., Baker, M. E., Rouch, D. A., *et al.* (1992). Membrane transport proteins: Implications of sequence comparisons. *Curr. Opin. Cell Biol.* **4,** 684–695.

Grim, E., and Smith, G. A. (1957). Water flux rates across dog gallbladder wall. *Am. J. Physiol.* **191,** 555–560.

Grinstein, S., and Foskett, J. K. (1990). Ionic mechanisms of cell volume regulation in leukocytes. *Annu. Rev. Physiol.* **52,** 399–414.

Grinstein, S., Furuya, W., and Bianchini, L. (1992). Protein kinases, phosphatases, and the control of cell volume. *NIPS* **7,** 232–237.

Guggino, W. B. (1986). Functional heterogeneity in the early distal tubule of the Amphiuma kidney: Evidence for two modes of Cl$^-$ and K$^+$ transport across the basolateral cell membrane. *Am. J. Physiol.* **250,** F430–F440.

Gullans, S. R., Avison, M. J., Ogino, T., Shulman, R. G., and Giebisch, G. (1986). Furosemide-sensitive K$^+$ efflux induced by glucose in the rabbit proximal tubule. *Kidney Int.* **29,** 396.

Ha, J.-H., Capp, M. W., Hohenwalter, M. D., Baskerville, M., and Record, M. T., Jr. (1992). Thermodynamic stoichiometries of participation of water, cations and anions in specific and non-specific binding of lac repressor to DNA. *J. Mol. Biol.* **228,** 252–264.

Haas, M. (1989). Properties and diversity of (Na-K-Cl) cotransporters. *Annu. Rev. Physiol.* **51,** 443–457.

Haas, M., and McBrayer, D. G. (1994). Na-K-Cl cotransport in nystatin-treated tracheal cells: Regulation by isoproterenol, apical UTP, and [Cl]$_i$. *Am. J. Physiol.* **266,** C1440–C1452.

Haberman, W. L., and Sayre, R. M. (1958). Motion of rigid and fluid spheres in stationary and moving liquids inside cylindrical tubes. *David Taylor Model Basin, U.S. Dep. Navy Rep.* **1143.**

Hakim, A., Lester, R. G., and Lifson, N. (1963). Absorption by an in vitro preparation of dog intestinal mucosa. *J. Appl. Physiol.* **18,** 409–413.

Hall, A. C., Ellory, J. C., and Klein, R. A. (1982). Pressure and temperature effects on human red cell cation transport. *J. Membr. Biol.* **68,** 47–56.

Hammel, H. T. (1979). Forum on osmosis. I. Osmosis: Diminished solvent activity or enhanced solvent tension? *Am. J. Physiol.* **237,** R95–R107.

Hansen, A. J., and Olsen, C. E. (1980). Brain extracellular space during spreading depression and ischemia. *Acta Physiol. Scand.* **108,** 355–365.

Hartley, G. S., and Crank, J. (1949). Some fundamental definitions and concepts in diffusion processes. *Trans. Faraday Soc.* **45**, 801–819.

Harvey, S. A., and Oxender, D. L. (1992). Amino acid transport in bacteria. *In* "Molecular Biology of Receptors and Transporters Bacterial and Glucose Transporters" (M. Friedlander and M. Mueckler, eds.), pp. 37–95. Academic Press, San Diego, CA.

Hasegawa, H., Zhang, R., Dohrman, A., and Verkman, A. S. (1993). Tissue-specific expression of mRNA encoding rat kidney water channel CHIP28k by in situ hybridization. *Am. J. Physiol.* **264**, C237–C245.

Hasegawa, H., Lian, S.-C., Finkbeiner, W. E., and Verkman, A. S. (1994a). Extrarenal tissue distribution of CHIP28 water channels by in situ hybridization and antibody staining. *Am. J. Physiol.* **266**, C893–C903.

Hasegawa, H., Ma, T., Skach, W., Matthay, M. A., and Verkman, A. S. (1994b). Molecular cloning of a mercurial-insensitive water channel expressed in selected water-transporting tissues. *J. Biol. Chem.* **269**, 5497–5500.

Häussinger, D., Lang, F., and Gerok, W. (1994). Regulation of cell function by the cellular hydration state. *Am. J. Physiol.* **267**, E343–E355.

Heeswijk, M. P. E., and Van Os, C. H. (1986). Osmotic water permeabilities of brush border and basolateral membrane vesicles from rat renal cortex and small intestine. *J. Membr. Biol.* **92**, 183–193.

Heisey, S. R., Held, D., and Pappenheimer, J. R. (1962). Bulk flow and diffusion in the cerebrospinal fluid system of the goat. *Am. J. Physiol.* **203**, 775–781.

Heremans, K. (1982). High presure effects on proteins and other biomolecules. *Annu. Rev. Biophys. Bioeng.* **11**, 1–21.

Hermansson, K., and Spring, K. R. (1986). Potassium induced changes in cell volume of gallbladder epithelium. *Pfluegers Arch.* **407**, S90–S99.

Higgins, C. F. (1992). ABC transporters: From microorganisms to man. *Annu. Rev. Cell Biol.* **8**, 67–113.

Hill, A. E. (1979). Osmosis. *Q. Rev. Biophys.* **12**, 67–99.

Hill, A. E. (1980). Salt-water coupling in leaky epithelia. *J. Membr. Biol.* **56**, 177–182.

Hill, A. E. (1982). Osmosis: A bimodal theory with implications for symmetry. *Proc. R. Soc. London* **215**, 155–174.

Hill, A. E. (1989a). Osmosis in leaky pores: The role of pressure. *Proc. R. Soc. London, Ser. B* **237**, 363–367.

Hill, A. E. (1989b). Osmotic flow equations for leaky porous membranes. *Proc. R. Soc. London* **237**, 369–377.

Hill, A. E. (1994). Osmotic flow in membrane pores of molecular size. *J. Membr. Biol.* **137**, 197–203.

Hill, A. E., and Hill, B. S. (1987). Steady-state analysis of ion fluxes in necturus gall-bladder epithelial cells. *J. Physiol.* (*London*) **382**, 15–34.

Hill, B. S., and Findlay, G. P. (1981). The power of movement in plants: The role of osmotic machines. *Q. Rev. Biophys.* **14**, 173–222.

Hill, B. S., and Hill, A. E. (1978). Fluid transfer by necturus gall bladder epithelium as a function of osmolarity. *Proc. R. Soc. London, Ser. B* **200**, 151–162.

Hille, B. (1992). "Ionic Channels of Excitable Membranes," 2nd ed. Sinauer Assoc., Sunderland, MA.

Hodgkin, A. L., and Keynes, R. D. (1953). The mobility and diffusion coefficient of potassium in giant axons from sepia. *J. Physiol.* (*London*) **119**, 513–528.

Horwitz, J., and Bok, D. (1987). Conformational properties of the main intrinsic polypeptide (MIP26) isolated from lens plasma membranes. *Biochemistry* **26**, 8092–8098.

House, C. R. (1974). "Water Transport in Cells and Tissues." Edward Arnold, London.

Johanson, C. E., and Preston, J. E. (1994). Potassium efflux from infant and adult rat choroid plexuses: Effects of CSF anion substitution, N-ethylmaleimide and Cl transport inhibitors. *Neurosci. Lett.* **169,** 207–211.

Johanson, C. E., Parandoosh, Z., and Smith, Q. R. (1985). Cl-HCO$_3$ exchange in choroid plexus: Analysis by the DMO method for cell pH. *Am. J. Physiol.* **249,** F478–F484.

Johanson, C. E., Sweeney, S. M., Parmelee, J. T., and Epstein, M. H. (1990). Cotransport of sodium and chloride by the adult mammalian choroid plexus. *Am. J. Physiol.* **258,** C211–C216.

Johanson, C. E., Murphy, V. A., and Dyas, M. (1992). Ethacrynic acid and furosemide alter Cl, K and Na distribution between blood, choroid plexus, CSF and brain. *Neurochem. Res.* **17,** 1079–1085.

Johanson, C. E., Palm, D. E., Dyas, M. L., and Knuckey, N. W. (1994). Microdialysis analysis of effects of loop diuretics and acetazolamide on chloride transport from blood to CSF. *Brain Res.* **641,** 121–126.

Kaback, H. R. (1992). In and out and up and down with lac permease. In "Molecular Biology of Receptors and Transporters Bacterial and Glucose Transporters" (M. Friedlander and M. Mueckler, eds.), pp. 97–125. Academic Press, San Diego, CA.

Kaplan, J. H. (1993). Molecular biology of carrier proteins. *Cell (Cambridge, Mass.)* **72,** 13–18.

Kawahara, K. M., Hunter, M., and Giebisch, G. (1987). Potassium channels in necturus proximal tubule. *Am. J. Physiol.* **253,** F488–F494.

Kersting, U., Wojnowski, L., Steigner, W., and Oberleithner, H. (1991). Hypotonic stress-induced release of KHCO$_3$ in fused renal epithelioid (MDCK) cells. *Kidney Int.* **39,** 891–900.

Kim, Y. K., Illsley, N. P., and Verkman, A. S. (1988). Rapid fluorescence assay of glucose and neutral solute transport using an entrapped volume indicator. *Anal. Biochem.* **172,** 403–409.

Kirk, K. L., Schafer, J. A., and DiBona, D. R. (1987). Cell volume regulation in rabbit proximal straight tubule perfused in vitro. *Am. J. Physiol.* **252,** F922–F932.

Knepper, M. A., and Nielsen, S. (1993). Kinetic model of water and urea permeability regulation by vasopressin in collecting duct. *Am. J. Physiol.* **265,** F214–F224.

Kornblatt, J. A., and Hoa, G. H. B. (1990). A nontraditional role for water in the cytochrome c oxidase reaction. *Biochemistry* **29,** 9370–9376.

laCour, M., and Zeuthen, T. (1992). Osmotic properties of the frog retinal pigment epithelium. *Exp. Eye Res.* **56,** 521–530.

Lang, F., Ritter, M., Völkl, H., and Häussinger, D. (1993). The biological significance of cell volume. *Renal Physiol. Biochem.* **16,** 48–65.

Larson, M., and Spring, K. R. (1983). Bumetanide inhibition of NaCl transport by necturus gallbladder. *J. Membr. Biol.* **74,** 123–129.

Latorre, R., and Miller, C. (1983). Conduction and selectivity in potassium channels. *J. Membr. Biol.* **71,** 11–30.

Lauf, P. K. (1985). K$^+$: Cl$^-$ cotransport: Sulfhydryls, divalent cations, and the mechanism of volume activation in a red cell. *J. Membr. Biol.* **88,** 1–13.

Lauf, P. K., McManus, T. J., Haas, M., et al. (1987). Physiology and biophysics of chloride and cation cotransport across cell membranes. *Fed. Proc., Fed. Am. Soc. Exp. Biol.* **46,** 2377–2394.

Lauf, P. K., Bauer, J., Adragna, N. C., et al. (1992). Erythrocyte K-Cl cotransport: Properties and regulation. *Am. J. Physiol.* **263,** C917–C932.

Leikin, S., Parsegian, V. A., and Rau, D. C. (1993). Hydration forces. *Annu. Rev. Phys. Chem.* **44,** 369–395.

Leirmo, S., Harrison, C., Cayley, D. S., Burgess, R. R., and Record, M. T., Jr. (1987). Replacement of potassium chloride by potassium glutamate dramatically enhances protein-DNA interactions in vitro. *Biochemistry* **26,** 2095–2101.

Le Neveu, S. M., Rand, R. P., and Parsegian, V. A. (1976). Measurement of forces between lecithin bilayers. *Nature (London)* **259,** 601–603.

Li, J., and Tooth, P. (1987). Size and shape of the *Escherichia coli* lactose permease measured in filamentous arrays. *Biochemistry* **26,** 4816–4823.

Lodish, H. F. (1988). Multi-spanning membrane proteins: How accurate are the models? *Trends Biochem. Sci.* **13,** 332–334.

Loike, J. D., Cao, L., Kuang, K., Vera, J. C., Silverstein, S. C., and Fischbarg, J. (1993). Role of facilitative glucose transporters in diffusional water permeability through J774 cells. *J. Gen. Physiol.* **102,** 897–906.

Loo, D. D. F., Brown, P. D., and Wright, E. M. (1988). Ca2+-activated K+ currents in necturus choroid plexus. *J. Membr. Biol.* **105,** 221–231.

Loo, D. D. F., Hazama, A., Supplisson, S., Turk, E., and Wright, E. M. (1993). Relaxation kinetics of the Na^+/glucose cotransporter. *Proc. Natl. Acad. Sci. U.S.A.* **90,** 5767–5771.

Lopez, A. G., Amzel, L. M., Markakis, D., and Guggino, W. B. (1988). Cell volume regulation by the thin descending limb of Henle's loop. *Proc. Natl. Acad. Sci. U.S.A.* **85,** 2873–2877.

Luby-Phelps, K., Castle, P. E., Taylor, D. L., and Lanni, F. (1987). Hindered diffusion of inert tracer particles in the cytoplasm of mouse 3T3 cells. *Proc. Natl. Acad. Sci. U.S.A.* **84,** 4910–4913.

Luby-Phelps, K., Mujumdar, S., Mujumdar, R. B., Ernst, L. A., Galbraith, W., and Waggoner, A. S. (1993). A novel fluorescence ratiometric method confirms the low solvent viscosity of the cytoplasm. *Biophys. J.* **65,** 236–242.

Lytle, C., Xu, J.-C., Biemesderfer, D., Haas, M., and Forbush, B., III (1992). The Na-K-Cl cotransport protein of shark rectal gland. *J. Biol. Chem.* **267,** 25428–25437.

Ma, T., Frigeri, A., Skach, W., and Verkman, A. S. (1993a). Cloning of a novel rat kidney cDNA homologous to CHIP28 and WCH-CD water channels. *Biochem. Biophys. Res. Commun.* **197,** 654–659.

Ma, T., Frigeri, A., Tsai, S.-T., Verbavatz, J.-M., and Verkman, A. S. (1993b). Localization and functional analysis of CHIP28k water channels in stably transfected Chinese hamster ovary cells. *J. Biol. Chem.* **268,** 22756–22764.

Ma, T., Hasegawa, H., Skach, W. R., Frigeri, A., and Verkman, A. S. (1994). Expression, functional analysis, and in situ hybridization of a cloned rat kidney collecting duct water channel. *Am. J. Physiol.* **266,** C189–C197.

Macey, R. I. (1984). Transport of water and urea in red blood cells. *Am. J. Physiol.* **246,** C195–C203.

MacLeod, R. J., and Hamilton, J. R. (1990). Regulatory volume increase in mammalian jejunal villus cells is due to bumetanide-sensitive $NaKCl_2$. *Am. J. Physiol.* **258,** G665–G674.

MacLeod, R. J., and Hamilton, J. R. (1991). Volume regulation initiated by Na^+-nutrient cotransport in isolated mammalian villus enterocytes. *Am. J. Physiol.* **260,** G26–G33.

MacRobbie, E. A. C., and Ussing, H. H. (1961). Osmotic behaviour of the epithelial cells of frog skin. *Acta Physiol. Scand.* **53,** 348–365.

Maffly, R. H., and Leaf, A. (1959). The potential of water in mammalian tissues. *J. Gen. Physiol.* **42,** 1257–1275.

Manning, G. S. (1968). Binary diffusion and bulk flow through a potential-energy profile: A kinetic basis for the thermodynamic equations of flow through membranes. *J. Chem. Phys.* **49,** 2668–2675.

Mao, B., Pear, M. R., and McCammon, J. A. (1982). Hinge-bending in L-arabinose-binding protein. *J. Biol. Chem.* **257,** 1131–1133.

Marčelja, S. (1976). Repulsion of interfaces due to boundary water. *Chem. Phys. Lett.* **42,** 129–131.

Marcus, D. C., Marcus, N. Y., and Greger, R. (1987). Sidedness of action of loop diuretics and ouabain on nonsensory cells of utricle: A micro-Ussing chamber for inner ear tissues. *Hear. Res.* **30**, 55–64.

Marcus, N. Y., and Marcus, D. C. (1987). Potassium secretion by nonsensory region of gerbil utricle in vitro. *Am. J. Physiol.* **253**, F613–F621.

Maurel, C., Reizer, J., Schroeder, J. I., and Chrispeels, M. J. (1993). The vacuolar membrane protein -TIP creates water specific channels in Xenopus oocytes. *EMBO J.* **12**, 2241–2247.

Mauro, A. (1957). Nature of solvent transfer in osmosis. *Science* **126**, 252–253.

Mauro, A. (1979). Forum on osmosis. III. Comments on Hammel and Scholander's solvent tension theory and its application to the phenomenon of osmotic flow. *Am J. Physiol.* **237**, R110–R113.

Mazzoni, M. R., Lundgren, E., Arfors, K.-E., and Intaglietta, M. (1989). Volume changes of an endothelial cell monolayer on exposure to anisotonic media. *J. Cell. Physiol.* **140**, 272–280.

Meyer, M. M., and Verkman, A. S. (1987). Evidence for water channels in renal proximal tubule cell membranes. *J. Membr. Biol.* **96**, 107–119.

Minton, A. P. (1983). The effect of volume occupancy upon the thermodynamic activity of proteins: Some biochemical consequences. *Mol. Cell. Biochem.* **55**, 119–140.

Mitchell, P. (1957). A general theory of membrane transport from studies of bacteria. *Nature (London)* **180**, 134–136.

Mitchell, P. (1990). Osmochemistry of solute translocation. *Res. Microbiol.* **141**, 286–289.

Moon, C., Preston, G. M., Griffin, C. A., Jabs, E. W., and Agre, P. (1993). The human aquaporin CHIP gene structure, organization, and chromosomal localization. *J. Biol. Chem.* **268**, 15772–15778.

Moreno, J. H., and Diamond, J. M. (1972). Cation permeation mechanisms and cation selectivity in "tight junctions" of gallbladder epithelium. *In* "Membranes" (G. Eisenman, ed.), pp. 383–515. Dekker, New York.

Muallem, S., Zhang, B.-X., Loessberg, P. A., and Star, R. A. (1992). Simultaneous recording of cell volume changes and intracellular pH or Ca^{2+} concentration in single osteosarcoma cells UMR-106-01*. *J. Biol. Chem.* **267**, 17658–17664.

Nicholson, C. (1993). Measurement of extracellular space. *In* "Practical Electrophysiological Methods" (H. Kettenmann and R. Grantyn, eds.), pp. 367–372. Wiley, New York.

Nicholson, C., Phillips, J. M., and Gardner-Medwin, A. R. (1979). Diffusion from an iontophoretic point source in the brain: Role of tortuosity and volume fraction. *Brain Res.* **169**, 580–584.

Nielsen, S., and Knepper, M. A. (1993). Vasopressin activates collecting duct urea transporters and water channels by distinct physical processes. *Am. J. Physiol.* **265**, F204–F213.

Nielsen, S., Smith, B. L., Christensen, E. I., Knepper, M. A., and Agre, P. (1993a). CHIP28 water channels are localized in constitutively water-permeable segments of the Nephron. *J. Cell Biol.* **120**, 371–383.

Nielsen, S., Smith, B. L., Christensen, E. I., and Agre, P. (1993b). Distribution of the aquaporin CHIP in secretory and resorptive epithelia and capillary endothelia. *Proc. Natl. Acad. Sci. U.S.A.* **90**, 7275–7279.

Nikaido, H., and Saier, M. H., Jr. (1992). Transport proteins in bacteria: Common themes in their design. *Science* **258**, 936–942.

O'Brien, J. A., Walters, R. J., Valverde, M. A., and Sepúlveda, F. V. (1993). Regulatory volume increase after hypertonicity- or vasoactive-intestinal-peptide-induced cell-volume decrease in small-intestinal crypts is dependent on Na^+-K^+-$2Cl^-$ cotransport. *Pflüegers Arch.* **423**, 67–73.

O'Grady, S. M., Palfrey, H. C., and Field, M. (1987). Characteristics and functions of Na-K-Cl cotransport in epithelial tissues. *Am. J. Physiol.* **253**, C177–C192.

Orengo, C. A., Jones, D. T., and Thornton, J. M. (1994). Protein superfamilies and domain superfolds. *Nature (London)* **372,** 631–634.

Otting, G., Liepinsh, E., and Wüthrich, K. (1991). Protein hydration in aqueous solution. *Science* **254,** 974–980.

Paganelli, C. V., and Solomon, A. K., (1957). The rate of exchange of tritiated water across the human red cell membrane. *J. Gen. Physiol.* **41,** 259–277.

Palfrey, C., and Cossins, A. (1994). Fishy tales of kidney function. *Nature (London)* **371,** 377–378.

Pao, G. M., Wu, L.-F., Johnson, K. D., *et al.* (1991). Evolution of the MIP family of integral membrane transport proteins. *Mol. Microbiol.* **5,** 33–37.

Pappenheimer, J. R. (1993). On the coupling of membrane digestion with intestinal absorption of sugars and amino acids. *Am. J. Physiol.* **265,** G409–G417.

Pappenheimer, J. R., and Reis, K. Z. (1987). Contribution of solvent drag through intercellular junctions to absorption of nutrients by the small intestine of the rat. *J. Membr. Biol.* **100,** 123–136.

Parent, L., Cardinal, J., and Sauve, R. (1986). Single channel investigation of the basolateral membrane in the rabbit proximal convoluted tubule. *Biophys. J.* **49,** 159.

Parker, J. C. (1993). In defense of cell volume? *Am. J. Physiol.* **265,** C1191–C1200.

Parr, C. E., and Finn, A. L. (1989). High serosal (S) K^+ induces cell swelling in Necturus gall-bladder by inhibiting KCl cotransport. *FASEB J.* **2A,** 1284 (abstr.).

Parsegian, V. A., and Rau, D. C. (1984). Water near intracellular surfaces. *J. Cell Biol.* **99,** 196s–200s.

Parsons, D. S., and Wingate, D. L. (1961). The effect of osmotic gradients on fluid transfer across rat intestine in vitro. *Biochim. Biophys. Acta* **46,** 170–183.

Pearce, D., and Verkman, A. S. (1989). NaCl reflection coefficients in proximal tubule apical and basolateral membrane vesicles. *Biophys. J.* **55,** 1251–1259.

Persson, B.-E., and Spring, K. R. (1982). Gallbladder epithelial cell hydraulic water permeability and volume regulation. *J. Gen. Physiol.* **79,** 481–505.

Persson, B.-E., and Spring, K. R. (1984). Permeability properties of the subepithelial tissues of Necturus gallbladder. *Biochim. Biophys. Acta* **772,** 135–139.

Petersen, O. H. (1988). The control of ion channels and pumps in exocrine acinar cells. *Comp. Biochem. Physiol.* **90A,** 717–721.

Preston, G. M., and Agre, P. (1991). Isolation of the cDNA for erythrocyte integral membrane protein of 28 kilodaltons: Member of an ancient channel family. *Proc. Natl. Acad. Sci. U.S.A.* **88,** 11110–11114.

Preston, G. M., Carroll, T. P., Guggino, W. B., and Agre, P. (1992). Appearance of water channels in zenopus oocytes expressing red cell CHIP28 protein. *Science* **256,** 385–389.

Preston, G. M., Jung, J. S., Guggino, W. B., and Agre, P. (1994). Membrane topology of aquaporin CHIP. *J. Biol. Chem.* **269,** 1–6.

Preston, G. M., Smith, B. L., Zeidel, M. L., Moulds, J. J., and Agre, P. (1994a). Mutations in aquaporin-1 in phenotypically normal humans without functional CHIP water channels. *Science* **265,** 1585–1587.

Rand, R. P., and Fuller, N. L. (1992). Water as an inhibiting ligand in yeast hexokinase. *Biophys. J.* **61,** A345.

Rand, R. P., and Parsegian, V. A. (1989). Hydration forces between phospholipid bilayers. *Biochim. Biophys. Acta* **988,** 351–376.

Rand, R. P., Fuller, N. L., Butko, P., Francis, G., and Nicholls, P. (1993). Measured change in protein solvation with substrate binding and turnover. *Biochemistry* **32,** 5925–5929.

Rapoport, S. I. (1976). "Blood-Brain Barrier in Physiology and Medicine." Raven Press, New York.

Rayner, M. D., Starkus, J. G., Ruben, P. C., and Alicata, D. A. (1992). Voltage-sensitive and solvent-sensitive processes in ion channel grating Kinetic effects of hypersmolar media on activation and deactivation of sodium channels. *Biophys. J.* **61,** 96–108.

Record, M. T., Jr., Anderson, C. F., and Lohman, T. M. (1978). Thermodynamic analysis of ion effects on the binding and conformational equilibria of proteins and nucleic acids: The roles of ion association or release, screening, and ion effects on water activity. *Q. Rev. Biophys.* **2,** 103–178.

Reuss, L. (1983). Basolateral KCl co-transport in a NaCl-absorbing epithelium. *Nature (London)* **305,** 723–726.

Reuss, L. (1985). Changes in cell volume measured with an electrophysiologic technique. *Proc. Natl. Acad. Sci. U.S.A.* **82,** 6014–6018.

Rosenberg, P. A., and Finkelstein, A. (1978). Water permeability of gramicidin A-treated bilayer membranes. *J. Gen. Physiol.* **72,** 341–350.

Sackin, H., and Palmer, L. G. (1987). Basolateral potassium channels in renal proximal tubule. *Am. J. Physiol.* **253,** F476–F487.

Sasaki, S. *et al.* (1988). KCl co-transport across the basolateral membrane of rabbit renal proximal straight tubules. *J. Clin. Invest.* **81,** 194–199.

Schafer, J. A. (1990). Transepithelial osmolality differences, hydraulic conductivities, and volume absorption in the proximal tubule. *Annu. Rev. Physiol.* **52,** 709–726.

Schild, L., Aronson, P. S., and Giebisch, G. (1991). Basolateral transport pathways for K^+ and Cl in rabbit proximal tubule: Effects on cell volume. *Am. J. Physiol.* **260,** F101–F109.

Schultz, S. G. (1989). Volume preservation: Then and now. *NIPS* **4,** 169–173.

Sepúlveda, F. V., Pargon, F., and McNaughton, P. (1991). K^+ and Cl^- currents in enterocytes isolated from guinea-pig small intestinal villi. *J. Physiol. (London)* **434,** 351–367.

Sheppard, D. N., Giraldez, F., and Sepúlveda, F. V. (1988a). K^+ channels activated by L-alanine transport in isolated necturus enterocytes. *FEBS Lett.* **234,** 446–448.

Sheppard, D. N., Giraldez, F., and Sepúlveda, F. V. (1988b). Kinetics of voltage- and Ca^{2+} activation and Ba^{2+} blockade of a large-conductance K^+ channel from necturus enterocytes. *J. Membr. Biol.* **105,** 65–75.

Sheppard, D. N., Valverde, M. A., Giraldez, F., and Sepúlveda, F. V. (1991). Potassium currents of isolated necturus enterocytes: A whole-cell patch-clamp study. *J. Physiol. (London)* **433,** 663–676.

Shi, L.-B., Fushimi, K., and Verkman, A. S. (1991). Solvent drag measurement of transcellular and basolateal membrane NaCl reflection coefficient in kidney proximal tubule. *J. Gen. Physiol.* **98,** 379–398.

Siegenbeek Van Heukelom, J. *et al.* (1981). Microscopical determination of the filtration permeability of the mucosal surface of the goldfish intestinal epithelium. *J. Membr. Biol.* **63,** 31–39.

Skach, W. R., Shi, L.-B., Calayag, M. C., Frigeri, A., Lingappa, V. R., and Verkman, A. S. (1994). Biogenesis and transmembrane topology of the CHIP28 water channel at the endoplasmic reticulum. *J. Cell Biol.* **125,** 803–815.

Smith, B. L., and Agre, P. (1991). Erythrocyte Mr 28,000 transmembrane protein exists as a multisubunit oligomer similar to channel proteins. *J. Biol. Chem.* **266,** 6407–6415.

Smith, B. L., Baumgarten, R., Nielsen, S., Raben, D., Zeidel, M. L., and Agre, P. (1993). Concurrent expression of erythroid and renal aquaporin CHIP and appearance of water channel activity in perinatal rats. *J. Clin. Invest.* **92,** 2035–2041.

Somero, G. N. (1986). Protons, osmolytes, and fitness of internal milieu for protein function. *Am. J. Physiol.* **251,** R197–R213.

Soodak, H., and Iberall, A. (1979). Forum on osmosis. IV. More on osmosis and diffusion. *Am. J. Physiol.* **237,** R114–R122.

Spring, K. R., and Ericson, A.-C. (1982). Epithelial cell volume modulation and regulation. *J. Membr. Biol.* **69,** 167–176.

Spring, K. R., and Hope, A. (1978). Size and shape of the lateral intercellular spaces in a living epithelium. *Science* **200**, 54–58.

Spring, K. R., and Hope, A. (1979). Fluid transport and the dimensions of cells and interspaces of living necturus gallbladder. *J. Gen. Physiol.* **73**, 287–305.

Stein, W. D. (1986). "Transport and Diffusion Across Cell Membranes." Academic Press, Orlando, FL.

Stein, W. D. (1990). "Channels, Carriers and Pumps." Academic Press, San Diego, CA.

Steitz, T. A., Shoham, M., and Bennett, W. S., Jr. (1981). Structural dynamics of yeast hexokinase during catalysis. *Philos. Trans. R. Soc. London* **293**, 43–52.

Sten-Knudsen, O. (1978). Passive transport processes. *In* "Membrane Transport in Biology" (G. Giebish, D. C. Tosteson, and H. H. Ussing, eds.), pp. 1–113. Springer-Verlag, Berlin.

Stoddard, J. S., and Reuss, L. (1988). Dependence of cell membrane conductances on bathing solution HCO3-/CO2 in necturus gallbladder. *J. Membr. Biol.* **102**, 163–174.

Stoddard, J. S., and Reuss, L. (1989). Electrophysiological effects of mucosal Cl- removal in necturus gallbladder epithelium. *Am. J. Physiol.* **257**, C568–C578.

Strange, K. (1988). RVD in principal and intercalated cells of rabbit cortical collecting tubule. *Am. J. Physiol.* **255**, C612–C621.

Tanford, C. (1973). Biological lipids. *In* "The Hydrophobic Effect: Formation of Micelles and Biological Membranes" (C. Tanford, ed.), pp. 94–99. Wiley, New York.

Tanford, C. (1983). Mechanism of free energy coupling in active transport. *Annu. Rev. Biochem.* **52**, 379–409.

Tauc, M., LeMaout, S., and Poujeol, P. (1990). Fluorescent video-microscopy study of regulatory volume decrease in primary culture of rabbit proximal convoluted tubule. *Biochim. Biophys. Acta* **1051**, 278–284.

Teeter, M. M. (1991). Water-protein interactions: Theory and experiment. *Annu. Rev. Biophys. Biophys. Chem.* **20**, 577–600.

Timasheff, S. N. (1993). The control of protein stability and association by weak interactions with water: How do solvents affect these processes? *Annu. Rev. Biophys. Biomol. Struct.* **22**, 67–97.

Tripathi, S., and Boulpaep, E. L. (1989). Mechanisms of water transport by epithelial cells. *Q. J. Exp. Physiol. Cogn. Med. Sci.* **74**, 385–417.

Ussing, H. H. (1985). Volume regulation and basolateral co-transport of sodium, potassium and chloride ions in frog skin epithelium. *Pflüegers Arch.* **405**, S2–S7.

Ussing, H. H., and Eskesen, K. (1989). Mechanism of isotonic water transport in glands. *Acta Physiol. Scand.* **136**, 443–454.

Valverde, M. A., Sheppard, D. N., Giraldez, F., and Sepúlveda, F. V. (1991). Two types of potassium currents seen in isolated necturus enterocytes with the single-electrode voltage-clamp technique. *J. Physiol. (London)* **433**, 645–661.

van der Goot, F. G., Podevin, R.-A., and Corman, B. J. (1989). Water permeabilities and salt reflection coefficients of luminal, basolateral and intracellular membrane vesicles isolated from rabbit kidney proximal tubule. *Biochim. Biophys. Acta* **986**, 332–340.

Van Hoek, A. N., Hom, M. L., Luthjens, L. H., de Jong, M. D., Dempster, J. A., and Van Os, C. H. (1991). Functional unit of 30 kDa for proximal tubule water channels as revealed by radiation inactivation. *J. Cell Biol.* **266**, 16633–16635.

Van Hoek, A. N., and Verkman, A. S. (1992). Functional reconstitution of the isolated erythrocyte water channel CHIP28. *J. Biol. Chem.* **267**, 18267–18269.

Van Hoek, A. N., Wiener, M., Bicknese, S., Miercke, L., Biwersi, J., and Verkman, A. S. (1993). Secondary structure analysis of purified functional CHIP28 water channels by CD and FTIR spectroscopy. *Biochemistry* **32**, 11847–11856.

Van Os, C. H., Deen, P. M. T., and Dempster, J. A. (1994). Aquaporins: water selective channels in biological membranes. Molecular structure and tissue distribution. *Biochim. Biophys. Acta.* **1197**, 291–309.

Vegard, L. (1908). On the free pressure in osmosis. *Proc. Cambridge Philos. Soc.* **15,** 13–23.

Verbavatz, J.-M., Brown, D., Sabolic, I., *et al.* (1993). Tetrameric assembly of CHIP28 water channels in liposomes and cell membranes: A freeze-fracture study. *J. Cell Biol.* **123,** 605–618.

Verkman, A. S. (1993). "Water Channels." R. G. Landes Company, Austin, TX.

Vodyanoy, I., Bezrukov, S. M., and Parsegian, V. A. (1993). Probing alamethicin channels with water-soluble polymers. Size-modulated osmotic action. *Biophys. J.* **65,** 2097–2105.

Walz, T., Smith, B. L., Zeidel, M. L., Engel, A., and Agre, P. (1994). Biologically active two-dimensional crystals of aquaporin CHIP. *J. Biol. Chem.* **269,** 1583–1586.

Walz, W. (1992). Mechanism of rapid $Ke11+$-induced swelling of mouse astrocytes. *Neurosci. Lett.* **135,** 243–246.

Wang, H., and Skalak, R. (1969). Viscous flow in a cylindrical tube containing a line of spherical tube containing a line of spherical particles. *J. Fluid Mech.* **38,** 75–96.

Wangemann, P., and Marcus, D. C. (1990). K^+-induced swelling of vestibular dark cells is dependent on Na^+ and Cl^- and inhibited by piretanide. *Pflüegers Arch.* **416,** 262–269.

Wangemann, P., and Shiga, N. (1994). Cell volume control in vestibular dark cells during and after a hyposmotic challenge. *Am. J. Physiol.* **266,** C1046–C1060.

Wangemann, P., Shiga, N., Welch, C., and Marcus, D. C. (1992). Evidence for the involvement of a K^+ channel in isosmotic cell shrinking in vestibular dark cells. *Am. J. Physiol.* **32,** C616–C622.

Weber, L. A., Hickey, E. D., Maroney, P. A., and Baglioni, C. (1977). Inhibition of protein synthesis by Cl^-. *J. Biol. Chem.* **252,** 4001–4010.

Weinstein, A. M., and Stephenson, J. L. (1981a). Coupled water transport in standing gradient models of the lateral intercellular space. *Biophys. J.* **35,** 167–191.

Weinstein, A. M., and Stephenson, J. L. (1981b). Models of coupled salt and water transport across leaky epithelia. *J. Membr. Biol.* **660,** 1–20.

Welling, L. W., Grantham, J. J., and Qualizza, P. (1972). Physical properties of isolated perfused renal tubules and tubular basement membranes. *J. Clin. Invest.* **51,** 1063–1073.

Welling, L. W., Welling, D. J., and Ochs, T. J. (1983). Video measurement of basolateral membrane hydraulic conductivity in the proximal tubule. *Am. J. Physiol.* **245,** F123–F129.

Welling, L. W., Welling, D. J., and Ochs, T. J. (1987). Video measurement of basolateral NaCl reflection coefficient in proximal tubule. *Am. J. Physiol.* **253,** F290–F298.

Whittembury, G., and Hill, B. S. (1982). Fluid reabsorption by Necturus proximal tubule perfused with solutions of normal and reduced osmolarity. *Proc. R. Soc.London* **215,** 411–431.

Whittembury, G., and Reuss, L. (1992). Mechanisms of coupling of solute and solvent transport in epithelia. *In* "The Kidney: Physiology and Pathophysiology" (D. W. Seldin and G. Giebish, eds.), pp. 317–360. Raven Press, New York.

Whittembury, G., Lindemann, B., Carpi-Medina, P., González, E., and Linares, H. (1986). Continuous measurements of cell volume changes in single kidney tubules. *Kidney Int.* **30,** 187–191.

Whittembury, G., Echevarria, M., Gutierrez, A., and González, E. (1993). Absorption of salt and water in the proximal tubule-revisited. *In* "Isotonic Transport in Leaky Epithelia" (H. H. Ussing, J. Fischbarg, O. Sten-Knudsen, E. H. Larsen, and N. J. Willumsen, eds.), pp. 37–48. Munksgaard, Copenhagen.

Wistow, G. J., Pisano, M. M., and Chepelinsky, A. B. (1991). Tandem sequence repeats in transmembrane channel proteins. *Trends Biochem. Sci.* **16,** 169–171.

Wright, E. M., and Diamond, J. M. (1968). Effects of pH and polyvalent cations on the selective permeability of gall-bladder epithelium to monovalent ions. *Biochim. Biophys. Acta* **163,** 57–74.

Wright, E. M., Widener, G., and Rumrich, G. (1977). Fluid secretion by the frog choroid plexus. *Exp. Eye Res., Suppl.* **25,** 149–155.

Wright, E. M., Hager, K. M., and Turk, E. (1992). Sodium cotransport proteins. *Curr. Opin. Cell Biol.* **4**, 696–702.

Xu, J.-C., Lytle, C., Zhu, T. T., Payne, J. A., Benz, E., Jr., and Forbush, B., III (1994). Molecular cloning and functional expression of the bumetanide-sensitive Na-K-Cl cotransporter. *Proc. Natl. Acad. Sci. U.S.A.* **91**, 2201–2205.

Yancey, P. H., Clark, M. E., Hand, S. C., Bowlus, R. D., and Somero, G. N. (1982). Living with water stress: Evolution of osmolyte systems. *Science* **217**, 1214–1223.

Zampighi, G. A., Hall, J. E., Ehring, G. R., and Simon, S. A. (1989). The structural organization and protein composition of lens fiber junctions. *J. Cell Biol.* **108**, 2255–2275.

Zeidel, M. L., Albalak, A., Grossman, E., and Carruthers, A. (1992a). Role of glucose carrier in human erythrocyte water permeability. *Biochemistry* **31**, 589–596.

Zeidel, M. L., Ambudkar, S. V., Smith, B. L., and Agre, P. (1992b). Reconstitution of functional water channels in liposomes containing purified red cell CHIP28 protein. *Biochemistry* **31**, 7436–7440.

Zeidel, M. L., Nielsen, S., Smith, B. L., Ambudkar, S. V., Maunsbach, A. B., and Agre, P. (1994). Ultrastructure, pharmacologic inhibition, and transport selectivity of aquaporin CHIP in proteoliposomes. *Biochemistry* **33**, 1606–1615.

Zeuthen, T. (1981). Transport and intracellular osmolarity in the necturus gall-bladder epithelium in water transport across epithelia. *In* "Water Transport across Epithelia" (H. H. Ussing, N. Bindslev, N. A. Lassen, and O. Sten-Knudsen, eds.), pp. 313–331. Munksgaard, Copenhagen.

Zeuthen, T. (1982). Relations between intracellular ion activities and extracellular osmolarity in necturus gallbladder epithelium. *J. Membr. Biol.* **66**, 109–121.

Zeuthen, T. (1983). Ion activities in the lateral intercellular spaces of gallbladder epithelium transporting at low external osmolarities. *J. Membr. Biol.* **76**, 113–122.

Zeuthen, T. (1987). The effects of chloride ions on electrodiffusion in the membrane of a leaky epithelium. *Pfluegers Arch.* **408**, 267–274.

Zeuthen, T. (1991a). Water permeability of ventricular cell membrane in choroid plexus epithelium from necturus maculosus. *J. Physiol. (London)* **444**, 133–151.

Zeuthen, T. (1991b). Secondary active transport of water across ventricular cell membrane of choroid plexus epithelium of necturus maculosus. *J. Physiol. (London)* **444**, 153–173.

Zeuthen, T. (1992). From contractile vacuole to leaky epithelia. *Biochim. Biophys. Acta* **1113**, 229–258.

Zeuthen, T. (1993a). Low reflection coefficient for KCl in an epithelial membrane. *In* "Isotonic Transport in Leaky Epithelia" (H. H. Ussing, J. Fischbarg, O. Sten-Knudsen, E. H. Larsen, and N. J. Willumsen, eds.), pp. 298–307. Munksgaard, Copenhagen.

Zeuthen, T. (1993b). Co-transport of K^+, Cl^- and H_2O in membrane proteins from the ventricular cell membrane of the choroid plexus epithelum in *Necturus maculosus*. *Proc. Congr. Int. Union Physiol. Sci.* **32**, 16.3 (abstr.).

Zeuthen, T. (1994). Cotransport of K^+, Cl^- and H_2O by membrane proteins from choroid plexus epithelium of Necturus maculosus. *J. Physiol. (London)* **478**, 203–219.

Zeuthen, T., and Stein, W. D. (1994). Co-transport of salt and water in membrane proteins: Membrane proteins as osmotic engines. *J. Membr. Biol.* **137**, 179–195.

Zeuthen, T., and Wright, E. M. (1981). Epithelial potassium transport: Tracer and electrophysiological studies in choroid plexus. *J. Membr. Biol.* **60**, 105–128.

Zeuthen, T., Christensen, O., Bærentsen, J. H., and laCour, M. (1987a). The mechanism of electrodiffusive K + transport in leaky epithelia and some of its consequences for anion transport. *Pfluegers Arch.* **408**, 260–266.

Zeuthen, T., Christensen, O., and Cherksey, B. (1987b). Electrodiffusion of Cl^- and K^+ in epithelial membranes reconstituted into planar lipid bilayers. *Pfluegers Arch.* **408**, 275–281.

Zhang, R., Skach, W., Hasegawa, H., Van Hoek, A. N., and Verkman, A. S. (1993). Cloning, functional analysis and cell localization of a kidney proximal tubule water transporter homologous to CHIP28. *J. Cell Biol.* **120,** 359–369.

Zimmerberg, J., and Parsegian, V. A. (1986). Polymer inaccessible volume changes during opening and closing of a voltage-dependent ionic channel. *Nature (London)* **323,** 36.

Zimmberberg, J., Bezanilla, F., and Parsegian, V. A. (1990). Solute inaccessible aqueous volume changes during opening of the potassium channel of the squid giant axon. *Biophys. J.* **57,** 1049–1064.

Zimmerman, S. B., and Harrison, B. (1987). Macromolecular crowding increases binding of DNA polymerase to DNA: An adaptive effect. *Proc. Natl. Acad. Sci. U.S.A.* **84,** 1871–1875.

Zimmerman, S. B., and Trach, S. O. (1991). Estimation of macromolecule concentrations and excluded volume effects for the cytoplasm of Escherichia coli. *J. Mol. Biol.* **222,** 599–620.

The Comparative Cell Biology of Accessory Somatic (or Sertoli) Cells in the Animal Testis

Sardul S. Guraya

Department of Zoology, Punjab Agricultural University, Ludhiana-141004, India

A comparative account is given of recent advances in the cell biology of testicular accessory somatic (or Sertoli) cells in mammals, nonmammalian vertebrates, and invertebrates by comparing and contrasting their structure and function. Their structure is discussed in relation to the nucleus, cytoplasmic organelles, and inclusions (lipids, the cytoskeleton, junctional complexes, and blood–testis barrier, which show great diversity and a variable testicular architecture), and mode of spermatogenesis. A very limited somatic cell–germinal association or its complete absence is observed in some groups of invertebrates. Wherever the somatic accessory cells are present, their comparative functions are discussed in relation to (1) mechanical support and nutrition; (2) translocation of germ cells; (3) paracrine regulation and a combination of male germ cell proliferation and differentiation by secretion of regulatory proteins, including peptide growth factors and hormones; (4) phagocytosis; (5) steroid hormone synthesis and metabolism; and (6) spermiation. Comparative cellular and molecular aspects of Sertoli cell–germ cell and peritubular cell interactions and the regulatory (hormonal) mechanisms involved as well as gaps in our knowledge about the molecular aspects of these interactions are emphasized for a better understanding of diversity in the patterns and regulation of spermatogenesis in animals.

KEY WORDS: Activin; Androgen-binding protein; Anti-Müllerian hormone; Blood–testis barrier, Ceruplasmin, Germ cells, Inhibin, Junctional complexes, Myoid cells, Phagocytosis, Sertoli cells, Spermiation, Transferrin, Tubulobulbar complexes.

I. Introduction

It is well known that the male germ cells in the testes of animals proliferate, differentiate, and mature in close morphological association with accessory somatic cells, which, because of this nurse cell function, have generally been designated as nurse, cyst, follicle, nutrient, sustentacular, supporting, or Sertoli cells (Roosen-Runge, 1977; Adiyodi and Adiyodi, 1983; Guraya, 1976, 1987; Hinsch, 1993; Grier, 1993; Pudney, 1993; Russell and Griswold, 1993). In this review, somatic cells, which show great diversity in their origin, proliferation, development, distribution, abundance, structural organization, and functional life, will be referred to as Sertoli cells. Our knowledge of these cells at the cellular and molecular levels is still meager, especially from a comparative point of view (Russell and Griswold, 1993). Although the reviews in Russell and Griswold discuss isolated aspects of mammalian Sertoli cells in much detail, very little attempt or none is made to give a consolidated account of their cell biology by comparing and contrasting their cellular, histochemical, and biochemical (or molecular) aspects throughout the animal kingdom; this will be covered here. The gaps in our knowledge about their comparative cell biology will also be emphasized for further investigations. The cellular and molecular aspects of mammalian Sertoli cells will be described briefly by discussing recent reviews and papers; however, the emphasis will be on the comparative cell biology of accessory somatic cells in other animal groups, which have received little attention in this regard. Such comparisons at the cellular and molecular levels will help us better understand the diversity in the regulation and coordination of proliferative, developmental, maturational, and degenerative processes of male germ cells as well as the evolutionary steps in the development of different patterns of contacts (or environmental, nutritional, and regulatory interactions) between male germ cells and somatic cells during the evolution of different groups of animals.

II. Characteristics of Sertoli Cells

A. Morphological Organization and Distribution

Sertoli cells and developing germinal cells provide one of the most complex examples of environmental cell–cell interactions because there is great diversity in morphological contacts or interrelationships between somatic cells and germ cells as a result of variable patterns of spermatogenesis in different groups of vertebrates and invertebrates.

1. Vertebrates

Sertoli cells form an important component of the seminiferous epithelia of all vertebrates, including amniotes (mammals, birds, and reptiles) and anamniotes (cyclostomes, cartilaginous fishes, teleosts, lungfishes, and amphibians) (Guraya, 1976, 1987; Lofts, 1987; Pudney, 1993; Grier, 1992, 1993; Russell, 1993a,c; Schulze and Holstein, 1993b). Sertoli cells are generally attached to the basal lamina by rudimentary hemidesmosomes and extend between germ cells to reach the tubular lumen, thus creating a suitable environment for spermatogenesis or interactions between Sertoli cells and germ cells (Skinner, 1991, 1993a,b; Russell and Griswold, 1993).

Sertoli cells continually change their shape to accommodate the structural modifications and mobilization of germ cells from the base to the free surface of the seminiferous epithelium. Meanwhile, the lateral membranes of the Sertoli cells form numerous veil-like processes that extend between spermatocytes and spermatids and develop morphological contacts with them. In the stage 5 Sertoli cell of the rat, these processes include (1) conical processes extending from the lateral surface near its base, (2) cup-shaped, sheet-like processes partially enveloping round germ cells, and (3) tapered luminal extensions of the sheet-like, cylindrical processes generally called apical processes (Russell, 1993a). These complex lateral processes are suggestive of the environmental interactions at the level of plasma membranes of Sertoli cells and differentiating male germ cells.

The cyclic alterations in Sertoli cell morphology along the length of the seminiferous tubule indicate that these cells function, along with germinal cells, in forming the spermatogenic cycle (Morales and Clermont, 1993; Russell, 1993a,c; Vogl et al., 1993) as well as in the extensive metabolic exchanges between Sertoli and germ cells, thus ensuring the development and differentiation of the latter. The Sertoli cell plasma membrane is now known to be involved with receptor-mediated endocytosis (Morales and Clermont, 1993).

As in mammals, the seminiferous epithelium in the testes of reptiles and birds also forms a permanent structure consisting of Sertoli cells and germ cells which undergo conspicuous morphological and biochemical changes during seasonal spermatogenic activity (Guraya, 1976; Lofts, 1987; Pudney, 1993). Sertoli cells of Cheloria (*Chelydra serpentina;* Mahmoud *et al.,* 1985; *Chrysemys picta;* Dubois *et al.,* 1988) and lizards (*Lacerta muralis* and *L. sicula;* Baccetti *et al.,* 1983; *L. vivipara;* Dufaure, 1971) undergo seasonal changes which correlate with spermatogenic activity. Sertoli cell processes interdigitate and invaginate neighboring Sertoli cells. With the initiation of spermatogenic activity, such invaginations and interdigitations become less extensive, possibly to meet the demand of surrounding more maturing germ cells, leading to the unfolding of these cytoplasmic processes to accommodate this increase in surface area (Pudney, 1993).

Complex morphological relationships between Sertoli cells and elongating spermatids are observed in some lizards, such as *Cnemidophorus lemniscatus lemniscatus* (Del Conte, 1976), *Anolis carolinensis* (Clark, 1967), and *Pseudemeys scripta* (Sprando and Russell, 1988) as shown by the formation of an accessory cap on a projection of Sertoli cytoplasm during spermiogenesis. In *P. scripta* the Sertoli cell processes invade the cytoplasm of elongating spermatids (Sprando and Russell, 1988), suggesting their possible role in sequestering and removing portions of spermatid cytoplasm; this process may function as a mechanism for decreasing the cytoplasmic volume of spermatids during their maturation.

The ultrastructure of Sertoli cells has been investigated during spermatogenic activity in a relatively few species of birds, such as the Barbary duck, *Cairine moschata* (Marchand, 1973) and domestic fowl (*Gallus domesticus*) (Cooksey and Rothwell, 1973); during sexual regression in quail (Brown *et al.*, 1975) and throughout the breeding cycles of the mute swan, *Cygnus olar* (Breuker, 1982), Pekin duck (Garnier *et al.*, 1973), budgerigar (Humphreys, 1975), Japanese quail (Scheib, 1974), and mallard, *Anas platyrynchus arratidae* (Pelletier, 1990). It is basically similar in all the avian species. Sertoli cells develop processes and complex interdigitations which undergo conspicuous changes with spermatogenic activity and regression, suggesting environmental interactions between Sertoli cells and spermatogenic cells to ensure the production of fertile spermatozoa during the breeding period.

The morphological organization of germinal tissue in the testes of anamniotes (Cyclostomata, Chondrichthyes, Teleostei, and Amphibia) differs greatly from that described in the amniotes (reptiles, birds, and mammals) as a cystic form of spermatogenesis is found in anamniotes (Pudney, 1993; Grier, 1992, 1993). Because of the basic differences in the relationship between somatic cells and germ cells in the testes of anamniotes and amniotes, there has been controversy whether the term ''Sertoli cell'' can be accurately used to describe somatic cells in the anamniote testis (Grier *et al.*, 1989; Grier, 1992, 1993; Pudney, 1993). In spite of this controversy, the great similarity in the cytological features of somatic cells in the testes of amniotes and anamniotes suggests an analogy between the two cells; the somatic cell of the anamniote testis can, therefore, be referred to as a Sertoli cell, in spite of whether this is morphologically right.

The testis of Cyclostomata, including lampreys (Petromyzonidae) and hagfishes (Myxinoidae) consists of aggregations of cysts, i.e., it is polyspermatocystic (Grier, 1992, 1993; Pudney, 1993). The cysts (follicles) of myxinoid testis show a single layer of Sertoli cells which form fluid, the liquor folliculi, during early stages of spermatogenesis (Dodd and Sumpter,

1984). Sertoli cells also develop highly complex processes that interdigitate and extend between germ cells (Grier, 1993). The peritoneal cells derived from the peritoneal epithelium covering the testis of the brook lamprey (*Entosphenus wilderi*) migrate into the testis to envelope individual germ cells (Okkelberg, 1921). Sertoli cells of the sea lamprey (*Petromyzon marinus*) isolate germ cells and envelope them to constitute nascent cysts (Hardisty, 1965). In the early upstream-migrating river lamprey (*Lampetra flaviatilis*), the Sertoli cells constituting the walls of the cyst (having spermatogonia) and developing complex interdigitations are fibroblast-like, with spindle-shaped nuclei (Barnes and Hardisty, 1972). With the spermatogenic activity, Sertoli cells undergo rapid and synchronous differentiation. Further study of the ultrastructural organization of Sertoli cells in cyclostomes may produce information valuable from an evolutionary and comparative point of view because the cyclostomes form the sole extant representatives of the most primitive jewelfishes (Agnatha).

Chondrichthyes include contemporary cartilaginous fishes, which are generally divided into two unequal classes such as Elasmobranchii (sharks, dogfish, skates, and rays) and Holocephali (ratfish). Their Sertoli cells and germ cells undergo conspicuous cytological changes during the development and maturation of spermatocytes (Pudney, 1993; Grier, 1992, 1993). Individual Sertoli cells surround spermatogonia within cytoplasmic processes, thus isolating them from each other. The cyst consists of spermatoblasts, each of which contains a single Sertoli cell and its cohort of isogenic germ cells. The Sertoli cell cytoplasm extends toward the lumen of the cyst to form the walls of the spermatoblast and encloses a central cavity (Jégou *et al.,* 1988; Pudney, 1993; Grier, 1992, 1993). The heads of developing spermatids forming a single layer are oriented toward the wall of the spermatoblast and their tails project into the cavity. The germ cells at this stage are not completely enveloped by Sertoli cytoplasm but reside in its shallow recesses. The latter, which house the immature germ cells, become effaced, freeing the spermatids within the spermatoblasts; only their heads are associated with the plasma membrane of the Sertoli cell. With the completion of spermiogenesis, the spermatids form tightly packed bundles which get enclosed within a pocket formed in the apical cytoplasm of the Sertoli cell. At this stage, the hypertrophied Sertoli cells are called "glandular" or "nurse" cells. Sertoli cells in the holocephan testis show basically the same cytological changes in relation to spermatogenic activity (Stanley and Lambert, 1985) as elasmobranchs.

In teleosts, Sertoli cells form spermatocysts lying within anastomosing tubules, lobules, or tubules (Billard, 1990; Grier, 1992, 1993). Each spermatocyst consists of spermatogenic cells and a Sertoli cell, with cytoplasmic processes extending between germ cells and constituting the border of the spermatocyst, which shows synchronous spermatogenesis (Billard,

1990; van Huren and Soley, 1990; Grier, 1992, 1993; Pudney, 1993). Primary spermatogonia are individually surrounded by Sertoli cell processes and do not lie within spermatocysts which, however, contain secondary spermatogonia and all subsequent stages of spermatogenesis.

Spermatozeugmata are formed within spermatocysts of some teleosts (Grier, 1993; Pudney, 1993). With the maturation of germ cells, Sertoli cells hypertrophy and gradually become columnar in shape. Meanwhile developing spermatids align themselves with their heads abutting the apical Sertoli cytoplasm to constitute the spermatozeugmata. The finger-like processes of Sertoli cells insinuate themselves between the heads of spermatozoa and anchor them at the lumenal margin of the cyst.

The morphology of the lungfish testis is similar to that described in the teleosts. Spermatogenesis occurs in cysts located in a seminiferous compartment, forming lobules (S. Guraya, personal unpublished observations). The heads of tightly packed bundles of spermatozoa lie in association with the Sertoli cells that rest on the basement membrane of the germinal epithelium.

The morphology of the testis in amphibians, including the Anura (frogs and toads), Urodela (newts and salamanders) and Apoda (caecilians) is also similar to that found in the teleosts because spermatogenesis occurs in spermatocysts placed in seminiferous compartments called lobules in urodeles but generally designated as tubules in anurans (Grier, 1992; Pudney, 1993). Both the Sertoli cells and germ cells undergo seasonal changes with spermatogenic activity in the cysts. The spermatocyst formation in urodeles starts with a fibroblast-like Sertoli cell. Sertoli cells remain relatively undifferentiated, forming the wall of the spermatocyst, which contains the developing germ cells. They generally conform to the circular outline of the spermatogonia and develop thin cytoplasmic processes surrounding immature germ cells. Spermatogonia develop small, blunt, cytoplasmic projections which invaginate adjacent Sertoli cell cytoplasm and are larger and more common on primary than secondary spermatogonia (Pudney, 1993). With the completion of spermiogenesis, Sertoli cells hypertrophy and bundles of spermatozoa get their heads embedded in large sheets of Sertoli cytoplasm, with their tails lying in the lumen. Each spermatid occupies an individual recess in the apical Sertoli cytoplasm, which contains an abundance of microtubules. Complex interdigitations and infoldings develop between neighboring Sertoli cells, which surround the maturing germ cells and do not send processes between them.

As in urodeles, the Sertoli cells and germ cells of anurans undergo conspicuous seasonal changes (Grier, 1992, 1993; Pudney, 1993). Sertoli cells form a layer of highly differentiated cells having spermatids embedded in their cytoplasm, but they basically undergo the same morphological changes during spermatogenesis as those described for urodelan Sertoli cells (Grier, 1992, 1993; Pudney, 1993). However, developing spermatids of the yellow-bellied toad (*Bombina variegata*) do not form bundles (Obert, 1976) as in other amphibians. It remains to be determined with an electron microscope whether Sertoli cells are present or absent in its testis.

2. Invertebrates

Relatively few species of invertebrates have been investigated for the ultrastructure of their gonads or testes, with special reference to accessory somatic cells (Roosen-Runge, 1977; Hinsch, 1993). The structural and functional interrelationships between accessory somatic cells and developing germ cells show great diversity in invertebrates. Testicular structures of Porifera (sponges) form masses of germ cells in various stages of spermatogenesis which are surrounded by a covering layer of cells (Adiyodi and Adiyodi, 1983). The sperm masses do not show any cells other than the sperm cells, which apparently derive from a single cell. In Cnidaria (coelenterates) spermatogenesis occurs in cysts of cells lying in the mesoglea. The cyst cell does not appear to function in any way comparable to a Sertoli cell because in the testes of the sea anemones *Bunodosoma cavernata* (Dewel and Clark, 1972) and *Ceriantheopsis americanus* (Hinsch and Moore, 1992); no cells comparable to somatic cells are observed. This appears to be a common feature in a variety of coelentcrates. However, in the highly differentiated gonads of the medusa jellyfish *Phialidium leuckhart*, the somatic epithelial cells extend throughout the germinal epithelium with pillar-like cell bodies up to 200 μm (Roosen-Runge and Szöllösi, 1965). Each pillar forms many fine lateral processes which develop contact with all germ cells without enveloping them. These supporting cells may form columns of more than one cell and are regularly spaced throughout the testis. Their basal region forms a foot-like association with the mesogleal lamella. Sertoli (supporting) cells in the medusa jellyfish get surrounded by zones of spermatogenic cells. After their migration from the gastrodecmis, the spermatogonia are placed adjacent to the mesoglea where they get separated from the mesogleal lamella by the processes of supporting cells. Thin processes of Sertoli cells occur between spermatogenic cells of the ctenophore, combjelly (*Beroe ovata;* Franc, 1973). The somatic or parietal cells penetrate deeper into the testicular follicles of the turbellarian *Polycelis,* and surround the spermatogonia and spermatocytes completely (Franquinet and Lender, 1973; Hendelberg, 1983). Cellular extensions of adjacent parietal cells overlap.

The testicular structures of Eucestoda and nematodes do not show any evidence of somatic tissues. However, a cytophore formed as a result of incomplete cytokinesis during meiotic division forms the characteristic feature of their spermatogenesis (Guraya and Gupta, 1970a,b; Gupta *et al.,* 1986; Davis and Roberts, 1983). In the nematode testes, the germ cells are attached by fine threads to a central anucleate protoplasmic mass, the so-called "rachis" which, with the proliferation of germ cells, branches dichotomously and finally its tertiary branches get attached to the germ cells in the zones of mitosis and meiosis, thus enhancing the environmental interactions between rachis and germ cells (Wright and Sommerville, 1984, 1985). As the spermatocytes approach meiosis, the rachis finally becomes the site of the breakdown of the residual cytoplasm. The rachis is absent in the testis of the marine nematode *Deontostoma californicum,* which shows epithelial cells with ameboid processes lying under the tunica propria (Hope, 1974).

A single tubular testis of the nematode *Dipetaloma dessetae* shows free germ cells at its apical end and more mature cells placed caudally (Marcaillou and Szöllösi, 1980). Somatic cells constitute a discontinuous layer of widely separated cells in the apical region, where germ cells appear to lie in contact with the surrounding sheath. In the caudal region, somatic cells form a continuous sheet.

A close association is observed between the numerous processes of sustentacular cells and spermatocytes in the testes of *Trichuris muris* (Jenkins *et al.,* 1979; Jenkins and Larkman, 1981). An accessory role in the regulation and synchronization of spermatogonial development is suggested for the testis of *T. muris.*

The cytoplasm of large Sertoli cells in the acanthocephalan (*Polymorphus minutus*) testis extends in various directions and their long processes penetrate between the gonocytes, forming contact with them (Whitfield, 1969; Guraya, 1971; Crompton, 1983). Sertoli cells are present among the germ cells of Pentastomida (Self, 1983). They lie along the basal lamina and envelope groups of spermatocytes. The pseudoacrosome of the sperm in *Argulus foliaceus* L. makes contact deep within the Sertoli cell cytoplasm (Wingstrand, 1972). The primitive germ cells lining the testis of pentastomida arise in the coelomic pouches of the early embryo and get enclosed within a cavity lined with epithelium (Self, 1983). The clumps of primordial germ cells enclosed in epithelial cysts multiply to produce a given number of spermatozoa; meanwhile the cyst is pinched off and floats free in the testicular cavity and its somatic cells are believed to function as nutritive and phagocytic cells. However, such functions remain to be defined by *in vitro* biochemical methods.

In the freshwater bryozoan *Pectinatella gelatinosa,* the Sertoli cells and sustentacular cells develop a close morphological relationship with synchronously developing spermatids and thus appear to play supporting and nutritive roles like the Sertoli cells of other animals (Tazima *et al.,* 1993). The masses of developing germ cells lie in projections or swellings of the peritoneum in different zones of segments in some species of polychaete annelids (marine worms) whereas in others they form discrete gonads (Adiyodi and Adiyodi, 1983; Sawada, 1984). Spermatogonia or primary spermatocytes are shed into the coelome, where they mature and undergo incomplete cytokinesis to form the cytophore as described for cestodes and trematodes. Masses of spermatogenic cells remain attached to the cytophore throughout spermatogenesis, suggesting that the cytophore is a kind of supporting element for the developing spermatogenic cells.

In oligochaetes (earthworms) also, spermatogonia are released into the coelome, where they mature. Maturing spermatogenic cells are attached to a cytophore by thin cytoplasmic bridges covered with an external fuzzy coat (Anderson et al., 1967; Block and Goodnight, 1980). Sperm also develop in association with the cytophore in the branchiobdellids (Hinsch, 1993). In Hirudinea (leeches), spermatogenesis occurs in the testicular lumen lined with a unicellular layer of parietal cells showing no specialized occluding junctions. Spermatogenic cells mature within the lumen attached to the cytophore (O'Donovan and Abraham, 1987).

The large accessory somatic cells in the crustacean testis are called by different names, such as giant cells in the Cladocera (Delavault and Bernard, 1974), follicular cells in the Isopoda and the Mysidacea (Kasaoka, 1974), nutritive or nurse cells in shrimps (Pillai, 1960; Arsenault et al., 1980), interstitial tissue (Fahrenbach, 1962; Manier et al., 1977) or sustentacular cells (Gupta, 1964; Wingstrand, 1978) in the copepods, intercalary cells in the decapod Reptantia (Pochon-Massion, 1968a,b), and Sertoli cells (Hinsch, 1993). Various electron microscope studies by Hinsch and co-workers and various other workers have clearly shown that in their organizational, morphological, and junctional features, crustacean somatic cells closely resemble vertebrate Sertoli cells (see the details in an excellent review by Hinsch, 1993).

An insect testis consists of follicles which are divided into distinct compartments (Szöllösi, 1982; Hinsch, 1993). Generally the greatest part of sperm development in insects occurs in spermatocysts in which germ cells differentiate, more or less synchronously surrounded by a capsule of somatic cells (Szöllösi, 1982), interstitial cells (Saxena and Tikku, 1989), or sustentacular cells (Smith, 1968). The cysts and their Sertoli (cyst) cells show a great diversity in their development, structure, and function in different groups of insects, as reviewed by Hinsch (1993). The long, slender, cytoplasmic processes of Sertoli cells in insects extend from the perinuclear region to surround the germ cells at various stages of development (Szöllösi, 1982; Saxena and Tikku, 1989, 1991; Hinsch, 1993).

It has been suggested that the processes of Sertoli cells in the lepidopteran testis extend between the spermatids, possibly to play a role not only in regulating their development and differentiation but also in providing nutritive substances. In all insects, the spermatids are arranged in bundles while they elongate (Hinsch, 1993) and meanwhile their heads are directed toward one of the Sertoli cells. Our knowledge is still very poor about the factors that cause the orientation of the spermatids toward one of the Sertoli cells. However, Szöllösi (1974, 1975) observed that in locusts, it is in the cyst and during spermiogenesis that the acrosomes of spermatids get tightly linked by a cap of glycoprotein, resulting in a firm association or adhesion of the sperm in the original bundle, which are transported to the female in the form of spermatodesms (Hinsch, 1993). It still remains to be determined whether the material of the cap is produced or transformed by the Sertoli cell.

Very divergent views exist about the origin, development, occurrence, structure, and function of apical cells or the apical complex in different insect groups (Roosen-Runge, 1977; Szöllösi, 1982; Hinsch, 1993). Primary spermatogonia surround the apical cell and their slender processes develop contact with its surface and interdigitate with radiating processes from the apical cell (Hinsch, 1993), suggesting environmental interactions between them. In *Locusta,* spermatogonia proliferate in cysts of cells surrounded by perifollicular cells in the apical region (Szöllösi, 1982). With the shifting of germ cells basally in the follicle, the perifollicular layer gets enveloped by the inner parietal cell layer. These cells develop many overlapping processes connected by septate desmosomes. Meiosis and spermiogenesis occur in this basal compartment. Sertoli cells in the germinal epithelium of the onychophoran testis develop an elaborate system of the processes that extend among the early spermatogenic cells (genocytes) (Storch and Ruhberg, 1983), suggesting environmental interrelationships.

Different classes of the phylum Mollusca show considerable variation in their reproductive systems because they are primitively dioecious but others are hermaphrodites; the latter are either simultaneous or protandric hermaphrodites. The gonad of the hermaphrodites is an ovotestis. A variety of terms such as nurse cells, nutritive cells, follicle cells, free cells, basal cells, and Sertoli cells have been used to describe the accessory somatic cells of mollusks, which form a close morphological association with developing spermatogenic cells as their cytoplasmic processes penetrate between the germ cells (Roosen-Runge, 1977; Hinsch, 1993), suggesting environmental interaction between them. They also form a barrier between the testicular and ovarian regions of the ovotestis. However, the term ''Sertoli cells'' has been frequently used for them because of their structural and possibly functional (phagocytic ability) resemblance to the corresponding Sertoli cells of the mammalian testis (de Jong-Brink *et al.,* 1977, 1981, 1984; Bergmann *et al.,*1984; Anelli *et al.,* 1985; Buckland-Nicks and Chia, 1986; Hinsch, 1993). In the mesogastropod snail *Cipango-paludina,* elongate processes extend from the Sertoli cell toward the lumen and are closely connected to the head and midpiece of the typical spermatid (Yasuzumi *et al.*, 1960; Yasuzumi, 1962). With further development of the spermatid, the processes become more and more numerous, develop contact with each other and finally coalesce to form a continuous thin sheet which surrounds each spermatid in a mantle. A typical spermatid gets lodged in a deep invagination of the nurse cell similar to that described for vertebrate Sertoli cells.

Sertoli cells showing two types of morphology are reported in the pulmonate *Arion ater* (Parivar, 1980). The first active type appears after gonial differentiation and sends processes between spermatocytes, spermatids, and sperm. The second type, or atrophic cells, are formed from the active Sertoli cells following spermiation and their cytoplasmic processes are highly interdigitated. Clusters of synchronously developing germ cells connected to large Sertoli cells rich in alkaline phosphatase and having polymorphic nuclei are observed in a slug, *Philomycus carolinianus* (Kugler, 1965), in which they are detached from the acinar wall but a cluster of 200 or more spermatozoa continue to be connected with them, thus providing environmental interaction between Sertoli cells and germ cells. The spermatozoa become free as the whole mass leaves the ovotestis and the Sertoli cell then disintegrates. In some mollusks such as *Planorbis, Halisoma, Vaginulus,* etc., the basal auxiliary somatic cells carry the developing spermatogenic cells from the distal to the proximal end of the lobule, where they are released into the lumen. However, de Jong-Brink *et al.* (1977) suggested that in *Bromphalaris glabrata* developing germ cells and Sertoli cells migrate in the opposite direction, toward the distal end of the acinus and this discrepancy remains to be clarified.

Buckland-Nicks and Chia (1986) provided details of the ultrastructure of Sertoli cells of three marine prosobranchs such as the hairy Oregon whelk *Fusitriton oregonensis*, the periwinkle *Lottorine sitkana*, and the leaf horn mouth, *Ceratostoma foliolatum*, and then discussed their functional morphology after comparing and contrasting them with the Sertoli cells of mammals and other vertebrates. Sertoli cells of prosobranchs are modified columnar epithelial cells and maintain continuous contact with the basal lamina from where they extend to the lumen of a testicular tubule. Cytoplasmic processes of Sertoli cells in some prosobranchs showing sperm polymorphism apparently cooperate to bring together a close eupyrene sperm and a carrier sperm at a particular time in development.

Most cephalopods, which include nautili, cuttlefish, squids, and octopods, are dioecious and have a gonad lying in the posterior part of their body. The large Sertoli cells of *Nautilus compilius* are branching cells forming pouches which surround synchronously developing spermatids (Arnold, 1978), suggesting environmental interactions between them.

Testicular structure varies among the different classes (asteroids, echinoids, ophiuroids, and holothuroids) of echinoderms and hence the structural organization of the germinal layer also varies, depending upon the species, season, and state of maturation of the gonad. It consists of somatic accessory cells (also called interstitial cells) and germinal cells (Chia and Bickell, 1983; Walker and Larochelle, 1984; Hinsch, 1993). Both types of cells often show an annual pattern in their abundance and distribution and in their morphological or ultrastructural characteristics (Nicotra and Serafino, 1988; Hinsch, 1993). The germinal epithelium in asteroids forms many distinct columns consisting of one Sertoli cell surrounded by approximately 400 spermatocytes which are apparently displaced from the basal regions toward the tips of these columns (Walker, 1980). With the initiation of each spermatogenic cycle, the Sertoli cells form a close association with the amitotic spermatogonia in ophiuroids (Hinsch, 1993). Filopodia terminating on the basal lamina as flattened discs support several Sertoli cells; the discs may overlap or touch those of adjacent cells and additional filopodia interdigitate laterally (Buckland-Nicks *et al.,* 1984). The filopod from the Sertoli cell also surrounds the developing germ cells, which move toward the testicular lumen where the filopodia are ultimately withdrawn and sperm is released into the lumen (Buckland-Nicks *et al.,* 1984). Sertoli cells in echinoids apparently function in the structural organization of spermatogenic cells and their movement.

The large and well-differentiated yolk cells occupy all the spaces between the developing male germ cells of the hemichordate *Ptychodera flava* but no morphological connections are observed between germ cells and yolk cells at any stage of spermatogenesis (Guraya, 1986). In Cephalochordata, commonly called *Amphioxus,* each testis at the height of spermatogenic activity consists of a tubular wall, Sertoli cells, and spermatogenic cells at various stages of growth, differentiation, and maturation (Guraya, 1983). The large Sertoli cells of irregular shape show a prominent vesicular nucleus. They send long cytoplasmic processes among the aggregates of developing spermatogenic cells. Although no electron microscopic study has been made, at the height of spermatogenic activity each testis of *Amphioxus* is comparable in its structural organization to the seminiferous tubule of the amniote testis, suggesting close morphological interrelationships or environmental interactions between Sertoli cells and germ cells.

In conclusion, it can be stated that the morphological organization of accessory somatic cells or Sertoli cells shows a great diversity in relation to developing spermatogenic cells in different groups of animals, as evidenced by numerous variations in their development and distribution. In some groups of animals such as poriferans, some coelenterates, trematodes, cestodes, nematodes, and annelids, accessory somatic cells do not occur and thus their spermatogenic cells proliferate and develop without any morphological association (environmental interaction) with accessory somatic cells. However, their developing male germ cells form a cyto-plasmic mass called a cytophore or rachis (present in nematodes) with which they remain connected until they become mature for release as sperm. In hemichordates, the accessory somatic cells are present but they do not develop any morphological association with developing spermato-genic cells. In the testes of arthropods, mollusks, and echinoderms, the accessory somatic cells are highly developed and closely resemble the Sertoli cells of reptiles, birds, and mammals in their morphological organi-zation or characteristics relative to developing spermatogenic cells. Al-though accessory somatic cells comparable to Sertoli cells are present in the testes of other invertebrate groups, they remain simple since they do not show much morphological differentiation. This is shown by the poor development of their cytoplasmic processes, which are highly developed in arthropods, mollusks, echinoderms, and chordates, including *Amphioxus*. Even among chordates, the degree of development of cytoplasmic pro-cesses varies greatly, depending upon the species and seasonal spermato-genic activity. In general, the complexity of morphological organization of Sertoli cells is closely related to the complexity of the testicular architec-ture in animals.

B. Nuclear and Cytoplasmic Components

In their morphology (including ultrastructure), histochemistry, and bio-chemistry, as well as in their distribution, the nuclear and cytoplasmic components of Sertoli cells show great variation not only in the cyclic activity of spermatogenesis but also with the species and seasons as re-vealed by recent morphometric analysis (Ueno and Mori, 1990; Russell *et al.*, 1990; Ye *et al.*, 1993; De Franca *et al.*, 1993; Russell, 1993a).

1. Nucleus

In its morphology and location, the mature Sertoli cell nucleus has varia-tions that depend on the species and stages of the spermatogenic cycle in vertebrates (Kurohmaru *et al.*, 1992; Russell, 1993a; Schulze and Holstein, 1993b; Morales and Clermont, 1993; Pudney, 1993) and invertebrates (Hinsch, 1993). It is interesting that Sertoli cells in the lizards *Anolis carolinensis, Sceloporus occidentalis,* and *Uta stansburiana stejneger* are binucleate (Pearson and Licht, 1990; Pudney, 1993), suggesting doubling of the amount of DNA preparatory to their growth and expanded function during spermatogenesis. The Sertoli cell nucleus shows an infolded nuclear envelope with pores, a relatively homogeneous nucleoplasm, and a single tripartite nucleolus, whose functional significance remains to be determined more precisely. The presence of primarily homogeneously distributed chromatin having a fine fibrillar granular texture is consistent with the postulate that Sertoli cells express a large portion of their genome in accordance with their highly versatile functions, especially in relation to secretion of many proteins (Bardin *et al.*, 1994). The tripartite nucleolus proper having a reticular structure is associated with two smaller, electron-dense, heterochromatin bodies (de Kretser and Kerr, 1994). Small heterochromatin masses termed heteropycnotic bodies, juxtanuclear bodies, perinuclear spheres, or satellite karyosomes but now known to be centromeric heterochromatin may also be seen elsewhere within the nucleoplasm or forming an association with the nuclear envelope (Guraya, 1987). The functional significance of these perinuclear bodies remains to be determined.

An unusual nucleolar fine structure is observed in the Sertoli cells of the dogfish *Scyliorhinus canicula* (Moyne and Collenot, 1982) as evidenced by the development of intranucleolar granules, similar to the perichromatin granules often found in mammals and lamella bodies; the latter appear to lie in contact with the nucleolus and consist of groups of parallel lamella containing DNA and ribonucleoprotein. The coiled body and new inclusion, the "mykaryon," form normal constituents of the Sertoli nucleus in the adult rat (Schultz, 1989). Their functional significance also remains to be determined. The Sertoli cell nucleolus of ruminants shows an aggregation of membrane-limited vesicles (Pawar and Wrobel, 1991), the functional significance of which is still not known. As seen by molecular immunochemistry, the fibrillar center within the nucleolus of human Sertoli cells is the only site where DNA and RNA are visualized together (Thiry, 1993). The nucleolus is now known to constitute a very original nuclear territory that reflects the compartmentation of nuclear functions which implicate specific RNA polymerases, specific RNAs, and specific proteins (Hernandez-Verdun, 1991). It forms the site of an intense exchange between the nucleus and cytoplasm that involves specific shuttled proteins and specific relationships with the nuclear envelope. Nucleoli are involved in the transport of certain messenger RNAs.

2. Cytoplasmic Organelles and Other Components

Sertoli cells have various organelles, including mitochondria, being long and slender with transverse tubular foliate cristae, a large Golgi apparatus (consisting of multiple separate Golgi elements scattered throughout the cytoplasm), a profuse endoplasmic reticulum (both rough and smooth, varying among different groups of animals in their relative proportion and location within the cytoplasm), numerous lysosomes of heterogeneous appearance, multivesicular bodies, residual bodies, phagosomes, but no secretory granules in mammals (Ueno and Mori, 1990; Ueno *et al.*, 1991; Kurohmaru *et al.*, 1992; Russell, 1993a; Vogl *et al.*, 1993; Morales and Clermont, 1993; Pudney, 1993; de Kretser and Kerr, 1994; Bardin *et al.*, 1994), human (Schulze and Holstein, 1993b), nonmammalian vertebrates (Pudney, 1993), and invertebrates (Hinsch, 1993).

Various cytoplasmic components showing species and cyclic or seasonal variations generally show a polarized distribution as evidenced from their abundance in the basal and trunk regions of Sertoli cells whose apical extensions usually show a paucity of organelles. Lipid inclusions, lipofuschin pigment, and glycogen granules also occur in Sertoli cells. A membrane body consisting of a complex array of thin, electron-dense membranous structures resembling endoplasmic reticulum is observed in the Sertoli cell of the skink *Eumeces laticeps* Schneider (Okia, 1992). Lysosomes, mitochondria, and microfilaments occur among the elements of the membranous body. Intralobular Sertoli cells show numerous lipid droplets in the fish species *Esox lucius* but not in *E. niger* (Grier *et al.*, 1989). Lipid droplets generally surrounded by cisternae of smooth endoplasmic reticulum lie in the basal regions of Sertoli cell.

Ye *et al.* (1993), using a sampling technique at the electron microscope level, have observed that among the many parameters investigated, only the surface area of the cell, the volume of lipid, and the volume and surface area of the rough endoplasmic reticulum vary cyclically, as demonstrated by statistical analysis. The parameters of rough endoplasmic reticulum generally showed a correlation with known patterns of protein secretion within the tubule and with the secretion of specific proteins as well as the factors important in regulating protein secretions. Methodological differences form potential sources of error for various discrepancies in the quantification of the structural changes during the cycle of the seminiferous epithelium of the rat in different investigations.

Cyclic changes in organelles appear to be correlated with specific generations of germ cells (Ueno and Mori, 1990). The cyclic changes in the Golgi apparatus appear to be related to synthesis of glycoproteins in response to the needs of the plasma membrane and lysosomes as well as to its probable role in the production and discharge of material substances (Ueno *et al.*, 1991). The cytological features of multiple Golgi bodies coupled with (1) the absence of large vacuoles or membrane-enclosed secretory granules in morphological association with the Golgi element, (2) the scarcity of cytological evidence for exocytosis, and (3) the small number of vesicles opening into the lateral surface of Sertoli cells indicate that the Sertoli cells do not release an appreciable amount of protein into the lumen of seminiferous tubules, lobules or cysts (Bardin *et al.*, 1994).

Smooth or agranular endoplasmic reticulum is generally reported as vesicular, tubular, cisternal, fenestrated, or lamellar, but it forms concentric whorls in the bull (Wrobel and Schimmel, 1989) and buffalo (Pawar and Wrobel, 1991). Large, compacted masses of agranular endoplasmic reticulum envelope the developing heads of elongating spermatids in some vertebrates (Russell, 1993a,c; Pudney, 1993). Just prior to sperm release, the spermatid head in some mammals is retained by the Sertoli cell through its contact with very large bulbous projections of Sertoli cell cytoplasm filled with agranular endoplasmic (Russell, 1993c). After the release of sperm from the Sertoli cell, these membranous masses are moved toward the base of each Sertoli cell. The functional significance of smooth endoplasmic reticulum still remains to be determined more precisely since its involvement in steroidogenesis is still not established in different groups of vertebrates (Dorrington and Khan, 1993).

Mitochondria adopting slender, spherical, and cup-shaped forms generally show a great diversity in form during the spermatogenic cycle. The precise functional significance of this diversity remains to be determined. The large mitochondria in Sertoli cells of the teleost, *Abudefduf marginatus,* are filled with lipid droplets (Mattei *et al.,* 1982). Numerous primary lysosomes, autophagic vacuoles, and heterophagic vacuoles are seen throughout the cytoplasm and appear to be involved in the phagocytosis and digestion of degenerating germ cells and residual bodies released by the spermatids during spermiation and engulfed by the Sertoli cells in vertebrates (Russell, 1993b; Morales and Clermont, 1993; Grier, 1993; Pudney, 1993) and invertebrates (Hinsch, 1993). They also play an important role in the natural autolytic process of degenerating germ cell cytoplasmic organelles.

The distribution of smooth endoplasmic reticulum around secondary lysosomes and pleomorphic lipofuschin pigment in the human Sertoli cell appears to be indicative of some enzymatic digestive process of the Sertoli cell (Schulze and Holstein, 1993a). Histochemical investigations have shown intense hydrolytic enzyme activity in the Sertoli cells (Guraya, 1976; Lofts, 1987), which primarily lies within membrane-bound, densebody lysosomes (Gunawerdana, 1992), supporting the participation of Sertoli cells in the phagocytosis and digestion of residual cytoplasm and degenerating germ cells; this has also been reported for nonmammalian vertebrates (Pudney, 1993). The development of lysosomes and their participation in the dissolution of residual bodies appear to provide a link between the endocytic and phagocytic activities of the Sertoli cell (Clermont *et al.,* 1987; Morales and Clermont, 1993).

The Sertoli cell cytoplasm may contain various crystals and vesicles as in man (Schulze and Holstein, 1993b) and some vertebrates (Pudney, 1993). Needle- or spindle-shaped crystaloids varying in length are also found in the basal part of the Sertoli cells in the three-toed sloth *Bradypus tridactylus* (Toyama *et al.,* 1990). The lipid content of Sertoli cells varies in amount, distribution, and chemistry from species to species as well as with season or the spermatogenic cycle in vertebrates (Guraya, 1976, 1987; Nagahama, 1986; Lofts, 1987; Russell, 1993a; Morales and Clermont, 1993; Pudney, 1993) and invertebrates (Hinsch, 1993). The physiological significance of various seasonal variations and species differences in the abundance of lipid remains to be precisely determined.

3. Cytoskeletal Elements

The cytoplasmic filaments and microtubules that form a remarkably elaborate cytoskeleton vary greatly in amount and distribution in different regions of the same Sertoli cells in vertebrates, especially in amniotes (Guraya, 1987; Pfeiffer and Vogl, 1992; Vogl *et al.*, 1993; Russell, 1993a,c; Morales and Clermont, 1993; Pudney, 1993) as well as in myoid cells and the tunica albuginea of rat and mouse testis; these cell types also show a variable distribution of actin filament bundles (Maekawa *et al.*, 1991). Very little attention or none has been paid to cytoskeletal elements in the Sertoli cells of lower vertebrates (Pudney, 1993) and invertebrates (Hinsch, 1993).

Cytoplasmic filaments of various sizes and forms are generally distributed around the Sertoli cell nucleus at the base of the cell in the apical cytoplasm and in association with junctions. Microfilaments or actin filaments are mainly localized in cortical regions and form aggregations in ectoplasmic specializations which are relatively more highly developed in amniotes than in other animal groups (Vogl *et al.*, 1991a,b, 1993; Pudney, 1993; Hinsch, 1993). Intermediate filaments are mainly of the vimentin type and form a network around the nucleus, as a layer or thin carpet along the base of the cell. They also distribute toward and are associated with desmosome-like junctions with early germ cells. Intermediate filaments, together with microtubules, also occur in abundance in portions of the cell involved with the transport of smooth endoplasmic reticulum, in cytoplasm associated with elongate spermatids, and in processes that extend into the residual cytoplasm of spermatogenic cells.

The formation of various complex cytoskeletal and contractile elements can be expected for the Sertoli cell because it undergoes a radical change in shape in conforming to the ever-changing events within the seminiferous or germinal epithelium (Russell, 1993a,c; Vogl *et al.*, 1993; Morales and Clermont, 1993; Pudney, 1993; Hinsch, 1993). The degree of development is closely related to the morphological complexity of seminiferous epithelium since relatively less differentiated somatic accessory cells of invertebrates with simple germinal epithelium do not show much development of cytoplasmic contractile elements (Hinsch, 1993). The dense filamentous networks appear to perform a major function as a structural support of the Sertoli cell when rigidity is required and at other times may engineer alterations in cell matrix viscosity, permitting the variations in plasticity that are essential to accommodate the constant mobility of the germ cells. The precise function of the intermediate filaments that are randomly distributed throughout the Sertoli cell cytoplasm is still not known. However, Ca^{2+}-calmodulin and cyclic adenosine monophosphate (cAMP) are believed to regulate the cytoskeleton in the Sertoli cells (Vogl et al., 1993). A high ATPase activity found near the filaments appears to be involved in providing an energy source for filament contractility.

Microtubules run predominantly parallel to the long axis of the Sertoli cell in thick bundles and thus to the long axis of apical crypts. (Vogl et al., 1993; Pudney, 1993). The dynamic nature of Sertoli cell microtubules emphasizes the stage-dependent role they may play in spermatogenesis. Microtubules associated with apically placed maturing spermatids in some species constitute structures like hoops that conform to the form of the spermatid head (Vogl et al., 1993). Sertoli cell microtubules are apparently needed for the normal development and translocation of spermatids in the seminiferous epithelium (Redenbach and Vogl, 1991; Redenbach et al., 1992) and are related to positional changes in the smooth endoplasmic reticulum of the Sertoli cell.

Redenbach and Vogl (1991) have indicated that Sertoli cell microtubules are of uniform polarity and are oriented with their minus ends toward the cell periphery. These observations are of great relevance to their proposed model of microtubule-based transport of spermatids through the seminiferous epithelium. Ectoplasmic specializations are believed to function as vehicles and microtubules as tracks (Vogl et al., 1993). Microtubule polarity provides the basis for the direction of force generated by available mechanisms. Experimental studies carried out on the binding between microtubules and spermatid ectoplasmic specializations (Redenbach et al., 1992) also support the proposal that elements in the cytoplasmic domain are a functional part of ectoplasmic specialization and are consistent with a microtubule-based model of spermatid translocation.

Vogl et al. (1993) have further proposed the involvement of mechanoenzymes (kinesin and cytoplasmic dynein) in the movement of the endoplasmic reticulum component of ectoplasmic specializations along adjacent microtubule tracts. With the movement of the endoplasmic reticulum, the whole adhesion complex is moved and the attached spermatid is shifted along; links are formed between the microtubules and endoplasmic reticulum. However, experimental studies are required to test the model of microtubule-based spermatid translocation in the testis as proposed by Vogl and co-workers (Vogl et al., 1993).

Microtubules may also play a role in shaping the spermatid head and acrosome (Allard *et al.*, 1993). Sertoli cell microtubules are also believed to be involved in protein secretion and in the formation of seminiferous tubule fluid (Johnson *et al.*, 1991). Allard *et al.* (1993) have recently proposed that the stage dependence of colchicine-induced effects reflects the dynamic and stage-dependent role of microtubules in spermatogenesis. Furthermore, cellular structures other than microtubules, such as vimentin filaments, may be required for maintaining the structural integrity of the seminiferous epithelium. Amselgruber *et al.* (1992) have localized S-100 protein immunohistochemically in the bovine Sertoli cell, which is believed to be involved in the microtubule assembly-disassembly system. The specificity of the immunolabelling observed should enable the antigen and/or antibody to S-100 to be used as an investigative and diagnostic tool in the study of Sertoli cell function. In the absence of actin filaments in prosobranch (mollusk) Sertoli cells, the changes in their shape are attributed to the abundant microtubules (Buckland-Nicks and Chia, 1986). However, de Jong Brink *et al.* (1977) believe in passive transportation of Sertoli cells involving myoepithelial cells and muscle cells in the acinar wall of pulmonates, rather than a mechanism involving microtubules or microfilaments.

Our knowledge about the regulation of the development of cytoskeletal elements in the Sertoli cells of invertebrates (Hinsch, 1993) and vertebrates (Pudney, 1993) is still meager. Kurohmaru *et al.* (1992) suggested a minimal influence of germ cells on Sertoli cell cytology and cytoskeleton. However, in a variety of physiological, pathological, and experimental (after hypophysectomy) situations, the evolution of the cytoskeleton (microfilaments, intermediate-sized filaments and microtubules) appears to be regulated directly or indirectly (through testosterone production) by pituitary hormones (Ghosh *et al.*, 1992; Morales and Clermont, 1993; Schulze and Holstein, 1993a,b).

Muffly *et al.* (1993) have shown that testosterone can maintain the normal spermiogenesis, the normal structure of association of ectoplasmic specialization and spermatids, and the peripheral distribution of Sertoli cell f-actin and vinculin in hypophysectomized rats if hormone replacement is begun shortly after removal of the pituitary. Testosterone can maintain the binding competency of Sertoli cells but cannot induce or restore this structural/functional characteristic of the cell. Seasonal changes in the fine structure of Sertoli cells are reported for the seasonal breeders, including mammals (Bartke *et al.*, 1993), various nonmammalian vertebrates (Pudney, 1993), and invertebrates (Hinsch, 1993); these may be the result of seasonal fluctuations in production of hormones and other factors which remain to be investigated more precisely.

III. Development of Junctional Complexes, the Blood–Testis Barrier, and Their Functional Significance

In any epithelium in the body, neighboring cells have structurally defined organelles, desmosomes that appear to hold the cells together (Larsen and Wert, 1988; Robards *et al.,* 1990). This is also true for the accessory somatic or Sertoli cells of the testis in vertebrates (Byers *et al.,* 1993a; Russell, 1993c; Pudney, 1993) and invertebrates (Hinsch, 1993). The multiplication, growth, and differentiation of Sertoli cells and germ cells in their germinal or seminiferous epithelium are closely accompanied by the development of surface specializations or complex junctional specializations which maintain within the seminiferous or germinal epithelium a highly specialized microenvironment for the development and differentiation of spermatogenic cells. This microenvironment is created by the blood–testis barrier as the Sertoli cells separate young and more mature germ cells into two compartments within the germinal or seminiferous epithelium of mammalian (Russell, 1993c; Hinton and Setchell, 1993) and nonmammalian vertebrates (Pudney, 1993; Grier, 1993) and some invertebrates (Hinsch, 1993).

A. Inter-Sertoli Cell Junctions and Blood–Testis Barrier

The formation of inter-Sertoli cell junctions parallels the cessation of Sertoli cell multiplication in the immature testis (Pelliniemi *et al.,* 1993; Gondos and Berndtson, 1993) and junctions form as spermatogenic cells proceed through the zygotene to pachytene steps of meiotic maturation in vertebrates (Gondos and Berndtson, 1993; Russell, 1993c; Grier, 1993; Pudney, 1993) and some invertebrates (Hinsch, 1993). The junctional complexes increase in numbers with the start of meiosis in spermatogonia and the formation of a tubule or cyst lumen for creating the blood–testis barrier in mammals (Russell, 1993c; Byers *et al.,* 1993a), nonmammalian vertebrates (including teleosts, amphibians, reptiles, and birds) (Pudney, 1993; Grier, 1993), and in some groups of invertebrates (Hinsch, 1993). The developmental phase at which spermatogenic cells get isolated from the systemic milieu also varies among vertebrate groups, i.e., postmeiotic spermatogenic cells for teleosts and amphibians, premeiotic spermatogenic cells for mammals. If the blood–testis barrier in the teleost and amphibian cyst is formed in response to sequestering autoantigens from the systemic circulation, this may reflect species differences in the presence and development of specific autoantigens which may influence the time of formation of the blood–testis barrier. Various morphological observations are complemented by physiological data showing that the barrier is absent at birth (Pelliniemi *et al.*, 1993; Gondos and Berndtson, 1993) or in seasonally quiescent testis (Bartke *et al.*, 1993), but with sexual maturation or with seasonal spermatogenic activity, the previously un-restricted transport of acriflavine dyes is prevented with meiotic matura-tion of the primary spermatocytes.

A great diversity is observed in the morphology, distribution, and num-bers of Sertoli cell junctions in the seminiferous or germinal epithelium of different groups of animals, including mammals (Russell, 1993c; Byers *et al.*, 1993a; Schulze and Holstein, 1993b), nonmammalian vertebrates (Pudney, 1993; Grier, 1993), and invertebrates (Hinsch, 1993), revealing the complexity of cell-to-cell contacts or communication (Enders, 1993; Skinner, 1991, 1993a). Depending upon the species, inter-Sertoli cell junc-tions may lack a subplasmalemmal layer of filaments and small tubules of smooth endoplasmic reticulum. Cytochalasin-D treatment causes alter-ations in the functional integrity of the Sertoli cell barrier and this is believed to be the result of disruption of microfilaments associated with the blood-testis barrier (Weber *et al.*, 1988).

The unique junctional specializations formed between Sertoli cells, which maintain the structural integrity of the tubular or germinal epithe-lium, create low-resistance pathways for electrical coupling of adjacent cells, and form the epithelial component of the blood–testis barrier that develops within the seminiferous epithelium, demonstrating spermato-genic activity. Sertoli cell–Sertoli cell junctions include (1) tight or occlud-ing junctions, (2) gap junctions, (3) septate-like junctions, (4) desmosomes, (5) close junctions, (6) Sertoli–Sertoli tubulobulbar complexes, and (7) ectoplasmic specializations (which are strictly not junctions). They show a great diversity in structure and distribution in mammals (Russell, 1993c; Vogl *et al.*, 1993), nonmammalian vertebrates (Pudney, 1993; Grier, 1993), and invertebrates (Hinsch, 1993). Various studies on invertebrates do not identify a particular variety of such junctions according to the classification of invertebrate intercellular junctions (Green and Bergquist, 1982) or a given preferential location between specific cell types of the intricate cell association of the seminiferous epithelium. However, by providing more details on the ultrastructural development and physiologi-cal aspects of the blood–testis barrier in nonmammalian vertebrates (Pud-ney, 1993) and invertebrates (Szöllösi, 1982; de Jong-Brink *et al.*, 1984; Bergmann *et al.*, 1984; Anelli *et al.*, 1985; Buckland-Nicks and Chia, 1986; Chia and Buckland-Nicks, 1987; Hinsch, 1993), various observations have revealed different situations in regard to structural and functional aspects of the blood–testis barrier.

The occluding tight junctions, fusing the membranes of adjacent Sertoli cells, form the structural component of the blood–testis barrier in mam-mals (Russell, 1993c; Byers *et al.*, 1993a; Hinton and Setchell, 1993), nonmammalian vertebrates (Pudney, 1993; Grier, 1993), and some groups of invertebrates such as arthropods and mollusks (Hinsch, 1992, 1993), and thereby separate the seminiferous or germinal epithelium into two compartments, namely the basal (exterior or peripheral) compartment having generally spermatogonia and preleptotene spermatocytes, and the adluminal (interior or central) compartment having meiotic spermatocytes and developing spermatids in amniotes.

It still remains to be determined at what phase of spermatogenesis Sertoli–Sertoli junctions develop in elasmobranch testis and whether physiologically they represent a blood–testis barrier. The blood–testis barrier in insects and mollusks effectively isolates the germ cells once they have entered the meiotic division (Hinsch, 1993), as clearly demonstrated for mammals. The basal compartment is contiguous with the basal lamina, myoid cells, and the blood, as well as with lymphatic vessels in the interstitial space in amniotes. The luminal space is continuous with the rete testis, efferent ducts, and epididymis in mammals. The substances observed in the basal compartment generally enter the adluminal and luminal compartments by traversing Sertoli cell cytoplasm and vice versa as judged by the physical separation between the exterior and interior compartments of the seminiferous epithelium through the development of the Sertoli–Sertoli tight junctional complexes (Setchell et al., 1994). This suggests that every substance of nutritional or regulatory nature must pass through the Sertoli cell cytoplasm to be available for germ cells placed within the adluminal compartment. This "nursing" nature of the Sertoli cells is further strengthened by the fact that late spermatid processes probe into deep recesses of the Sertoli cell cytoplasm.

Sertoli–Sertoli and Sertoli–germ cell configurational relationships in mammals have been recently determined using morphometric techniques and direct measurement (Ren and Russell, 1992; Russell, 1993a,c). Ye et al. (1993) demonstrated cyclic differences in volumes and surface areas of Sertoli cells by an electron microscope technique that proportionally sampled the Sertoli cells within the seminiferous tubule. The surface area of the Sertoli cell varies cyclically. Specific Sertoli–Sertoli and Sertoli–germ cell contacts and/or junctions, their types, numbers, and position relative to the reconstructed cell and also to adjacent Sertoli cells are revealed. The Sertoli–Sertoli junctional contact areas show a belt-like arrangement near the base of the cell. Each Sertoli cell forms a contact with the basal lamina, other Sertoli cells, and also germ cells (Russell, 1993a,c; Schulze and Holstein, 1993b; Byers et al., 1993a; Pudney, 1993; Grier, 1993; Hinsch, 1993). All of these junctional contacts serve as cohesive elements to hold the epithelium together, facilitating the interaction between different cell types. Yazama et al. (1991) have done a deep-etch visualization of the Sertoli cell (blood–testis) barrier in the boar;

ectoplasmic specializations of the Sertoli cell barrier are held together by three kinds of cross-bridging structures, providing important morphological evidence that implicates ectoplasmic specializations in the dynamic function of the microfilament bundles of the Sertoli cell barrier. The peculiar occluding zonules that form the blood–testis barrier in the nonseasonal breeder birds actually encircle the apices of the Sertoli cells although they occur in the basal third of the seminiferous epithelium (Pudney, 1993). Their functional significance remains to be determined.

Recent studies of the permeability properties of inter-Sertoli cell tight junctions in mammals have demonstrated a progressive loss of impermeability from the base to the lumen of the seminiferous tubule that is also correlated with an increasing incidence of disintegration of the junctional complexes (Russell, 1993c; Hinton and Setchell, 1993; Setchell et al., 1994). A belt-like arrangement of the Sertoli–Sertoli junctional contact areas near the base of the Sertoli cell is in agreement with limiting the passage of substances toward the tubular lumen (Russell, 1993c; Hinton and Setchell, 1993; Byers et al., 1993a; Setchell et al., 1994). The presence of a blood–testis barrier (or Sertoli cell barrier) is further supported by the fact that a number of endogenous or exogenous compounds in serum readily enter the lymphatics and yet are not present in the rete testis of mammals. The impermeable character of the Sertoli junctions is verified by their ability to exclude electron-dense compounds such as horseradish peroxidase and lanthanum and sucrose (Hinton and Setchell, 1993; Setchell et al., 1994). The blood–testis barrier, which has also been demonstrated in different groups of nonmammalian vertebrates (Grier, 1993; Pudney, 1993) and invertebrates—especially insects, mollusks, and some echinoderms (Hinsch, 1993)—is responsible for maintaining the differences in chemical composition of the tubule fluid and the blood plasma or lymph, or hemolymph. This provides a structural basis for the maintenance of a different environment in each of two compartments, which may be required during specific stages of germ cell development.

The compartmentation of the seminiferous epithelium as a result of the formation of a blood–testis barrier appears to have important consequences in relation to spermatogenesis that are yet to be understood. Compartmentation does not occur in the testes of lower chordates (Pudney, 1993; Grier, 1993) and some groups of invertebrates (Hinsch, 1993), but spermatogenesis still occurs. The factors involved in the formation and maintenance of junctional complexes of the blood–testis barrier, therefore, remain to be determined at the molecular level (Byers et al., 1993b). However, Muffly et al. (1993) have suggested that testosterone can maintain the binding competency of Sertoli cells and normal Sertoli-spermatid interaction but cannot restore them. In the rat, Sertoli cell binding competency is required for maximizing spermiogenesis by max-

imizing spermatid binding to Sertoli cells at Stage VIII of the spermatogenic cycle. Byers *et al.* (1991) have studied the development of Sertoli cell
junctional specializations and distribution of the tight junction-associated
protein ZO-1 in the mouse testis.

B. Junctional Complexes between Sertoli Cells and
Developing Germ Cells

Besides Sertoli cell interactions, the supporting cell develops specialized
junctional complexes of several kinds in the plasma membrane, which
adhere to developing germ cells in mammals (Russell, 1993c; Byers *et al.*,
1993a; Enders, 1993), nonmammalian vertebrates (Pudney, 1993; Grier,
1993), and invertebrates (Hinsch, 1993), suggesting that these environmental interactions may be required for cell attachment, association, and cell-
to-cell communication. The junctional complexes between Sertoli cells
and developing germ cells include desmosomes and gap junctions which
in their structure and distribution vary greatly with the stages of spermatogenesis. In mammals, the desmosome-like structures appear between Sertoli cells and all types of germ cells, as is also reported for reptiles and
birds (Pudney, 1993). Typical nonoccluding gap junctions and nexuses
occur between Sertoli cells and spermatids. Owing to their frequent association with desmosomes, they are occasionally called desmosome-gap-
junctional complexes (Ren and Russell, 1992; Russell, 1993c; Byers *et
al.*, 1993a). Ren and Russell (1992) have studied Sertoli cell–germ cell
desmosome gap junctions in relation to meiotic divisions in the male rat.
The results obtained have suggested that these junctions may relate more
to the period of initiation of meiosis than to its continuance. Furthermore,
a third kind of Sertoli cell–spermatid cell contact was observed: projections of spermatid processes extending deep into the cytoplasm of the
Sertoli cells (Russell, 1993c; Enders, 1993). However, such structures
are absent at the border between Sertoli cells and primary or secondary
spermatocytes of mammals. Therefore, the exchange of signals between
the two cell types appears to depend on mechanisms that do not require
direct membrane contact, but rather transport of chemical substances that
traverse the cell membranes or bind to cell surface receptors.

A "ball-and-socket-like" junction between branches of Sertoli cells and
developing spermatids has been observed in the cephalopod *Nautilus*
(Arnold, 1978); a cytoplasmic extension of a Sertoli cell fits into a pocket
in the spermatid and has a constricted neck region. The frequent occurrence of multivesicular bodies in the Sertoli cell extension and of small
vesicles in the spermatid cytoplasm in the area of the ball-and-socket-like
junction has suggested the possibility of communication materials between

the somatic and germ cells. Arnold (1978) has suggested that the junctions more likely function in the transport of nutrients and other substances from somatic (Sertoli) cells to the germ cells rather than in coordination of developmental process of spermatogenic cells.

No structural evidence has been produced to show pinocytotic activity in germ cells in previous studies, but Sertoli cells show a pronounced endocytotic activity, which is supported by the phagocytosis of residual cytoplasmic bodies and degenerating germ cells in mammals (Clermont, 1993; Morales and Clermont, 1993; Russell, 1993b), nonmammalian vertebrates (Grier, 1993; Pudney, 1993) and invertebrates (Hinsch, 1993). However, a recent study by Segretain *et al.* (1992) has demonstrated receptor-mediated and adsorptive endocytosis by male germ cells of different mammalian species, which is believed to allow germ cells to take up from the extracellular space and intercellular space within the seminiferous tubules important factors required for their division, differentiation, and metabolism. This suggestion is also supported by receptor-mediated endocytosis of testicular transferrin by germinal cells of the rat testis (Petrie and Morales, 1992). Mannose 6-phosphate-containing glycoproteins secreted by Sertoli cells are also endocytosed by spermatogenic cells (O'Brien *et al.*, 1993).

C. Ectoplasmic Specializations

Ectoplasmic specializations are reported in the Sertoli cells of mammals and some nonmammalian vertebrates (reviewed by Vogl *et al.*, 1993; Pudney, 1993). Vogl *et al.* (1993) have discussed the details of ultrastructure, chemistry, and possible functions of ectoplasmic specializations and thus a very brief account of these structures will be given here. Ectoplasmic specializations are composed of a dense band of actin-rich filaments sandwiched between the Sertoli cell plasma membrane and a cistern of endoplasmic reticulum, resembling one half of the paired ectoplasmic specializations that constitute the inter-Sertoli cell tight junctions at the base of the Sertoli cells. Their structure at Sertoli–Sertoli and Sertoli–spermatid contacts is the same except that (1) the cytoplasmic surface of the endoplasmic reticulum in basal sites shows ribosomes attached to it whereas that in apical regions generally does not; (2) the intercellular contact area with a narrowing of the intercellular space to between 70 and 100 Å shows prominent gap junctions at basal sites whereas such junctions generally do not occur at the apical locations; and (3) ectoplasmic specializations in basal zones develop on both sides of the intercellular contact area whereas those in relation to spermatids are formed only on the Sertoli cell side. The absence of ectoplasmic specializations facing pachytene cells suggests that they are needed later in spermatogenesis (Vogl *et al.*, 1993; Russell, 1993a,c).

The ectoplasmic specializations of nonmammalian vertebrates differ from those of mammalian Sertoli cells in some aspects (Vogl *et al.*, 1993; Pfeiffer and Vogl, 1993). The ectoplasmic specializations of nonmammalian vertebrates apparently occupy only the Sertoli cell zones associated with spermatids, as evidenced by their absence in basal sites related to Sertoli–Sertoli cell junctions as is found in mammals. Actin filaments do not form paracrystalline arrays similar to those of mammals because they are distributed more loosely in bundles or form networks (Pudney, 1993). The ectoplasmic specializations of nonmammalian vertebrates appear to contain myosin II (Pfeiffer and Vogl, 1990, 1993), which is absent in those of mammals (Vogl *et al.*, 1993). This is further supported by the contraction of detergent-extracted ectoplasmic specializations of the turtle but not of mammals.

Pfeiffer and Vogl (1993) have suggested that ectoplasmic specializations in eutherian mammals appear to have evolved from actin-associated adhesion junctions in which actin bundles were initially contractile but from which myosin II was lost. The endoplasmic reticulum does not form a consistent characteristic of ectoplasmic specializations in nonmammalian vertebrates (Pfeiffer and Vogl, 1993; Vogl *et al.*, 1993), but it forms a specific feature in mammals. The features of the actin complexes in ectoplasmic specializations of nonmammalian vertebrates are believed to be similar to those of comparable complexes associated with intracellular adhesion junctions occurring generally in epithelial cells (Vogl *et al.*, 1993; Pfeiffer and Vogl, 1993). Ectoplasmic specializations between developing spermatids and Sertoli cells have not yet been described in the testis of teleosts and urodeles (Pudney, 1993), suggesting that these are not required because spermatids do not form intimate physical relationships with Sertoli cells. However, a homogeneous, low-density material is deposited between plasma membranes of the spermatids and Sertoli cells.

Ectoplasmic specializations are apparently involved in adhesion of Sertoli cells to germ cells, (2) in structural support for the Sertoli cell during germ cell mobilization, (3) in participation in sperm release and acrosome shaping, (4) in contraction, and (5) in the development of sites of intercellular attachment (Vogl *et al.*, 1993; Russell, 1993a,b,c; Morales and Clermont, 1993). Vogl *et al.* (1993), after discussing the evidence in regard to the possible functional significance of ectoplasmic specialization, have proposed that these facilitate intercellular attachment or adhesion and help in positioning spermatogenic cells within the seminiferous epithelium, relating them to "maintaining and regulating cell adhesion" and not to establishing permeability junctions. The presence of actin together with vinculin at these sites is also in agreement with this hypothesis. Extensive cross-linking and hexagonal packing of the actin filaments are believed to contribute to the reinforcement of membrane adhesion domains. In apical sites, the ectoplasmic specializations appear with the microtubules (Redenbach *et al.,* 1992) and intermediate filaments maintain the position of the spermatid in the seminiferous epithelium. Further investigations also remain to be carried out to clarify the origin and fate of ectoplasmic specializations. The factors regulating the differentiation and degradation of ectoplasmic specializations and their suggested functions in germ cell movement, sperm release, and structural support of the Sertoli cell remain to be determined more precisely at the molecular level (Vogl *et al.,* 1993; Russell, 1993b,c).

Very few attempts or none have been made to determine whether ectoplasmic specializations comparable to those of mammals and nonmammalian vertebrates also occur in invertebrates since some of these also show a complex nature for somatic (Sertoli)–germ cell contacts (Hinsch, 1993). From this comparative account of ectoplasmic specializations, it can be suggested that these specializations have evolved to solve the unusual mechanical problem of maintaining the functional integrity of the seminiferous epithelium in the face of constantly changing morphology of the Sertoli cells as well as to help in positioning spermatogenic cells within the epithelium.

D. Tubulobulbar Complexes and Other Specializations

Tubulobulbar complexes are known to form the characteristic feature of the mammalian seminiferous epithelium although species variations occur in their number, position, and morphology. For a detailed discussion of their development, structure, and function, see the excellent review by Vogl *et al.* (1993); here these are described very briefly.

Tubulobulbar complexes form specialized projections that penetrate into Sertoli cells from neighboring cells and consist of the invaginated Sertoli cell membrane and associated substructure. They develop apically in association with elongate spermatids before spermiation in mammals, thus attaching spermatids to the seminiferous epithelium (Russell, 1993b). Tubulobulbar complexes also develop basally between adjacent Sertoli cells at the level of the blood–testis barrier and occasionally constitute tight and gap junctions in the zone of interdigitation (Vogl *et al.,* 1993). The complexes at both the sites develop at specific stages of spermatogenesis in zones associated with or previously occupied by ectoplasmic specializations in regions of intercellular attachment. Two rows of tubulobulbar complexes develop along the edges of the concave surface of the hook-shaped spermatid head in the rat. Each complex forms an elongated tubular segment terminating in bulbous dilation (a structure in the form of a coated pit develops at the end of the structure, and a cistern of endoplasmic reticulum occurs immediately beneath the Sertoli cell plasma membrane in zones of the bulb). A dense network of actin filaments lying in the Sertoli cell cytoplasm surrounds the tubular parts of the complex. The structure of the tubulobulbar complexes that develop between neighboring Sertoli cells basically resembles that of complexes formed between elongate spermatids and Sertoli cells except that the apical complexes are longer than the basal ones, which also show profiles of tight and gap junctions between adjacent membranes.

Most of the basal complexes degrade one or more days before the upward movement of spermatocytes (Russell, 1993b,c). The development of several generations of tubulobulbar complexes in the apical regions and their degradation in succession by the Sertoli cells indicates that a proportion of spermatid cytoplasm resorbs through the degradation of numerous tubulobulbar complexes (Kojma, 1990; Vogl et al., 1993; Russell, 1993b,c). The flow of spermatid cytoplasm into these complexes also possibly triggers the sperm release mechanism, as discussed by Russell (1993b). The filaments and microtubules of Sertoli cell cytoplasmic projections that surround and invaginate the residual cytoplasm apparently pull the residual cytoplasm from the lumen to the apical trunk of the Sertoli cell (Vogl et al., 1993; Russell, 1993b; Morales and Clermont, 1993).

The development of tubulobulbar complexes between the Sertoli cell and spermatids is also suggestive of the transfer of some materials between them. They are also believed to serve the function of internalizing junctions prior to two major phenomena in the seminiferous epithelium—release of sperm and shifting of spermatocytes from basal to adluminal compartments, which is supported by complexes at both apical and basal sites (Vogl et al., 1993).

A complex morphological relationship between Sertoli cells and elongating spermatids is observed in certain lizards—*Cnemidophorus lemniscatus lemniscatus* and *Anolis carolinensis*—and the turtle *P. scripta* (Pudney, 1993). During their spermiogenesis, a projection of Sertoli cytoplasm develops an accessory cap which surrounds the developing acrosome and is in turn covered by spermatid cytoplasm. The Sertoli accessory cap or mantle lacks organelles except for, depending on the species, fine filaments or microtubules. In *P. scripta* the Sertoli cytoplasm constituting the accessory cap forms ectoplasmic specializations in areas adjacent to the spermatid (Sprando and Russell, 1988). Sertoli cell processes also invade the cytoplasm of elongating spermatids in the turtle and these are apparently involved in sequestering and removing portions of spermatid cytoplasm, thus forming a mechanism for decreasing the cytoplasmic volume of spermatids as they undergo maturation.

As in reptiles, the avian Sertoli cell also forms cytoplasmic processes which penetrate spermatid cytoplasm (Sprando and Russell, 1988). While these processes sequester portions of cytoplasm, unlike in reptiles, no evidence exists to show that these areas of spermatid cytoplasm are removed by Sertoli cells. A Sertoli accessory cap or mantle associated with mature spermatids is also reported. As in reptiles, the Sertoli cytoplasm of this cap lying adjacent to the spermatid forms ectoplasmic specializations. Tubulobulbar complexes and other specializations are not reported in invertebrates (Hinsch, 1993), suggesting that they originated in mammals with the evolution of a complex structural organization of seminiferous epithelium.

E. Translocation of Germ Cells from the Basal to the Adluminal Compartment

The multiplication and renewal of spermatogonia occur in the peripheral (basal) compartment of the mammalian seminiferous epithelium, while the spermatogonia and preleptotene spermatocytes lie close to the basal lamina and below the occluding, intercellular Sertoli junctions (Morales and Clermont, 1993; de Kretser and Kerr, 1994). The type B spermatogonia divide to produce preleptotene spermatocytes, which undergo a final replication of nuclear DNA before embarking on the prolonged meiotic prophase (Guraya, 1987). In the leptotene and zygotene stages of meiotic prophase, the spermatocytes translocate across the Sertoli junction into the adluminal (central) compartment of the seminiferous epithelium (Guraya, 1987; Morales and Clermont, 1993; Russell, 1993c; de Kretser and Kerr, 1994).

The precise way in which the germ cells ascend from the basal to the adluminal compartment may reveal species differences in the way the barrier maintained by the Sertoli cell tight junctions (Russell, 1993c; Byers et al., 1993a; de Kretser and Kerr, 1994) is broken. Various investigations have suggested that the Sertoli cell junctions do not form permanent structures but open from time to time, possibly in response to the action of some enzymes, to facilitate the passage of the preleptotene spermatocytes. The production of plasminogen activator and other protease activities by Sertoli cells (Fritz et al., 1993) is believed to be required in several aspects of tissue remodelling associated with translocation of spermatocytes and degradation of junctional complexes, and in release of mature spermatozoa into the lumen of the tubule. However, the function and regulation of the proteolytic activities in the seminiferous tubule remain to be determined precisely. In the early stages of the development of a tight junction beneath migrating germ cells, an increasing number of intramembranous junctional strands grow in length and start assuming parallel orientations, thus making a collective contribution to an increasing degree of continuity, culminating in a typical junctional complex. The progressive translocation of differentiating germ cells from the basal zone of the seminiferous epithelium to the tubular lumen appears to involve complex surface interactions between Sertoli and germ cells (Enders, 1993; Russell, 1993c) which still need to be more precisely understood at the molecular level, not only for mammals but also for nonmammalian vertebrates and invertebrates. The precise mechanisms regulating the passage of germ cells through the blood–testis barrier are yet to be resolved. It also remains to be determined by which mechanisms mature spermatozoa detach from Sertoli cells at the time of spermiation (Russell, 1993b).

Abundant actin filaments, found in the Sertoli cell cytoplasm between the subsurface cisternae and the junctional complexes (Vogl et al., 1993), may be involved in inducing conformational alterations in the Sertoli cell and thereby facilitate the intercellular centripetal translocation of spermatocytes toward the lumen of the tubule, which may also be accompanied by a constant breaking of old connections and a making of new ones. Vogl et al. (1993) have suggested that ectoplasmic specializations are not only involved in maintaining intercellular attachment per se, but play a direct role in the basic processes by which spermatocytes move through Sertoli cell junctions at the base of the seminiferous epithelium and by which mature spermatozoa are finally released from Sertoli cells at the end of spermiogenesis (Russell, 1993b). Cameron and Muffly (1991) have demonstrated that maximal spermatid binding to Sertoli cells in vitro requires follicle-stimulating hormone (FSH) and testosterone and is associated with peripheral distribution of actin and viniculin.

Developing germ cells show specific surface receptors, possibly glycoproteins, involved in cell–cell adhesion (Byers et al., 1993b; Enders, 1993). The multiple antigenic surface macromolecules observed on spermatogonia, spermatocytes, and spermatids (Guraya, 1987) may be of great significance in their centripetal translocation but whether any of these are involved in intercellular adhesion needs to be investigated. Bellvé (1979) believes that any array of "adhesive" surface receptors could greatly facilitate the orderly transfer of preleptotene spermatocytes to the adluminal compartment of the mammalian seminiferous epithelium and that this process certainly involves coordinate movement of large syncytial clusters of germ cells intercalated between numerous Sertoli cells (Morales and

Clermont, 1993). The cell adhesions are not static; rather, dynamic changes in cell adhesive interactions may occur. The balance between various modes of adhesion between Sertoli cells and germ cells may change from one moment to the next, governing the translocation of germ cells toward the center of the tubule. Pratt *et al.* (1993) have suggested that cell surface β1, 4 galactosyl transferase (Gal Tase) on primary spermatocytes facilitates their initial adhesion to Sertoli cells *in vitro,* possibly by binding the glycoside ligands on the Sertoli cells, which show surface-associated glycosyl transferase activities *in vitro* (Raychoudhury and Millette, 1993). It is possible that Gal Tase functions in the initial reversible adhesion of spermatocytes to Sertoli cells that permits pachytene spermatocytes to change positions along the Sertoli cell in order to assume new positions further toward the center of the tubule.

A depression of spermatogenesis in men of diverse fertility status is related to an increase in the incidence of tubules with apparently defective tight junctions (Schulze and Holstein, 1993a). The status of the blood–testis barrier and the integrity of Sertoli occluding junctions after experimental treatment or in animals with genetic disorders has been investigated by several workers (Russell and Peterson, 1985; Russell, 1993c). The functional integrity of Sertoli junctions *in vitro* is more difficult to resolve than their integrity in the intact animal (Steinberger and Jakubowiak, 1993). However, dissociated Sertoli cells from juvenile animals maintained in culture are found to develop tight junctions (Russell and Peterson, 1985; Russell, 1993c).

IV. Functions

The precise functions of the somatic accessory cells of Sertoli cells in relation to regulation of spermatogenesis still remain to be determined precisely since their initial discovery and description by Enrico Sertoli in 1885. However, recent investigations have suggested that they perform several functions in the physiology of the seminiferous or germinal epithelium, as indicated by their ultrastructural, histochemical, and biochemical features. These functions include (1) provision of nutrient substances to developing spermatogenic cells, (2) maintenance of ionic composition, (3) phagocytosis of residual cytoplasmic bodies and degenerated spermatogenic cells, (3) synthesis of steroid hormones, (4) secretion of proteins, and (5) spermiation. Some of the proposed functions of Sertoli cells are based on the concept of the blood–testis barrier.

A. Provision of Nutrient Materials to Developing Spermatogenic Cells

The close anatomical association of somatic accessory cells with developing spermatogenic cells as discussed in Sections II and III has suggested that they function as nurse cells by providing nutrient materials to developing germ cells; this is supported by cyclic (Morales and Clermont, 1993) and seasonal changes (Bartke *et al.*, 1991, 1993). For the more advanced germ cells, the blood–testis barrier makes it possible to transfer nutrient materials through the Sertoli cells. Most of the evidence produced to support the provision of nutrient materials to developing spermatogenic cells is based on morphological and histochemical data which have revealed the presence of glycogen, lipids, etc. in the Sertoli cells of vertebrates and invertebrates, as described in Section II,A. They appear to constitute the major nutrient substances for the proliferating and developing germ cells. The substances derived from phagocytized residual bodies and degenerated germ cells by Sertoli cells in vertebrates and invertebrates are also believed to be used to supply nutrients to spermatogenic cells for their proliferation and development and survival (Morales and Clermont, 1993; Pudney, 1993; Hinsch, 1993).

In response to FSH, insulin, and insulin-like growth factor stimulation, Sertoli cells of mammals produce considerable amounts of lactate from glucose (Bartke *et al.*, 1991, 1993; Newton *et al.*, 1992; Skinner, 1993a), a metabolic substrate preferred by advanced spermatogenic cells. Localization of metabolic enzymes such as ATPase in the tubule and the observation that Sertoli cells produce glucose metabolites such as inositol suggest that Sertoli cells are involved in the transport of energy metabolites to spermatogenic cells. Both pyruvate and lactate are well known to support germinal cell metabolism and provide an efficient energy source for germ cells (Skinner, 1991, 1993a). Demonstration of metabolic enzymes such as lactate and malate dehydrogenases and the actions of FSH on glucose transport by Sertoli cells also support this suggestion. Other cellular metabolites, such as coenzymes, nucleotides, and amino acids may also need a nutritional interaction between Sertoli cells and spermatogenic cells to maintain spermatogenesis (Skinner, 1991; Skinner *et al.*, 1991).

Besides transport of small, soluble, nutrient molecules through junctional complexes, a large number of substances need binding proteins to facilitate their transport owing to low solubilities or reactive chemical properties (Sylvester, 1993). Lipids also need a binding protein for transport and in this regard sulfated glycoprotein-1 (SPG-1) is of great significance because the transport of lipid precursors and specific fatty acids between Sertoli cells and germinal cells may be required to support the rapid expansion of the germ cell population during spermatogenesis. Further investigations remain to be made of lipid transport and the binding proteins involved (Skinner, 1991).

Developing spermatogenic cells also need vitamins (Kim and Wang, 1993) and Sertoli cells must develop a mechanism to transport them to germinal cells. Sertoli cells produce vitamin-binding proteins such as folate-binding protein, biotin-binding protein, and retinoid-binding proteins (Skinner, 1991, 1993a; Kim and Wang, 1993). Further investigations are required to define the precise mechanisms of transport of various types of nutrients from Sertoli cells to germinal cells, not only in mammals but also in other groups of animals that show a great diversity in the interrelationship of somatic accessory cells with germinal cells, as discussed in Section II.

The somatic accessory cells and germinal cells in some groups of invertebrates form associations similar to those of reptiles, birds, and mammals (amniotes), suggesting the strong possibility of carrying out several of the potential functions of amniote Sertoli cells, which include (1) synthesis and transport of nutrients and regulatory substances to the developing germ cells, (2) coordination, movement, and mechanical support of germ cell development, and (3) phagocytosis of residual cytoplasmic bodies and degenerated spermatogenic cells and their utilization as nutrient materials (Hinsch, 1993). However, the mechanisms of synthesis and release of nutrients from somatic cells and of their subsequent transport and utilization by germinal cells remain to be investigated at the cellular and molecular levels.

1. Maintenance of Ionic Composition of Tubular Fluid

Sertoli cells are strategically located to create through their metabolic secretory activities a special environment that promotes germ cell differentiation. However, the precise mechanisms involved remain to be defined. The special environment is the result of the secretion of fluid which starts being produced at about 18 days of age in the rat when lumen development is first observed. The blood–testis barrier described in Section III,A appears to regulate the molecular composition of the intercellular fluid in the adluminal compartment (Hinton and Setchell, 1993; Setchell et al., 1994). This may also be true for those groups of invertebrates and vertebrates who show the development of a blood–testis barrier (see Section III,A).

The maintenance of intercellular ionic composition is a function of the Sertoli cell cytoplasm and the occluding inter-Sertoli cell junctions (Hinton and Setchell, 1993; Setchell et al., 1994). Regulation of the secretion and composition of tubular fluid has not yet been studied in much detail, even in mammals (Hinton and Setchell, 1993; Setchell et al., 1994). Actually, more information is available on the characterization of fluid obtained from the rete testis of the rat than directly from its seminiferous tubules. About nine-tenths of the rete fluid is secreted directly through the epithelium of the rete testis. In its ionic composition the rete fluid more closely resembles blood plasma than the "primary" seminiferous tubule fluid. Therefore, the latter must be investigated in detail if the Sertoli cell contribution is to be understood more precisely (Hinton and Setchell, 1993; Setchell *et al.*, 1994).

When taking into account the vast differences in ionic composition of the fluids on the opposite sides of the seminiferous epithelium, it can be assumed that this specific intratubular milieu is required for normal maturation of the postmeiotic germ cells and that interference with this ionic milieu may damage spermatogenesis; our knowledge is meager in this regard (Hinton and Setchell, 1993; Setchell *et al.*, 1994). Studies also remain to be carried out on the comparative ionic composition of testicular fluid in different groups of animals, which show a great diversity in the structural organization of germinal epithelium in their testes or gonads, especially in regard to interrelationships between somatic accessory cells and spermatogenic cells (Section II) (Pudney, 1993; Hinsch, 1993; Grier, 1993). Comparative studies of its ionic composition will certainly help us to determine its precise functions in maintaining spermatogenesis in different groups of animals with diverse testicular architecture.

C. Phagocytic Activity and Its Functional Significance

The somatic accessory cells or Sertoli cells in the testes of various groups of animals that have been investigated so far are phagocytic, eliminating from the germinal or seminiferous epithelium residual bodies and degenerating germ cells and ingesting particulate matter (Morales and Clermont, 1993; Russell, 1993b; Pudney, 1993; Hinsch, 1993). The residual bodies, which consist of a variable amount of lipid droplets, ribosomes, mitochondria, and various cytoplasmic membranes, are phagocytized by the Sertoli cells into the cytoplasm and then transported from the luminal into a more peripheral position of germinal epithelium, especially in vertebrates.

Phagocytic activity, which forms the characteristic feature of Sertoli cells, is generally greatly accelerated following epithelial damage and massive degeneration of germ cells in different groups of animals and under variable physiological and pathological conditions (Boekelheide, 1993; Chakraborty, 1993; Schulze and Holstein, 1993a); this further supports the phagocytic function of their somatic accessory cells. The Sertoli cells of seasonally breeding mammals (Bartke *et al.*, 1991, 1993), nonmammalian vertebrates (Pudney, 1993), and invertebrates (Tikku and Saxena,

1990; Hinsch, 1993) show conspicuous variations in phagocytic activity, which is closely related to seasonal spermatogenic activity. Endocytotic activity is common in the cytoplasm of apical cells in moths and locusts (Szöllösi and Marcaillou, 1977, 1980).

The precise regulation of phagocytic activity of Sertoli cells remains to be determined (Clermont et al., 1987; Morales and Clermont, 1993). However, Stefanini et al. (1988) suggested that FSH substantially inhibits the phagocytic activity of these cells in a dose-dependent fashion in in vitro cultures of mammalian Sertoli cells and that the effect is mediated by increased levels of cAMP. Although Sertoli cells can phagocytize in the absence of testosterone stimulation, the possibility cannot be ruled out that testosterone or other hormones may modulate the phagocytic activity of Sertoli cells. The presence of Sertoli cell-conditioned media significantly increases the internalization of beads, suggesting the possible involvement of some Sertoli cell secretory products in this phenomenon; the nature and molecular mechanism of action of such substances remain to be determined (Clermont et al., 1987; Morales and Clermont, 1993). However, residual bodies must be freed from the released spermatozoa, remain attached to the Sertoli cells, and subsequently be phagocytized and metabolized by the Sertoli cells (Clermont et al., 1987; Morales and Clermont, 1993; Russell, 1993b).

Since phagocytic activity forms a specific feature of somatic accessory cells in diverse groups of animals showing different patterns of spermatogenesis (Pudney, 1993; Hinsch, 1993), the causative factors may, therefore, be variable and thus remain to be determined through in vitro experiments (Pineau et al., 1991). Co-culture models provide a useful system for studying phagocytic activity by Sertoli cells. Peritubular cells and FSH can modulate the interaction between Sertoli cells and residual bodies and different steps of residual body disposal can be reproduced by co-culture. Since phagocytic activity forms a specific feature of somatic accessory cells in the various groups of animals investigated, the nature and functions of products formed from this activity may be variable, depending upon the species; this remains to be determined.

D. Steroid Hormone Synthesis

Although the presence of organelles (smooth endoplasmic reticulum and mitochondria with some tubular cristae), enzyme systems, and cholesterol-positive lipid droplets, which are believed to be indicative of steroidogenesis, has been reported in Sertoli cells or somatic accessory cells of vertebrates (Guraya, 1968, 1976, 1987; Nagahama, 1986; Lofts, 1987; Pudney, 1993; Sourdaine and Garnier, 1993; Dorrington and Khan, 1993) and some invertebrates (Saxena and Tikku, 1989; Hinsch, 1993) (see Section II), so far no direct evidence has been produced to show that these organelles and cholesterol-positive lipids are utilized by their Sertoli cells for steroid biosynthesis. However, the mammalian Sertoli cells are observed to synthesize testosterone from exogenous progesterone, and to aromatize exogenous testosterone to estradiol-17β *in vitro* under the influence of FSH (Lambert *et al.,* 1991; Dorringon and Khan, 1993). This occurs only in young animals (i.e., before 20 days in the rat) but not in the adult (Guraya, 1980). Aromatase is also highest in young animals and is highly responsive to FSH (Rosselli and Skinner, 1992; Dorrington and Khan, 1993). Aromatase activity is found to be high and FSH-responsive in prepubertal Sertoli cells and to decline and be nonresponsive to FSH in late pubertal Sertoli cells (Rosselli and Skinner, 1992). FSH is the only hormone found to influence aromatase activity and estrogen production.

The paracrine factor (termed PMods and produced by testicular peritubular cells) alone does not affect aromatase activity at any point in the developmental stages (Rosselli and Skinner, 1992). However, it suppresses the ability of FSH to stimulate aromatase activity and estrogen production in midpubertal Sertoli cells, suggesting that PMods may promote Sertoli cell differentiation to a more adult stage of development that is less responsive to FSH in stimulating aromatase activity.

The estradiol production of cultured Sertoli cells in response to FSH constitutes a bioassay for this hormone (Dorrington and Khan, 1993; Griswold, 1993a). FSH-induced estradiol secretion in rat Sertoli cells is affected by protein kinase C activity (Lambert *et al.,* 1991). FSH-induced estradiol secretion from immature rat Sertoli cells is modulated by intra- and extracellular Ca^{2+} (Talbot *et al.,* 1991). The results of various investigations suggest that Sertoli cells do not make a significant contribution to the androgen pool in the blood of adult mammals since no convincing evidence for *de novo* testosterone synthesis in their testicular cell types other than the Leydig cells has been produced (Dorrington and Khan, 1993) and this suggestion still holds good for other vertebrates (Guraya, 1976, 1979, 1994; Lofts, 1987; Nagahama, 1986; Pudney, 1993) and invertebrates (de Jong-Brink *et al.,* 1981; Buckland-Nicks and Chia, 1986; Hinsch, 1993; Saxena and Tikku, 1989). However, this does not rule out a function for the seminiferous tubular cells of vertebrates—which contain dehydrogenases and other enzyme systems related directly or indirectly to steroid biosynthesis and metabolism—in utilizing steroid hormones produced in the Leydig cells for further steroid synthesis.

Recently Sourdaine and Garnier (1993) have produced convincing evidence by demonstrating stage-dependent modulation of Sertoli cell steroid production in the dogfish (*Scyliorhinus canicula*). Progesterone is the major steroid in seminiferous lobules at all stages of spermatogenesis except in lobules with spermatogonia, in contrast to the observations of Sourdaine *et al.* (1990), who found testosterone as the predominant steroid at all stages of spermatogenesis. Progesterone, 4-androstenedione, testosterone, and 17α-hydroxy, 20β-dihydroxy progesterone concentrations were highest in lobules with late spermatids, whereas dihydrotestosterone concentrations decreased during spermatogenesis. Isolated seminiferous lobules at all stages can synthesize steroids from hydroxycholesterol, and lobules with late spermatids showed the highest basal contents of androstenedione and testosterone. This was also suggested in several previous studies (Sourdaine and Garnier, 1993), indicating a relationship between these steroids and spermatogenesis. The responsiveness of the lobules to dibutyryl cAMP also varied according to the stage of spermatogenesis and to the steroid assayed.

Various studies have suggested that germ cells apparently markedly influence the steroidogenesis of Sertoli cells in the adult dogfish testis. However, the molecular aspects of this influence remain to be defined in future investigations. Sertoli cell steroid production in dogfish lobules is also in agreement with previous ultrastructural and histochemical data which demonstrated organelles and enzyme systems related to steroidogenesis in Sertoli cells (Sourdaine and Garnier, 1993). The presence of enzymes necessary for the biosynthesis of progesterone, androstenedione, and testosterone from cholesterol is in agreement with the observation of Dubois and Callard (1989) showing that 3β-hydroxysteroid dehydrogenase and hydroxylase/lyase activities are present in Sertoli cells of *Squalus acanthias* testis at all stages of spermatogenesis. In contrast to steroidogenesis of seminiferous lobules in the dogfish, the rat seminferous tubules (or Sertoli cells) cannot convert cholesterol to androgens and pregnenolone to progesterone (Dorrington and Khan, 1993).

The precise physiological meaning of possible testosterone synthesis and testosterone metabolism by the Sertoli cells of vertebrates remains to be determined. However, it is possible that either testosterone or its metabolites may influence some biological activities within the Sertoli cells (Sar *et al.*, 1993), indicating autocrine function; be made available to the developing germ cells, suggesting a paracrine role; or exert their effect outside the seminiferous tubules, performing an endocrine function, especially on the rete testis or epididymis. Various investigations have suggested that the pattern of testosterone metabolism in the seminiferous tubule differs greatly from that in the accessory sex organs (Dorrington and Khan, 1993). 5α-Reductase activity is higher in testes from younger animals and is stimulated by FSH and decreased by luteinizing hormone. It is observed in spermatocytes, spermatids, and Sertoli cells. Dihydroxytestosterone forms the major 5-reductase product of testosterone in spermatocytes but testosterone is further metatolized to 5α-androstene-3α,

17β-diol (Dorrington and Khan, 1993). In contrast to regulation of steroid conversions in the rat seminiferous epithelium by pituitary gonadotrophins (Dorrington and Khan, 1993), the production of androgens by the testis of elasmobranchs is independent of the pituitary to some extent (Sourdaine and Garnier, 1993), suggesting the possible effects of germ cells in the control of steroid production in their testicular lobules. This suggestion needs to be supplemented by further investigations.

Saxena and Tikku (1989) have suggested that the insect *Dysdercus testis* has apparently the capability to synthesize proteins and steroids by cyst or interstitial cells and spongy cells respectively, for an uninterrupted spermiogenesis. This suggestion, which was made on the basis of their ultrastructural features, has been supported by *in vitro* experiments in *Mamestra brassicae* which have indicated that both the proteins and ecdysteroids are needed for an uninterrupted spermiogenesis (Shimizu and Yagi, 1982; Shimizu *et al.*, 1985).

Since the steroid-synthesizing capacity of the gonad of the pulmonate mollusk *L. stagnalis* is rather low, it is believed that the steroid(s) produced by the Sertoli cells (has) have only a local hormonal effect in the gonad, e.g., on the maintenance and completion of spermatogenesis (paracrine function) (de Jong-Brink *et al.*, 1981). The stimulating (hormonal) effect of the lateral lobes, small ganglia attached to the cerebral ganglia, on spermatogenesis in maturing young snails may be exerted via an increase in the production of steroids in the Sertoli cells. Further comparative *in vitro* biochemical studies should be carried out on the somatic accessory cells or Sertoli cells in different groups of animals to determine their steroidogenic activity and its regulation and functional significance, as well as their role in steroid metabolism. The results of such comparative studies would also be very useful in defining the autocrine, paracrine, and endocrine roles of steroid hormones in maintenance of spermatogenesis in diverse groups of animals that have testes (or Sertoli cells) with variable morphological organizations (Section II) (Grier, 1993).

E. The Sertoli Cell Secretory Proteins and Their Regulation and Functions

The presence of variable amounts of granular endoplasmic reticulum, free ribosomes, and polysomes in the cytoplasm of somatic Sertoli cells as well as their nuclear characteristics in mammals, nonmammalian vertebrates, and invertebrates (Section II,B) are indicative of protein synthesis. The secretion of a variety of peptides and proteins by Sertoli cells is now well documented in *in vitro* biochemical studies for mammals (Griswold, 1988, 1993b; Griswold *et al.*, 1988; Skinner, 1991, 1993a,b; Skinner *et al.*,

1991; Sylvester, 1993; Spiteri-Grech and Nieschlag, 1993; Fritz *et al.*, 1993; Parvinen, 1993) and some nonmammalian vertebrates (Callard and Mak, 1988; LeGac and Loir, 1988; Foucher and LeGac, 1989) but not for comparable somatic cells of most nonmammalian vertebrates (Pudney, 1993; Grier, 1993) and invertebrates (Hinsch, 1993).

The glycoproteins secreted by the mammalian (rat) Sertoli cells have been grouped by their biochemical properties. The transport, binding, and bioprotective proteins include metal ion-binding and transport proteins (e.g., transferrin, ceruloplasmin, and SPARC (secreted protein that is acidic and rich in cysteine), lipid-binding and transport proteins (e.g., sulfated glycoprotein-2 and prosaposin), and hormone- and vitamin-binding proteins (e.g., androgen-binding protein, insulin-like growth factor-binding proteins, and vitamin-binding proteins). Secreted binding and transport proteins are responsible for the movement of nutrients, vitamins, hormones, and wastes to and from the dividing and developing germ cells (Davis and Ong, 1992). Proteases and protease inhibitors of Sertoli cell origin play an important role in tissue remodeling processes during spermiation and movement of preleptotene spermatocytes into the adluminal compartment (Fritz *et al.*, 1993). Müllerian-inhibiting substance (MIS), inhibin, and activin are the regulatory proteins which are secreted in very low abundance and will carry out their biological and biochemical roles. The basement membrane glycoproteins form the basal lamina between the Sertoli cells and peritubular cells which interact with each other through paracrine factors, including PMods (secreted by peritubular cells) and various growth factors (Skinner *et al.*, 1991; Skinner, 1993b).

In recent years there has been significant progress in the identification and characterization of novel regulatory polypeptides (generally called growth factors), and in the delineation of their molecular mechanism of action (Waterfield, 1990; Skinner *et al.*, 1991; Skinner, 1991, 1993a; Syed *et al.*, 1993; Spiteri-Grech and Nieschlag, 1993; Onoda and Suarez-Quían, 1994). Various growth factors, which function primarily through local paracrine and autocrine pathways, are also secreted by Sertoli cells. Growth factors are believed to regulate somatic and germ cell division, but recent investigations have shown that growth factors also influence differentiated cell function (Teerds and Dorrington, 1992).

In the mammalian testis, the growth factors, which are implicated in testicular paracrine regulation of spermatogenesis and Leydig cell function, include seminiferous growth factor (SGF), epidermal growth factor, acidic and basic fibroblast growth factors (αFGF and βFGF), somatomedin C/insulin-like growth factors I and II (IGF-I and II), Sertoli cell-secreted growth factor (SCSGF), transforming growth factor α and β (TGFα, TGFβ), β-nerve growth factor (β-NGF), platelet-derived growth factor (PDGF), and interleukin-1 (Il-I) (Bellvé and Zheng, 1989; Skinner *et al.*, 1991; Skinner, 1991, 1993a; Spiteri-Grech and Nieschlag, 1993). Their secretion is also attributed to the Sertoli cells as well as to the peritubular moyid cells based on direct biological assays, biochemical characterization and purification, and/or specific immunological and molecular probes. The most important prerequirements for the identification of growth factors involved in the regulation of the seminiferous epithelial compartment are (1) the demonstration of the synthesis and secretion of the peptide in the compartment, (2) demonstration of an effect or effects on differentiated cell function or somatic or germ cell division or on both, and (3) localization of receptors on the suggested target cells.

The cyclically produced proteins include plasminogen activator, transferrin, clusterin, growth factors, including the somatomedin-like molecule, and proteins S_2, S_5, T_8, T_9 (Griswold, 1988, 1993b; Griswold *et al.*, 1988; Parvinen, 1993; Sylvester, 1993; Zwain *et al.*, 1993). Sertoli cell proteins were and will continue to be employed as probes for studying Sertoli cells and their interactions with the multiple-germ cell types in the adluminal and basal compartments of the seminiferous epithelium (Skinner, 1991, 1993a; de Kretser *et al.*, 1991; Skinner *et al.*, 1991; Stahler *et al.*, 1991; Cheng *et al.*, 1992; Sylvester, 1993; Spiteri-Grech and Nieschlag, 1993). The molecular mechanisms of action of various secreted proteins, including growth factors, in relation to cell-to-cell interactions in the mammalian testis are of great current interest. Accessory somatic (or Sertoli) cell proteins (including growth factors) and their roles in cell-to-cell interactions remain to be investigated for the testes of nonmammalian vertebrates and invertebrates. The results of such comparative studies on the Sertoli cell secretory proteins and their autocrine and paracrine roles will be very useful for a better understanding of the regulation of different patterns of spermatogenesis in animals.

The details of regulation of protein secretion products will be omitted here because these are described in recent papers (Rosselli and Skinner, 1992; Kanganiemi *et al.*, 1992; Karzai and Wright, 1992) and reviews (Griswold, 1993b,c; Sar *et al.*, 1993; Kim and Wang, 1993; Bardin *et al.*, 1994). FSH, testosterone, insulin, and vitamins are the known regulators that affect the secretory function of Sertoli cells. These cells secrete many serum- and testis-specific proteins as a result of mRNA expression, which is of great current interest (Griswold *et al.*, 1988; Lönnerberg *et al.*, 1992; Choongkittaworn *et al.*, 1993). mRNA expression, protein synthesis, and secretion of proteins by Sertoli cells are highly regulated in relation to the stages in the seminiferous epithelium (Parvinen, 1993) cycle. Sertoli cell-secreted proteins are also believed to be involved in the attachment of Sertoli cells to spermatogenic cells. Membrane-associated proteoglycans produced by the Sertoli cells, which are not randomly distributed on the cell surface (Rodriguez and Minguell, 1992), may be involved in this attachment process.

FSH and testosterone are required for the maximal binding of spermatids to Sertoli cells (Cameron and Muffly, 1991). Sertoli cells also must develop firm adherence to the membrane of the residual bodies of released spermatozoa to stop their loss from the surface of the seminiferous epithelium (Russell, 1993b). FSH-induced cell–cell associations *in vitro* can be blocked by specific lactins, inhibitors of protein synthesis, inhibitors of RNA synthesis, and cell surface modification, supporting the suggestion that FSH enhances the formation of cell surface glycoproteins (Griswold, 1993a,b,c; Parvinen, 1993) which in turn serve the function of cell adhesion molecules (Edelman and Crossin, 1991; Byers *et al.*, 1993b).

Developing germ cells are also now known to regulate the secretion of Sertoli cell proteins. Grima *et al.* (1992) suggested that germ cells differentially regulate the biosynthesis of rat Sertoli cell clusterin, α_2-macroglobulin, and testins. Recently germ cell-conditioned medium has been found to contain multiple biological factors that modulate the secretion of testins, clusterin, and transferrin by Sertoli cells (Pineau *et al.*, 1993a). A germ cell factor that stimulates transferrin secretion by Sertoli cell has been identified and partially purified (Pineau *et al.*, 1993b). Germ cell control of testin production is found to be inverse to that of other Sertoli cell products (Jegou *et al.*, 1993). Germ cell factor(s) appear to regulate opioid gene expression in Sertoli cells (Fujisawa *et al.*, 1993). Sertoli germ cell interactions are also found to be determinants of bidirectional secretion of androgen binding protein (ABP) (Gunsalus and Bardin, 1991).

Various observations made in regard to the effects of germinal cells in their co-culture have been extended through analysis of the effects of conditioned medium obtained from spermatogenic cell cultures on Sertoli cells. Germ cell-conditioned medium is observed to stimulate phosphorylation of specific proteins and γ-glutamyl transpeptidase activity in Sertoli cells, to increase ABP production and decrease estradiol synthesis by Sertoli cells, to decrease RNA synthesis in Sertoli cells, to increase transferrin production and transferrin gene expression, and to influence vectorial secretion of proteins by Sertoli cells (Skinner, 1991; Skinner *et al.*, 1991). Conditioned medium from cultures of pachytene spermatocytes and early-stage round spermatids generally showed the most dramatic effect. Partial purification of the heat activity and trypsin-sensitive activity of the conditioned medium showed that it was present in a fraction containing three polypeptides between 10 and 30 kDa. Further investigations remain to be carried out on purification and characterization of the factors in germ cell-conditioned medium to elucidate the number of factors needed for biological activities and potential identification of the paracrine factor.

It also remains to be investigated whether the paracrine factor(s) is actively secreted by the germ cells or is simply present in germinal cell cytosol and derives from cell lysis.

F. Spermiation

The cellular and molecular mechanisms involved in the release of spermatozoa from the seminiferous epithelium or germinal epithelium are still poorly understood (Russell, 1993b). This process of release of spermatozoa, termed "spermiation," begins with the breakdown of membrane junctional specializations between Sertoli cells and maturing spermatids (Russell, 1993b). The associated portions of Sertoli cells with maturing spermatids, especially the ectoplasmic specializations and their actin filaments, spermatid tubulobulbar complexes, Sertoli–Sertoli–cell membrane processes, and junctional complexes, etc. undergo conspicuous ultrastructural and biochemical changes, resulting in the disengagement of spermatozoa (Russell, 1993b). Junctional contacts are eliminated via receptor-mediated endocytosis. With the elimination of junctional complexes, the spermatid becomes free of the adhesion molecules which link it to the Sertoli cell. The tubulobulbar complexes and cytoplasmic remnant residual body are finally phagocytized by the Sertoli cell and degraded through the activity of Sertoli lysosomes as the latter form the characteristic feature of somatic accessory cells or Sertoli cells in invertebrates and vertebrates (see Section IV,C). Various investigations have suggested the involvement of two mechanisms in the elimination of cytoplasm and its organelles, one taking place via the formation of organelle-rich residual bodies and the other occurring by way of numerous phagocytized, waterish-appearing pockets of cytoplasm (tubulobulbar complexes) (Morales and Clermont, 1993; Russell, 1993b).

In the ground squirrel, numerous Sertoli cell processes are formed in morphological association with the residual cytoplasm of elongate spermatids (Vogl et al., 1993). Following retraction into the epithelium, the residual cytoplasm condenses, Sertoli cell processes disappear, and after sperm release, the residual bodies are phagocytized and degraded by the Sertoli cells (Clermont et al., 1987; Section IV,C). In other mammals, Sertoli cell processes do not develop specialized contacts with germ cells (Russell, 1993b). According to Vogl et al. (1993), the Sertoli cell microtubules possibly facilitate the retraction of residual cytoplasm from germ cells. The disappearance of ectoplasmic specializations in Sertoli cells is related also to sperm release because these structures disappear from around the spermatid heads in a sequence that correlates well with the retraction of the Sertoli cell cytoplasm from the germ cells.

The precise regulation of the changes that occur during spermiation remains to be understood. However, LH and human chorionic gonadotropin (hCG) are observed to induce spermiation in some nonmammalian vertebrates, especially amphibians (Russell, 1993b). Some hydrolytic enzymes, especially proteases, may be involved in hydrolyzing the adhesion molecules present on the Sertoli cell surface as well as on the sperm head (Fritz *et al.*, 1993). As the spermatozoa start loosening from the Sertoli cells in mammals, the spermatid cytoplasm becomes lobulated, but continues to remain attached to the spermatid neck by a thin stalk which finally separates to release the spermatozoon (Russell, 1993b). Therefore, the Sertoli cells appear to play an important role in separating the cytoplasmic bleb from the spermatozoon. However, biophysical and molecular mechanisms involved in this process still remain to be defined (Russell, 1993b).

Divergent views have been expressed about the role of Sertoli cell microtubules in spermiation, as revealed by experiments with microtubule-disrupting agents (Vogl *et al.*, 1993; Russell, 1993b). Spermatozoa are not released if the morphological organization of microtubules is disrupted experimentally. Spermiation is possibly due to the physical segregation of two segments of the spermatid plasma membrane (Russell, 1993b) as a result of the activity of some hydrolytic enzymes, especially proteases (Fritz *et al.*, 1993). Further investigations are required to determine the mechanism(s) of spermiation at the molecular and physiological levels (Russell, 1993b). This author has put forth the hypothesis of receptor-mediated loss of junctional links between Sertoli cell and mature spermatids during spermiation in mammals, and this hypothesis needs further experimental testing. However, elimination of ectoplasmic specialization junctional contacts between spermatids and Sertoli cells in nonmammalian vertebrates (Sprando and Russell, 1987) occurs in spite of the absence of tubulobulbar complexes or similar structures.

The mechanism of spermiation and the factors involved in it also remain to be investigated for invertebrates (Hinsch, 1993). Basically these may be similar to those described for mammals and nonmammalian vertebrates, especially in those invertebrate groups where developing spermatids form a close morphological association with Sertoli cells (Section II,A). In different groups of invertebrates, a great diversity can be seen in morphological contacts (or environmental interactions) between somatic accessory cells (Sertoli cells) and developing spermatids. The roles of cell types in the process of spermiation may, therefore, be variable, depending upon the species. In some invertebrate species, the maturing spermatids may play a predominant role in spermiation, as illustrated by trematodes, cestodes, and annelids where the developing spermatids are connected to the cytophore (cytoplasmic mass) but not to Sertoli cells (see Section II). In such cases spermatozoa must play an active role to get themselves disengaged from the cytophore, which is left behind. Even in this type of spermiation process, some physicochemical changes must occur and these remain to be investigated at the molecular level. In hemichordates, the process of spermiation is absent because the developing spermatid and somatic cells do not develop morphological contact (Guraya, 1986).

V. General Conclusions and Future Perspectives

The testicular somatic accessory cells associated with developing male germ cells show various patterns of development, distribution, size, abundance, permanancy, and structure in mammals, nonmammalian vertebrates, and invertebrates, which appear to be related to the variable mode of their spermatogenesis, but their regulation remains to be determined. Wherever the somatic accessory cells are present, they play important roles in (1) providing mechanical support and nutrition to developing male germ cells; (2) translocation of developing germ cells within the seminiferous or germinal epithelium; (3) paracrine regulation and coordination of male germ cell differentiation by secretion of regulatory autocrine and paracrine molecules and proteins, polypeptide growth factors, and hormones; (4) phagocytosis of residual cytoplasm and organelles sloughed off during spermiogenesis, degenerate germ cells, and even excluded spermatozoa; (5) the synthesis and metabolism of steroid hormones; and (6) spermiation.

The diversity in the morphology and numbers of Sertoli cell junctions is closely related to the complexity of the structural organization of the testes (especially seminiferous or germinal epithelium) in animals. The ways in which Sertoli cells influence germ cell proliferation, development, differentiation, and metabolic activities in different groups of animals are possibly diverse, and need to be understood at the molecular level. A major goal in future investigations of the junctional complexes of testes in different groups of animals should be a convergent application of biochemical, immunological, molecular, biological, and physiological techniques to reveal the functional meaning of their diversity (Zong et al., 1992; Cheng et al., 1993), as is being done for other tissues and organs (Robards et al., 1990; Green, 1992). The occluding tight junctions, which fuse the membranes of adjacent Sertoli cells in the testes of vertebrates and some invertebrates, constitute the structural component of the blood–testis barrier, and sequester the meiotic and postmeiotic spermatogenic cells (depending upon the animal group or species) in an adluminal compartment which is inaccessible to macromolecules from the serum or lymph. Further investigations with recently developed and improved methods for immunolabelling and the production of specific probes (e.g., affinity-purified polyclonal antibodies and monoclonal antibodies) will contribute to new discoveries about the comparative molecular composition and morphogenesis of the blood–testis barrier in the germinal epithelium of testes in different groups of animals.

Immunocytochemical characterization of blood–testis barrier-related proteins (including enzymes as well as receptor–transporter complexes) is also required to determine which of the general processes (such as passive diffusion, facilitated diffusion, active transport, and receptor-mediated endocytosis) of substrate movement function most effectively in the occluding junctions of the blood–testis barrier in different animal groups. The movement of molecules across the Sertoli cell body is potentially a more selective process involving receptors, pumps, exchange mechanisms, enzyme systems, transport and internalization systems, which also remain to be defined more precisely in future investigations. The precise nature and mechanisms of transport of nutritive substances (lipids, carbohydrates, proteins, amino acids, vitamins, etc.) from the Sertoli cells to germ cells also remain to be determined, but binding proteins and endocytosis are being identified.

The germ cells are now shown to provide signals to somatic cells (Sertoli cells) to coordinate their various stage-specific functions, suggesting a mutual functional interrelationship, which remains to be understood at the molecular level. The study of cell adhesion molecules, substrate adhesion molecules, and cell junctional molecules, as well as enzymes will be very helpful in revealing the nature of the molecules involved in the adhesion, communication, and transport processes between Sertoli cells and germ cells. Advances in the identification and characterization of cell adhesion molecules have largely paralleled progress in protein chemistry, immunology, and molecular genetics (Grunwald, 1992). Cadherins and integrins are of great current interest in relation to cell adhesion and recognition in the seminiferous tubules of the mammalian testis.

The complexity of testicular structure and the multiple intratesticular interactions occurring at cellular levels make the experimental design of investigations and interpretation of data difficult. Therefore, the use of a combination of traditional methods and new molecular biology probes *in vitro* and *in vivo* in comparative studies of the synthesis, chemical characterization, and biological actions of proteins, including peptide growth factors, in the testes of animals that have a variable testicular architecture (or structural complexity in somatic cell–germ cell contacts) holds promise for future investigations because our knowledge in these areas is very poor, especially for the testes of invertebrates and lower vertebrates.

For answering questions about the local regulation of testicular functions, it is, therefore, important to study the differential regulation of gene expression throughout development and then its correlation with the development of specific functions or characteristics. Although immunocytochemistry, mRNA expression, and *in situ* hybridization indicate synthesis of a peptide, a final demonstration of the specific effects of factors isolated from testicular tissue and subsequently characterized is necessary. Comparative studies on the identification of the secretory products, including proteins and steroid hormones, and their biological roles in the "nurse cell" function of Sertoli cells in different groups of animals would be very useful in improving our understanding of the precise mechanisms involved in spermatogenesis.

However, a very limited somatic cell–germ cell association or its complete absence in the testes of some animals (Section II) clearly suggests that the information needed for germ cell development is present in the genome of the spermatogenic cell itself. An alternative type of cell–cell interaction may be a more passive process which does not require regulatory (i.e., signal transduction type) interactions. The somatic cell may be simply involved in providing a proper environment, metabolites, and nutrients to permit spermatogenesis to proceed, suggesting that the primary Sertoli cell–spermatogenic cell interactions are environmental and nutritional rather than regulatory. It also remains to be determined whether somatic cell function in the testes of the mature (or adult) animal is simply optimally maintained or regulated (i.e., changing).

Acknowledgments

This chapter was prepared and supported under the program of the Council of Scientific and Industrial Research (New Delhi), Emeritus Scientist Scheme. It was updated while the author was a visiting scientist in the Department of Biochemistry, University of Georgia, Athens, Georgia for which funds were provided by Dr. P. N. Srivastava from his research grant, as well as by Dr. David Puett, chairman, from a departmental grant. I am very grateful to both of them for making my stay in the department very productive and comfortable. The help of Mr. Don Roberts in collecting the references is also acknowledged.

References

Adiyodi, K. G., and Adiyodi, R. G., eds. (1983). "Reproductive Biology of Invertebrates." Wiley, New York.

Allard, E. K., Johnson, K. J., and Boekelheide, K. (1993). Colchicine disrupts the cytoskeleton of the rat testis seminiferous epithelium in a stage-dependent manner. *Biol. Reprod.* **48,** 143–153.

Amselgruber, W., Sinowatz, F., Schams, D., and Lehmann, M. (1992). S-100 protein immunoreactivity in bovine testis. *Andrologia* **24,** 231–235.

Anderson, W. A., Ellis, R. A., and Weissman, A. (1967). Cytodifferentation during spermio-genesis in *Lumbricus terrestris*. *J. Cell Biol.* **32**, 11–26.

Anelli, G., Franchi, F., and Camatini, M. (1985). Structure and possible functional role of septate junctions in the ovotestis of a pond snail: Inter-Sertoli junctions. *J. Submicrosc. Cytol.* **17**, 213–222.

Arnold, J. M. (1978). Spermiogenesis in *Nautilus pompilius*. II. Sertoli cell-spermatid junctional complexes. *Anat. Rec.* **191**, 261–268.

Arsenault, A. L., Clattenburg, R. E., and Odense, P. H. (1980). Further observations on spermiogenesis in the shrimp, *Crangon septemspinosa*. A mechanism of cytoplasmic reduction. *Can. J. Zool.* **58**, 497–506.

Baccetti, B., Bigliardi, E., Talluri, M. V., and Burrini, A. G. (1983). The Sertoli cell in lizards. *J. Ultrastruct. Res.* **85**, 11–23.

Bardin, C. W., Cheng, C. Y., Mustow, N. A., and Gunsalus, G. L. (1994). The Sertoli cell. *In* "Physiology of Reproduction" (E. Knobil and J. D. Neil, eds.), 2nd ed., Vol. 1, pp. 1291–1334. Plenum, New York.

Barnes, K., and Hardisty, M. W. (1972). Ultrastructural and histochemical studies on the testis of the river lamprey *Lampetra flaviatilis* (L.) *J. Endocrinol.* **53**, 59–69.

Bartke, A., Mayerhofer, A., Newton, S., Mayerhofer, D., Majumdar, S., and Chandra-shekar, V. (1991). Alterations in the control and functions of somatic cells in the testis associated with suppression of spermatogenesis in seasonal breeders. *Ann. N. Y. Acad. Sci.* **637**, 143–151.

Bartke, A., Sinha Hikim, A. P., and Russell, L. D. (1993). Sertoli cell structure and function in seasonally breeding mammals. *In* "The Sertoli Cell" (L. D. Russell and M. D. Griswold, eds.), pp. 349–364. Cache River Press, Clearwater, FL.

Bellvé, A. R. (1979). The molecular biology of mammalian spermatogenesis. *Oxford Rev. Reprod. Biol.* **1**, 531–567.

Bellvé, A. R., and Zheng, W. (1989). Growth factors as autocrine and paracrine modulators of male gonadal function. *J. Reprod. Fertil.* **85**, 771–793.

Bergman, N. M., Greven, H., and Schindelmeise, J. (1984). Tight junctions in the ovotestis of the pond snail *Lymnaea stagnalis* (L.) (Gastropoda, Basommatophora). *Int. J. Invertebr. Reprod. Dev.* **7**, 291–296.

Billard, R. (1990). Spermatogenesis in teleost fish. *In* "Marshall's Physiology of Reproduction." (G. E. Lamming, ed.), 4th ed., Vol. 2, pp. 183–212. Churchhill-Livingstone, New York.

Block, E. M., and Goodnight, C. J. (1980). Spermatogenesis in *Limnodrilus hoffmeisteri* (Annelida, Tubificidae): A morphological study of the development of two sperm types. *Trans. Am. Microsc. Soc.* **99**, 368–384.

Boekelheide, K. (1993). Sertoli cell toxicants. *In* "The Sertoli Cell" (L. D. Russell and M. D. Griswold, eds.), pp. 551–576. Cache River Press, Clearwater, FL.

Breuker, H. (1982). Seasonal spermatogenesis in the mute swam (*Cygnus olor*). *Adv. Anat. Embryol. Cell Biol.* **72**, 1–90.

Brown, N. L., Bayle, J. D., Scanes, C. G., and Follett, B. K. (1975). Chicken gonadotro-phins: Their effects on the testis of immature and hypophysectomized Japanese quail. *Cell Tissue Res.* **156**, 499–520.

Buckland-Nicks, J., and Chia, F.-S. (1986). Fine structure of Sertoli cells in three marine snails with a discussion on the functional morphology of Sertoli cells in general. *Cell Tissue Res.* **245**, 305–313.

Buckland-Nicks, J., Walrer, C. W., and Chia, F.-S. (1984). Ultrastructure of the male reproductive system of spermatogenesis in the viviparous brittle star, *Amphipholis lividus*. *J. Morphol.* **179**, 243–262.

Byers, S., Graham, R., Dai, H. N., and Hoxter, B. I. (1991). Development of Sertoli cell junctional specializations and the distribution of the tight junction-associated protein ZO-I in the mouse testis. *Am. J. Anat.* **191**, 35–47.

Byers, S., Pelletier, R.-M., and Suarez-Quian, C. (1993a). Sertoli-Sertoli cell junctions and the seminiferous epithelium barrier. *In* "The Sertoli Cell" (L. D. Russell and M. D. Russell, eds.), pp. 431–446. Cache River Press, Clearwater, FL.

Byers, S., Degou, B., MacCalman, C., and Blaschuk, O. (1993b). Sertoli cell adhesion molecules and the collective organization of the testis. *In* "The Sertoli Cell" (L. D. Russell and M. D. Griswold, eds.), pp. 461–476. Cache River Press, Clearwater, FL.

Callard, G. V., and Mak, P. (1988). Characteristics and stage-dependent distribution of a novel steroid binding protein in the testis of *Squalus acanthias. Steroids* **52**, 359–360.

Cameron, D. F., and Muffly, K. E. (1991). Hormonal regulation of spermatid binding. *J. Cell Sci.* **100**, 623–633.

Chakraborty, J. (1993). Conditions adversely affecting Sertoli cells. *In* "The Sertoli Cell" (L. D. Russell and M. D. Griswold, eds.), pp. 577–596. Cache River Press, Clearwater, FL.

Cheng, C. Y., Grima, J., Pineau, C., Stahler, M. S., Jégou, B., and Bardin, C. W. (1992). α-Macroglobulin protects the seminiferous epithelium during stress. *In* "Stress and Reproduction" (K. E. Sheppard, J. H. Boublik, and J. W. Funder, eds.), Vol. 86, pp. 123–133. Raven Press, New York.

Cheng, C. Y., Morris, I., and Bardin, C. W. (1993). Testins are structurally related to the mouse cysteine proteinase precursor proregion but devoid of any protease/anti-protease activity. *Biochem. Biophys. Res. Commun.* **191**, 224–231.

Chia, F. S., and Bickell, L. R. (1983). Echinodermata. *In* "Reproductive Biology of Invertebrates" (K. G. Adiyodi and R. G. Adiyodi, eds.), Vol. 2, pp. 545–620. Wiley, New York.

Chia, F. S., and Buckland-Nicks, J. (1987). Sertoli-like interstitial cells in the echinoderm testis: A test of a permeability barrier. *J. Invertebr. Reprod. Dev.* **13**, 239–250.

Choongkittaworn, N. M., Kim, K. H., Danner D. B., and Griswold, M. D. (1993). Expression of proinhibin in rat seminiferous epithelium. *Biol. Reprod.* **49**, 300–310.

Clark, A. W. (1967). Some aspects of spermiogenesis in a lizard. *Am. J. Anat.* **121** 369–400.

Clermont, Y. (1993). Introduction to the Sertoli cells. *In* "The Sertoli Cell" (L. D. Russell and M. D. Griswold, eds.), pp. xi–xviii. Cache River Press, Clearwater, FL.

Clermont, Y., Morales, C., and Hermo, L. (1987). Endocytic activities of Sertoli cells in the rat. *Ann. N. Y. Acad. Sci.* **513**, 1–15.

Cooksey, E. J., and Rothwell, B. (1973). The ultrastructure of the Sertoli cell and its differentiation in the domestic fowl (*Gallus domesticus*). *J. Anat.* **114**, 329–345.

Crompton, D. W. (1983). Acanthocephala. *In* "Reproductive Biology of Invertebrates" (K. G. Adiyodi and R. G. Adiyodi, eds.), Vol. 2, pp. 477–490. Wiley, New York.

Davis, J. T., and Ong, D. E. (1992). Synthesis and secretion of retinol-binding protein by cultured rat Sertoli cell. *Biol. Reprod.* **47**, 528–533.

Davis, R. E., and Roberts, L. S. (1983). Eucestoda. *In* "Reproductive Biology of Invertebrates" (R. G. Adiyodi and R. G. Adiyodi, eds.), Vol. 2, pp. 131–149. Wiley, New York.

De Franca, L. R., Ghosh, S., Ye, S.-J., and Russell, L. D. (1993). Surface and surface-to-volume relationships of the Sertoli cell during the cycle of the seminiferous epithelium in the rat. *Biol. Reprod.* **49**, 1215–1228.

de Jong-Brink, M., Boer, H. H., Hommes, T. G., and Kodde, A. (1977). Spermatogenesis and the roles of Sertoli cells in the freshwater snail *Biomphalaria glabrata. Cell Tissue Res.* **181**, 37–58.

de Jong-Brink, M., Schot, L. P. C., Schoenmakers, H. J. N., and Bergamin-Sassen, A. (1981). A biochemical and quantitative electron microscopy study on steroidogenesis in ovotestis and digestive gland of the pulmonate snail *Lymnaea stagnalis. Gen. Comp. Endocrinol.* **45**, 30–38.

de Jong-Brink, M., de With, N. D., Hurkmans, P. J., and Bergamin, M. J. M. (1984). A morphological enzyme-cytochemical and physiological study of the blood–gonad barrier in the hermaphroditic snail *Lymnaea stagnalis*. *Cell Tissue Res.* **235**, 543–600.

de Kretser, D. M., and Kerr, J. B. (1994). The cytology of the testis. *In* "Physiology of Reproduction" (E. Knobil and J. D. Neil, eds.), IInded., Vol. 1, pp. 1177–1290. Plenum, New York.

de Kretser, D. M., Risbridger, G. P., Drummond, A. E., Gonzales, G., and Sun, Y. T. (1991). Paracrine mechanisms in the regulation of testicular function. *In* "Growth Factors in Fertility Regulation" (E. P. Haseltine and J. K. Findlay, eds.), pp. 143–156. Cambridge Univ. Press, Cambridge, UK.

Delavault, R., and Bernard, J. J. (1974). Etude ultrastructurale de la spermatogénèse chez *Daphnia magna* S. (Entomostraces, Brachiopodes, Cladoceres). *C. R. Hebd. Seances Acad. Sci.* **278**, 1589–1592.

Del Conte, E. (1976). The subacrosimal granule and its evolution during spermiogenesis in a lizard. Observations about the acrosomal fringe and the spermatid-Sertoli cell relationship. *Cell Tissue Res.* **171**, 483–498.

Dewel, W. C., and Clark, W. H., Jr. (1972). An ultrastructural investigation of spermatogenesis and the mature sperm in the Anthozoan *Bunodosoma cavernat* (Cnidaria). *J. Ultrastruct. Res.* **40**, 417–431.

Dodd, J. M., and Sumpter, J. P. (1984). Fishes, *In* "Marshall's Physiology of Reproduction" (G. E. Lamming, ed.), 4th ed. Vol. 1, pp. 1–126. Churchill-Livingstone, New York.

Dorrington, J. H., and Khan, S. A. (1993). Steroid production, metabolism and release by Sertoli cells. *In* "The Sertoli Cell" (L. D. Russell and M. D. Griswold, eds.), pp. 537–550. Cache River Press, Clearwater, FL.

Dubois, W., and Callard, G. V. (1989). Role of the Sertoli cell in spermatogenesis: The *Squalus* testis model. *Fish Physiol. Biochem.* **7**, 221–227.

Dubois, W., Pudney, J., and Callard, J. P. (1988). The annual testicular cycle in the turtle *Chrysemys picta:* A histochemical and electron microscopic study. *Gen. Comp. Endocrinol.* **71**, 191–204.

Dufaure, J. P. (1971). L'ultrastructure du testiculede lézard vivipare (Reptile, Lacertilien). II. Les cellules de Sertoli étude du glycogene. *Z. Zellforsch. Mikrosk. Anat.* **115**, 565–578.

Edelman, G. M., and Crossin, K. L. (1991). Cell adhesion molecules: Implications for a molecular history. *Annu. Rev. Biochem.* **60**, 155–190.

Enders, G. C. (1993). Sertoli-Sertoi and Sertoli-germ cell communication. *In* "The Sertoli Cell" (L. D. Russell and M. D. Russell, eds.), pp. 447–460. Cache River Press, Clearwater, FL.

Fahrenbach, H. (1962). The biology of a harpacticoide copepod. *Cellule* **62**, 303–376.

Foucher, J.-L., and LeGac, F. (1989). Evidence for an androgen binding protein in the testis of a teleost fish (*Salmo gairdneri* R): A potential marker of Sertoli cell function. *J. Steroid. Biochem.* **32**, 545–552.

Franc, J.-M. (1973). Etude ultrastructurale de la spermatogénèse du Ctenaire *Beroe ovata*. *J. Ultrastruct. Res.* **42**, 255–267.

Franquinet, R., and Lender (1973). Etude ultrastructurale des testicules de polycelis tenuis et poly nigra. *Z. Zellforsch. Mikrosk. Anat.* **37**, 4–22.

Fritz, I. B., Tung, P. S., and Ailenberg, M. (1993). Proteases and antiproteases in the seminiferous tubule. *In* "The Sertoli Cell" (L. D. Russell and M. D. Griswold, eds.), pp. 217–236. Cache River Press, Clearwater, FL.

Fujisawa, M., Bardin, C. W., and Morris, P. L. (1993). Male germ cell factor(s) regulate opioid gene expression in Sertoli cells. *Recent Prog. Horm. Res.* **48**, 497–503.

Garnier, D. H., Tixiev-Vidale, A., Gourdji, D., and Picart, R. (1973). Ultrastructure des cellules de Leydig et des cellules de Sertoli au cours du cycle testiculaire du Canard

Pekin. Correlation avec les données biochimique et cytoenzymologiques. *Z. Zellforsch. Mikrosk. Anat.* **144,** 369–394.

Ghosh, S., Bartke, A., Grasso, P., Reichert, L. E., Jr., and Russell, L. D. (1992). Structural response of the hamster Sertoli cell to hypophysectomy. A correlative morphometric and endocrine study. *Anat. Rec.* **234,** 513–529.

Gondos, B., and Berndtson, W. E. (1993). Postnatal and pubertal development. *In* "The Sertoli Cell" (L. D. Russell and M. D. Griswold, eds.), pp. 115–154. Cache River Press, Clearwater, FL.

Green, C. R. (1992). The role of gap junctions during development and patterning. *In* "Developmental Biology" (F. E. Bittar, ed.), pp. 133–150. JAI Press, Greenwich, CT.

Green, C. R., and Bergquist, P. R. (1982). Phylogenetic relationship within the invertebrate in relation to the structure of the septate junction and the development of occluding junctional types. *J. Cell Sci.* **53,** 279–305.

Grier, H. J. (1992). Chordate testis: The extracellular matrix hypothesis. *J. Exp. Zool.* **261,** 151–160.

Grier, H. J. (1993). Comparative organization of Sertoli cells including the Sertoli cell barrier. *In* "The Sertoli Cell" (L. D. Russell and M. D. Griswold, eds.), pp. 703–740. Cache River Press, Clearwater, FL.

Grier, H. J., van den Hurk, R., and Billard, R. (1989). Cytological identification of cell types in the testis of *Exox lucius* and *E. niger*. *Cell Tissue Res.* **257,** 491–496.

Grima, J., Pineau, C., Bardin, C. W., and Cheng, C. Y. (1992). Rat Sertoli cell clusterin, α-macroglobulin and testins: Biosynthesis and differential regulation by germ cells. *Mol. Cell. Endocrinol.* **89,** 127–140.

Griswold, M. D. (1988). Protein secretions of Sertoli cells. *Int. Rev. Cytol.* **110,** 133–156.

Griswold, M. D. (1993a). Protein secretion by Sertoli cells: General considerations. *In* "The Sertoli Cell" (L. D. Russell and M. D. Griswold, eds.), pp. 195–200. Cache River Press, Clearwater, FL.

Griswold, M. D. (1993b). Unique aspects of the biochemistry and metabolism of Sertoli cells. *In* "The Sertoli Cell" (L. D. Russell and M. D. Griswold, eds.), pp. 485–492. Cache River Press, Clearwater, FL.

Griswold, M. D. (1993c). Action of FSH on mammalian Sertoli cells. *In* "The Sertoli Cell" (L. D. Russell and M. D. Griswold, eds.), pp. 493–508. Cache River Press, Clearwater, FL.

Griswold, M. D., Morales, C., and Sylvester, S. (1988). Molecular biology of the Sertoli cell. *Oxford Rev. Reprod. Biol.* **10,** 124–161.

Grunwald, G. B. (1992). Cell adhesion and recognition in development. *In* "Developmental Biology" (E. E. Bittar, ed.), pp. 103–132. JAI Press, Greenwich, CT.

Gunawerdana, V. K. (1992). Lysosomes in the seminiferous tubules of the domestic fowl as revealed by acid phosphatase activity. *Acta Histochem. Cytochem.* **25,** 91–96.

Cunsalus, G. L., and Bardin, C. W. (1991). Sertoli-germ cell interactions as determinants of bidirectioal secretion of androgen-binding protein *Ann. N.Y. Acad. Sci.* **627,** 322–326.

Gupta, B. L. (1964). Cytological studies of the male germ cells in some freshwater ostracods and copepods. Ph.D Thesis, Cambridge University, Cambridge, UK.

Gupta, B. L., Parshad, V. R., and Guraya, S. S. (1986). Morphological and histochemical studies on the development of testis and spermatogenesis in the ruminal sheep trematode, *Paramphistomum cervi* (Digenea: Paramphistomatidae). *Folia Parasitol.* **33,** 131–144.

Guraya, S. S. (1968). A histochemical study of lipios in the goat testis. *Acta Morphol. Neerl.-Scand.* **7,** 15–27.

Guraya, S. S. (1971). Morphological and histochemical observations on the acanthocephalan spermatogenesis. *Acta Morphol. Neerl.-Scand.* **9,** 75–83.

Guraya, S. S. (1976). Recent advances in morphology, histochemistry and biochemistry of steroid-synthesizing cellular sites in the testes of non-mammalian vertebrates. *Int. Rev. Cytol.* **47**, 99–136.

Guraya, S. S. (1979). Recent advances in the morphology and histohemistry of steroid-synthesizing cellular sites in the gonads of fish. *Proc. Indian Nat. Sci. Acad., Part B* **45**, 401–452.

Guraya, S. S. (1980). A recent progress in morphology: Histochemistry, biochemistry and physiology of developing and maturing mammalian testis. *Int. Rev. Cytol.* **62**, 187–309.

Guraya, S. S. (1983). Cephalochordata. *In* "Reproductive Biology of Invertebrates" (K. G. Adiyodi and R. G. Adiyodi, eds.), Vol. 2, pp. 633–648. Wiley, New York.

Guraya, S. S. (1986). Morphological studies on the gonads and gametogenesis in the hemichordate *Ptychodera flava*. *Z. Mikrosk. Anat. Forsch.* **100**, 711–728.

Guraya, S. S. (1987). "Biology of Spermatogenesis and Spermatozoa in Mammals." Springer-Verlag, Heidelberg.

Guraya, S. S. (1994). Gonadal development and production of gametes in fish. *Proc. Indian Nat. Sci. Acad., Part B* **60**, 14–32.

Guraya, S. S., and Gupta, A. N. (1970a). Histochemical studies on the spermatogenesis of *Fasciola indica*. *Acta Morphol. Neerl.-Scand.* **8**, 9–18.

Guraya, S. S., and Gupta, A. N. (1970b). Histochemical observations on the spermatogenesis of the trematode *Paramphistomum* (Explanatum) *bathycotyle*. *Acta Morphol. Neerle-Scand.* **8**, 19–28.

Hardisty, M. W. (1965). Sex differentiation and gonadogenesis in lampreys. Part II. The ammocoete gonads of the land-locked sea lamprey *Petromyzon murinus*. *J. Zool.* **146**, 346–387.

Hendelberg, J. (1983). Platyhelminthes. *In* "Reproductive Biology of Invertebrates" (K. G. Adiyodi and R. G. Adiyodi, eds.), Vol. 2, pp. 75–104. Wiley, New York.

Hernandez-Verdun, D. (1991). The nucleolus today. Commentary. *J. Cell Sci.* **99**, 465–482.

Hinsch, G. W. (1992). Junctional complexes between Sertoli cells in the testis of the crayfish *Procambarus paeninsulanus*. *Tissue Cell* **24**, 379–386.

Hinsch, G. W. (1993). Comparative organization and cytology of Sertoli cells in invertebrates. *In* "The Sertoli Cell" (L. D. Russell and M. D. Griswold, eds.), pp. 659–684. Cache River Press, Clearwater, FL.

Hinsch, G. W., and Moore, J. A. (1992). The structure of the reproductive mesenteries of the sea anemone *Ceriantheopsis americanus*. *J. Invertebr. Reprod. Dev.* **21**, 25–32.

Hinton, B. T., and Setchell, B. P. (1993). Fluid secretion and movement *In* "The Sertoli Cell" (L. D. Russell and M. D. Griswold, eds.), pp. 249–268. Cache River Press, Clearwater, FL.

Hope, W. D. (1974). Nematoda. *In* "Reproduction in Marine Invertebrates" (A. C. Giese and J. S. Pearse, eds.), Chapter 3, pp. 391–469. Academic Press, New York.

Humphreys, P. N. (1975). Ultrastructure of the budgerigar testis during a photo-periodically induced cycle. *Cell Tissue Res.* **159**, 541–550.

Jégou, B., Le Magueresse, B., Sourdaine, P., Pineau, C., Velez de la Calle, J. F., Garnier, D. H., Guillou, F., and Boisseau, C. (1988). Germ cell-Sertoli cell interactions in vertebrates. *In* "Molecular and Cellular Endocrinology of the Testis" (B. A. Cooke and R. M. Sharpe, eds.). Serano Symposic Publications, vol. 30, pp. 255–270. Raven Press, New York.

Jégou, B., Pineau, C., Velez de la Calle, J. F., Touzulin, A.-M., Bardin, C. W., and Cheng, C. Y. (1993). Germ cell control of testin production is inverse to that of other Sertoli cell products. *Endocrinology (Baltimore)* **132**, 2557–2562.

Jenkins, T., and Larkman, A. (1981). Spermatogenesis in a trichuroid nematode. *Trichuris muris*. II. Fine structure of primary spermatocyte and first meiotic division. *Int. J. Invertebr. Reprod.* **3**, 257–273.

Jenkins, T., Larkman, A., and Funnell, M. (1979). Spermatogenesis in a trichuroid nematode, *Trichuris muris*. I. Fine structure of spermatogonia. *Int. J. Invertebr. Reprod.* **1**, 371–385.

Johnson, K. J., Hall, E. S., and Bokelheide, K. (1991). 2,5-Hexanedione exposure alters the rat Sertoli cell cytoskeleton. I. Microtubules and seminiferous tubule fluid secretion. *Toxicol. Appl. Pharmacol.* **111**, 432–442.

Kanganiemi, M., Cheng, C. Y., Toppari, J., Grima, J., Stahler, M., Bardin, C. W., and Parvinen, M. (1992). Basal and FSH-stimulated steady levels of SGP, α2-macroglobulin, and testibumin in culture media of rat seminiferous tubules at defined stages of the epithelial cycle. *J. Androl.* **13**, 208–213.

Karzai, A. W., and Wright, W. W. (1992). Regulation of the synthesis and secretion of transferrin and cyclic protein-2/cathepsin. *Biol. Reprod.* **47**, 823–831.

Kasaoka, L. D. (1974). The male genital system in two species of mysid Crustacea. *J. Morphol.* **143**, 259–283.

Kim, K. H., and Wang, Z. (1993). Action of vitamin A on the testis: Role of the Sertoli cell. In "The Sertoli Cell" (L. D. Russell and M. D. Griswold, eds.), pp. 517–536. Cache River Press, Clearwater, FL.

Kojma, Y. (1990). Ultrastructure of goat testes, tubulobulbar complexes between spermatids and Sertoli cells. *Jpn. J. Vet. Sci.* **52**, 781–786.

Kugler, O. E. (1965). A morphological and histochemical study of the reproductive system of the slug *Philomycus carolinianus* (Bosc.). *J. Morphol.* **116**, 117–132.

Kurohmaru, M., Kanai, Y., and Hayashi, Y. (1992). A cytological and cytoskeletal comparison of Sertoli cells without germ cells and those with germ cells using the w/w mutant mouse. *Tissue Cell* **24**, 895–904.

Lambert, A., Talbot, A., Mitchell, R., and Robertson, W. R. (1991). Inhibition of protein kinase C by staurosponine increases estrogen secretion by rat Sertoli cells. *Acta Endocrinol. (Copenhagen)* **125**, 286–290.

Larsen, W. J., and Wert, S. E. (1988). Roles of cell junctions in gametogenesis and early embryonic development. *Tissue Cell* **20**, 809–843.

LeGac, F., and Loir, M. (1988). Control of testis function in fish: *In vitro* studies of gonadotropic regulation in the trout (*Salmo gairdneri*). *Reprod. Nutr. Dev.* **28**, 1031–1046.

Lofts, B. (1987). Testicular function. In "Hormones and Reproduction in Fishes, Amphibians and Reptiles" (D. O. Norris and R. E. Jones, eds.), pp. 283–326. Plenum, New York.

Lönnerberg, P., Parvinen, M., Jahnsen, T., Hansson, V., and Persson, H. (1992). Stage- and cell-specific expression of cyclic adenosine 3′, 5′- monophosphate-dependent protein kinases in rat seminiferous epithelium. *Biol. Reprod.* **46**, 1057–1068.

Maekawa, M., Nagano, T., Kamimura, K., Murakami, T., Ishikawa, H., and Dezawa, M. (1991). Distribution of actin-filament bundles in myoid cells, Sertoli cells and tunica albuginea in rat and mouse testes. *Cell Tissue Res.* **266**, 295–300.

Mahmoud, I. Y., Cyrus, R. V., Bennett, T. M., Woller, M. J., and Montag, D. M. (1985). Ultrastructural changes in testes of the snapping turtle, *Chelydra serpentina* in relation to plasma testosterone, Δ5-3-hydroxysteroid dehydrogenase and cholesterol. *Gen. Comp. Endocrinol.* **57**, 454–464.

Manier, J.-F., Raibaut, A., Rousset, V., and Coste, F. (1977). Reproduction et sexualité des copepodes parasites de poissons. II. L'appareil genital male et la spermiogénèse de. *Naobranchia cygniformis* Hasse, 1863. *Ann. Sci. Nat., Zool. Biol. Anim.*

Marcaillou, C., and Szöllösi, A. (1980). The "blood-testis" barrier in a nematode and a fish: A generalizable concept. *J. Ultrastruct. Res.* **70**, 128–136.

Marchand, C. R. (1973). Ultrastructure des cellule de Leydig et des cellules de Sertoli du testicule de canard de Barbarie (*Cairina moschata* L.) en activité sexuelle. *C. R. Seances Soc. Biol. Ses Fil.* **167**, 933–937.

Mattei, X., Mattei, C., Marchand, B., and DLT, Kit (1982). Ultrastructure des cellules de Sertoli d'un poisson téléostéen: *Abudefdug marginatus*. *J. Ultrastruct Res.* **81**, 222–240.

Morales, C., and Clermont, Y. (1993). Structural changes of the Sertoli cell during the cycle of the seminiferous epithelium. *In* "The Sertoli Cell" (L. D. Russell and M. D. Griswold, eds.), pp. 305–330. Cache River Press, Clearwater, FL.

Moyne, G., and Collenot, G. (1982). Unusual nucleolar fine structure in the Sertoli cells of the dog-fish *Scyliorhinus canicula* (L). *Biol. Cell.* **44**, 239–248.

Muffly, K. E., Nazian, S. J., and Cameron, D. F. (1993). Junction on related Sertoli cell cytoskeleton in testosterone-treated hypophysectomized rats. *Biol. Reprod.* **49**, 1122–1132.

Nagahama, Y. (1986). Testis. *In* "Vertebrate Endocrinology: Fundamentals and Biochemical Implications" (P. K. T. Pang and M. P. Schreibman, eds.), Vol. 1, pp. 399–437. Academic Press, New York.

Newton, S. C., Walsh, M. J., and Bartke, A. (1992). Effects of age, photoperiod and follicle-stimulating hormone on lactate production by cultured Sertoli cells from prepubertal siberian hamsters. *(Phodopus sungorus)*. *J. Reprod. Fertil.* **95**, 87–95.

Nicotra, A., and Serafino, A. (1988). Ultrastructural observations on the interstitial cells of the testis of *Paracentrotus lividus. Int. J. Invertebr. Reprod. Dev.* **13**, 239–250.

Qbert, H. J. (1976). Die Spermatogenese bei der Gelbbauchunke (*Bombina variegata* L.) in Verlauf der jahrlichen Activitats periode und die korrelation zur Paarungsrufaktivitat (Discoglossidae, Anura). *Z. Mikrosk.-Anat. Forsch.* **90**, 908–924.

O'Brien, D. A., Gabel, C. A., and Eddy, E. M. (1993). Mouse Sertoli cells secrete mannose 6-phosphate containing glycoprotein that are endocytosed by spermatogenic cells. *Biol. Reprod.* **49**, 1055–1065.

O'Donovan, P., and Abraham, M. (1987). Somatic tissue-male germ cell barrier in three hermaphrodite invertebrates, *Dugesia biblica* (Platyhelminthes, *Placobdella costata* (Annelida) and *Levantina hierosoluma* (Mollusca). *J. Morphol.* **192**, 217–227.

Okia, N. O. (1992). The cell membrane body in the skink. *Tissue Cell* **24**, 283–289.

Qkkelberg, P. (1921). The early history of the germ cells in the brook lamprey, *Entosphenus wilderi* (Gage) up to and including the period of sex differentiation. *J. Morphol.* **35**, 1–151.

Onoda, M., and Suarez-Quian, C. A. (1994). Modulation of transferrin secretion by epidermal growth factor in immature rat Sertooi cells *in vitro. J. Reprod. Fertil.* **100**, 541–550.

Parivar, K. (1980). Differentiation of Sertoli cells and post-reproductive epithelial cells in the hermaphrodite gland of *Arion ater* (L.) Pulmonata). *J. Molluscan Stud.* **46**, 139–147.

Parvinen, M. (1993). Cyclic function of Sertoli cells. *In* "The Sertoli Cell" (L. D. Russell and M. D. Griswold, eds.), pp. 331–348. Cache River Press, Clearwater, FL.

Pawar, H. S., and Wrobel, K.-N. (1991). The Sertoli cell of the water buffalo (*Bubalus bubalis*) during the spermatogenic cycle. *Cell Tissue Res.* **265**, 43–50.

Pearson, A. K., and Licht, P. (1990). The double nucleus of the Sertoli cell in the lizard *Anolis Carolinensis. Tissue Cell* **22**, 221–229.

Pelletier, R.-M. (1990). A novel perspective. The occluding zonule encircles the apex of the Sertoli cell as observed in birds. *Am. J. Anat.* **188**, 87–108.

Pelliniemi. L. J., Frojdman, K., and Paranko, J. (1993). Embryological and prenatal development and function of Sertoli cells. *In* "The Sertoli Cell" (L. D. Russell and M. D. Griswold, eds.), pp. 87–114. Cache River Press, Clearwater, FL.

Petrie, R. G., Jr., and Morales, C. R. (1992). Receptor-mediated endocytosis of testicular transferrin by germinal cells of the rat testis. *Cell Tissue Res.* **267**, 45–55.

Pfeiffer, D. C., and Vogl, A. W. (1990). Ectoplasmic "junctional" specializations in Sertoli cells of the rooster and turtle. *Anat. Rec.* **226**, 80A.

Pfeiffer, D. C., and Vogl, A. W. (1992). Actin filaments associated with the basal Sertoli cell surface in the alligator and turtle. *Tissue Cell* **24**, 643–654.

Pfeiffer, D. C., and Vogl, A. W. (1993). Ectoplasmic ("junctional") specializations in Sertoli cells of the rooster and turtle: evolutionary implications. *Anat. Rec.* **235**, 33–50.

Pillai, R. S. (1960). Studies on the shrimp (*Caridina laevis* (Heller). II. The reproductive system. *J. Mar. Biol. Assoc. India* **2**, 226–236.

Pineau, C., Le Maguaesse, B., Courtens, J. L., and Jégou, B. (1991). Study *in vitro* of the phagocytotic function of Sertoli cells in the rat. *Cell Tissue Res.* **264**, 589–598.

Pineau, C., Syed, V., Bardin, C. W., Jégou, B., and Cheng, C. Y. (1993a). Germ cell-conditioned medium contains multiple biological factors that modulate the secretion of testins, clusterin and transferrin by Sertoli cells. *J. Androl.* **14**, 87–98.

Pineau, C., Syed, V., Bardin, C. W., Jégou, B., and Cheng, C. Y. (1993b). Identification and partial purification of a germ cell factor that stimulates transferrin secretion by Sertoli cells. *Recent Prog. Horm. Res.* **48**, 539–542.

Pochon-Massion, J. (1968a). L'ultrastructure des spermatozoide vesiculaires chez les Crustaces Decapodes avant et au cours de leur dévagination experimentale. I. Brachyeures et Anomoures. *Ann. Sci. Nat., Zool. Biol. Anim.* **10**, 1–98.

Pochon-Massion, J. (1968b). L'ultrastructure des spermatozoides vesiculaires chez les Crustaces Decapodes avant et au cours de leur devagination expérimentale. II. Macroures discussion et conclusions. *Ann. Sci. Nat., Zool. Biol. Anim.* **10**, 367–454.

Pratt, S. A., Scully, N. F., and Shur, B. D. (1993). Cell surface B1, 4 galactosyltransferase on primary spermatocytes facilitates their initial adhesion to Sertoli cells *in vitro*. *Biol. Reprod.* **49**, 470–482.

Pudney, J. (1993). Comparative cytology of the non-mammalian Sertoli cell. *In* "The Sertoli Cell" (L. D. Russell and M. D. Griswold, eds.), pp. 611–658. Cache River Press, Clearwater, FL.

Raychoudhury, S., and Millette, C. F. (1993). Surface-associated glycosyltransferase activities in rat Sertoli cells *in vitro*. *Mol. Reprod. Dev.* **36**, 195–202.

Redenbach, D. M., and Vogl, A. W. (1991). Microtubule polarity in Sertoli cells: A model for microtubule based spermatid transport. *Eur. J. Cell Biol.* **54**, 277–290.

Redenbach, D. M., Boekelheide, K., and Vogl, A. W. (1992). Binding between mammalian spermatid-ectoplasmic specialization complexes and microtubules. *Eur. J. Cell Biol.* **59**, 433–448.

Ren, H. P., and Russell, L. D. (1992). Quantitation of Sertoli cell-germ cell desmosome-gap junctions in relation to meiotic divisions in the male rat. *Tissue Cell* **24**, 565–574.

Robards, A. W., Lucas, W. J., Pitts, J. D., Jongsma, H. J., and Spray, D. C., eds. (1990). "Parallels in Cell to Cell Junctions in Plants and Animals." Springer-Verlag, Berlin.

Rodriguez, J. P., and Minguell, J. J. (1992). Membrane-associated proteoglycans produced by Sertoli cells are not randomly distributed on the cell surface. *Eur. J. Cell Biol.* **59**, 348–351.

Roosen-Runge, E. C. (1977). "The Process of Spermatogenesis in Animals." Cambridge Univ. Press, London.

Roosen-Runge, R. E., and Szöllösi, D. (1965). On biology and structure of the testis of *Phialidium leuckhart* (Leptomedusae). *Z. Zellforsch. Mikrosk. Anat.* **68**, 597–610.

Rosselli, M., and Skinner, M. K. (1992). Developmental regulation of Sertoli cell aromatase activity and plasminogen activator production by hormones, retinoids and the testicular paracrine factor, PModS. *Biol. Reprod.* **46**, 586–594.

Russell, L. D. (1993a). Form, dimensions, and cytology of mammalian Sertoli cells. *In* "The Sertoli Cell" (L. D. Russell and M. D. Griswold, eds.), pp. 1–38. Cache River Press, Clearwater, FL.

Russell, L. D. (1993b). Role in spermiation. *In* "The Sertoli Cell" (L. D. Russell and M. D. Griswold, eds.), pp. 269–304. Cache River Press, Clearwater, FL.

Russell, L. D. (1993c). Morphological and functional evidence for Sertoli-germ cell relationships. *In* "The Sertoli Cell" (L. D. Russell and M. D. Griswold, eds.), pp. 365–390. Cache River Press, Clearwater, FL.

Russell, L. D., and Griswold, M. D., eds. (1993). "The Sertoli Cell." Cache River Press, Clearwater, FL.

Russell, L. D., and Peterson, R. N. (1985). Sertoli cell junctions: Morphological and functional correlates. *Int. Rev. Cytol.* **94**, 177–211.

Russell, L. D., Ren, H. P., Hikim, I. S., Schulz, W., and Hikim, A. P. S. (1990). A comparative study in twelve mammalian species of volume densities, volumes and numerical densities of selected testis components emphasizing those related to the Sertoli cell. *Am. J. Anat.* **188**, 21–30.

Sar, M., Hall, S. II., Wilson, E. M., and French, F. S. (1993). Androgen regulation of Sertoli cells. In "The Sertoli Cell" (L. D. Russell and M. D. Griswold, eds.), pp. 509–516. Cache River Press, Clearwater, FL.

Sawada, N. (1984). Electron microscopical studies of spermatogenesis in polychaetes. *Forsch. Zool.* **29**, 100–114.

Saxena, B. P., and Tikku, K. (1989). Ultrastructural studies of two tissues in *Dysdercus koenigii* testis and their role in spermatogenesis. In "Regulation of Insect Reproduction IV" (M. Tonner, T. Soldan, and B. Bennettova, eds.), pp. 141–151. Czech. Publ. Acad. Sci., Prague.

Saxena, B. P., and Tikku, K. (1991). Scanning electron microscopical confirmation of spongy and interstitial cell structure in *Dysdercus koenigii* F. testis. *Invertebr. Reprod. Dev.* **20**, 115–119.

Scheib, D (1974). Sur quelques aspects cytologiques de la regression du testicle chez le poussin du Coeille sous tract a la photostimulation. *C. R. Itebd. Seances Acad. Sci., Ser.* D **279**, 1297–1300.

Schultz, M. C. (1989). Ultrastructural study of the coiled body and new inclusion, the "mykaryon" in the nucleus of the adult rat Sertoli cell. *Anat. Rec.* **225**, 21–25.

Schulze, C., and Holstein, A. F. (1993a). Human Sertoli cells under pathological conditions. In "The Sertoli Cell" (L. D. Russell and M. D. Russell, eds.), pp. 597–610. Cache River Press, Clearwater, FL.

Schulze, C., and Holstein, A. F. (1993b). Human Sertoli cell structure. In "The Sertoli Cell" (L. D. Russell and M. D. Griswold, eds.), pp. 685–702. Cache River Press, Clearwater, FL.

Segretain, D., Egloff, M., Gerard, N., Pinean, C., and Jégou, B. (1992). Receptor-mediated and absorptive endocytosis by male germ cells of different mammalian species. *Cell Tissue Res.* **268**, 471–478.

Self, J. T. (1983). Pentastomida. In "Reproductive Biology of Invertebrates" (K. G. Adiyodi and R. G. Adiyodi, eds.), Vol. 2, pp. 257–267. Wiley, New York.

Setchell, B. P., Maddocks, S., and Brooks, D. E. (1994). Anatomy vasculature and fluids of male reproduction tract. In "Physiology of Reproduction" (E. Knobil and J. D. Neil, eds.), 2nd ed. Vol. 1, pp. 1063–1176. Plenum, New York.

Shimizu, T., and Yagi, S. (1982). Hormonal effect on cultivated insect tissues. IV. The role of testis proteins in spermatogenesis of cabbage army worm, *Mamestra brassicae in vitro* (Lepidoptea: Nocktuidae) *Appl. Entomol. Zool.* **17**, 385–392.

Shimizu, T., Moribayashi, A., and Agai, N. (1985). *In vitro* analyses of spermatogenesis and testicular ecdysteroids in the cabbage army worms, *Mamestra brassicae* L (Lepidoptera: Nectuidae). *Appl. Entomol. Zool.* **20**, 56–61.

Skinner, M. K. (1991). Cell-cell interactions in the testis. *Endocr. Rev.* **12**, 45–77.

Skinner, M. K. (1993a). Secretion of growth factors and other regulatory factors. In "The Sertoli Cell" (L. D. Russell and M. D. Griswold, eds.), pp. 237–248. Cache River Press, Clearwater, FL.

Skinner, M. K. (1993b). Sertoli cell-peritubular myoid cell interactions. In "The Sertoli Cell" (L. D. Russell and M. D. Griswold, eds.), pp. 477–484. Cache River Press, Clearwater, FL.

Skinner, M. K., Norton, J. N., Mullaney, B. P., Rosselli, M., and Anthony, C. T. (1991).

Cell-cell interactions and the regulation of testis function. *Ann. N.Y. Acad. Sci.* **637**, 354–363.

Sourodaine, P., and Garnier, D. H. (1993). Stage-dependent modulation of Sertoli cell steroid production in dog fish (*Soyliorhinus canicula*). *J. Reprod. Fertil.* **97**, 133–142.

Sourdaine, P., Garnier, D. H., and Jégou, B. (1990). The adult dogfish (*Scyliorhinus canicula* L.) testis: A model to study stage-dependent changes in steroid levels during spermatogenesis. *J. Endocrinol.* **127**, 451–460.

Spiteri-Grech, J., and Nieschlag, E. (1993). Paracrine factors relevant to the regulation of spermatogenesis—a review. *J. Reprod. Fertil.* **98**, 1–14.

Sprando, R. L., and Russell, L. D. (1987). Comparative study of Sertoli cell ectoplasmic specializations in selected non-mammalian vertebrates. *Tissue Cell* **19**, 479–493.

Sprando, R. L., and Russell, L. D. (1988). Spermiogenesis in the red-ear turtle (*Pseudemeysscripta*) and the domestic fowl (*Gallus domestious*). A study of cytoplasmic events including cell volume changes and cytoplasmic elimination. *J. Morphol.* **198**, 95–118.

Stahler, M. S., Cheng, C. Y., Morris, P. L., Cailleau, G., Verhaeven, G., and Bardin, C. W. (1991). α-Macroglobulin: A multifunctional protein of the seminiferous tubule. *Ann. N.Y. Acad. Sci.* **626**, 73–80.

Stanely, H. P., and Lambert, C. C. (19850. The role of a Sertoli cell actinmyosin system in sperm bundle formation in the ratfish, *Hydrolagus colliei* (Chondrichthyes, Holocephali). *J. Morphol.* **186**, 223–236.

Stefanini, M., Russo, M. A., Filippini, A., Canipari, R., Palombi, F., Bertalot, G., Di Bonito, P., and Ziparo, E. (1988). Role of cell mitrations within the semineferous opithelium in the regulation of spermatogenesis. *In* "The Molecules and Cellular Endocrinology of the Testies" (B. A. Cooke, and R. M. Sharpe, eds.), pp. 281–296. Raven Press, New York.

Steinberger, A., and Jakubowiak, A. (1993). Sertoli cell culture: Historical perspective and reveiw of methods. *In* "The Sertoli Cell" (L. D. Russell and M. D. Griswold, eds.), pp. 155–180. Cache River Press, Clearwater, FL.

Storch, V., and Ruhberg, H. (1983). Onychophora. *In* "Reproductive Biology of Invertebrates" (K. G. Adiyodi and R. G. Adiyodi, eds.), pp. 397–405. Wiley, New York.

Syed, V., Gerard, N., Kaipia, A., Bardin, C. W., Parvinen, M., and Jégou, B. (1993). Identification, ontogeny, and regulation of an interleukin-6-like factor in the rat seminiferous tubule. *Endocrinology (Baltimore)* **132**, 293–299.

Sylvester, S. R. (1993). Secretion of transport and binding proteins. *In* "The Sertoli Cell" (L. D. Russell and M. D. Griswold, eds.), pp. 201–216. Cache River Press, Clearwater, FL.

Szöllösi, A. (1974). Ultrastructural study of spermatodesm in *Locusta migratoria* migratorioides (R. et F. J: Acrosome and cap formation. *Acrida* **3**, 175–192.

Szöllösi, A. (1975). Electron microscope study of spermiogenesis in *Locusta migratoria* (Insect Orthoptera). *J. Ultrastruct. Res.* **50**, 322–346.

Szöllösi, A. (1982). Relationships between germ and somatic cells in the testes of locusts and moths. *In* "Insect Ultrastructure" (R. C. King and H. Akai, eds.), Vol. 1, pp. 32–60. Plenum Press, New York.

Szöllösi, A., and Marcaillou, C. (1977). Electron microscope study of the blood-testis barrier in an insect, *Locusta migratoria*. *J. Ultrastruct. Res.* **59**, 158–172.

Szöllösi, A., and Marcaillou, C. (1980). Gap junctions between germ and somatic cells in the testis of the moth, *Angasta kuehniella* Insects: Lepiodptera. *Cell Tissue Res.* **213**, 137–148.

Talbot, J. A., Lambert, A., Mitchell, R., Grabinski, M., Anderson, D. C., Tsatroulis, A., Spalet, S. M., and Robertson, W. R. (1991). Follicle stimulating hormone dependent estrogen secretion by rat Sertoli cells *in vitro* Modulation by calcium. *Acta Endocrinol. (Capenhagen)* **125**, 280–285.

Tazima, I., Inoue, S., and Gopal Dutt, N. H. (1993). Spermatogenesis in fresh water bryzoan *Pectinatella gelatinosa:* a light microscopy. *Proc. Natl. Acad. Sci. India, Sect. B* **63**, 301–304.

Teerds, K. J., and Dorrington, J. H. (1992). Localization of TGF-B1 and TGF-B2 in the rat testis. *Proc. Eur. Workshop Mol. Cell. Endocrinol. Testis, 7th.* Castle Elmau, Germany, 1992.

Thiry, M. (1993). Ultrastructural distribution of DNA and RNA within the nucleolus of human Sertoli cells as seen by molecular immunocytochemistry. *J. Cell Sci.* **105**, 33–39.

Tikku, K., and Saxena, B. P. (1990). Ultrastructural spermatid and sperm morphology in *Poecilocerus pictus* (Fab) with a reference to spermiophagic cells in the testis and sperm duct. *Tissue Cell* **22**, 71–80.

Toyama, Y., Urena Calderon, F., and Quesada, R. (1990). Ultrastructural study of crystalloids in Sertoli cells in the three toad sloth (*Bradypus tridactylus*). *Cell Tissue Res.* **259**, 599–602.

Ueno, H., Nishimune, Y., and Mori, H. (1991). Cyclic localization change of Golgi apparatus in Sertoli cells induced by mature spermatids in rats. *Biol. Reprod.* **44**, 656–662.

Ueno, N., and Mori, H. (1990). Morphometric analysis of Sertoli cell ultrastructure during the seminiferous epithelial cycle in rats. *Biol. Reprod.* **43**, 769–776.

van Huren, J. H. J., and Soley, J. T. (1990). Some ultrastructural observations of the Leydig and Sertoli cells in the testis of *Tilapia rendalli* following induced testicular recrudescence. *J. Morphol.* **206**, 57–64.

Vogl, A. W., Pfeiffer, D. C., and Redenbach, D. M. (1991a). Ectoplasmic ("junctional") specializations in mammalian Sertoli cells. Influence on spermatogenic cells. *Ann. N. Y. Acad. Sci.* **637**, 175–202.

Vogl, A. W., Pfeiffer, D. C., and Redenbach, D. M. (1991b). Sertoli cell cytoskeleton. Influence on spermatogenic cells. *Proc. Serono Symp.* 704–715.

Vogl, A. W., Pfeiffer, C., Redenbach, D. M., and Grove, B. D. (1993). Sertoli cell cytoskeleton. *In* "The Sertoli Cell" (L. D. Russell and M. D. Griswold, eds.), pp. 39–86. Cache River Press, Clearwater, FL.

Walker, C. W. (1980). Spermatogenic columns, somatic cells and the microenvironment of germinal cells in the testes of asteroids. *J. Morphol.* **166**, 81–107.

Walker, C. W., and Larochelle, D. (1984). Interactions between germinal and somatic accessory (SA) cells of the spermatogenic epithelium of *Asterias vulgaris in vivo* and *in vitro*. *Adv. Invertebr. Reprod.* 41–52.

Waterfield, M. D., ed. (1990). Growth factors in cell and developmental biology. BSCB. *J. Cell Sci., Suppl.* **13**, 1–300.

Weber, J. E., Turner, T. T., Tung, K. S. K., and Russell, L. D. (1988). Effects of cytochalasin D on the integrity of the Sertoli cell (blood-testis) barrier. *Am. J. Anat.* **182**, 130–147.

Whitfield, P. J. (1969). Studies on the reproduction of Acanthocephala. Ph.D Dissertation, University of Cambridge, Cambridge, UK.

Wingstrand, K. G. (1972). Pentastomid, *Raillietiella hemidactyli. Biol. Skr.—Dan. Vidensk. Selsk.* **19**, 1–72.

Wingstrand, K. G. (1978). Comparative spermatology of the crustacea Entomostracea I. Subclass Brachiopoda. *Biol Sk. Kgl. Dan. Vidnsk. Selsk.* **22**, 1–66.

Wright, E. J., and Sommerville, R. I. (1984). Structure and development of the spermatozoon of the parasitic nematode, *Nematospiroides dubius. Parasitology* **90**, 179–191.

Wright, E. J., and Sommerville, R. I. (1985). Spermatogenesis in a nematode *Nippostrongylus bresilensis. Int. J. Parasitol.* **15**, 283–299.

Wrobel, K.-N., and Schimmel, M. (1989). Morphology of bovine Sertoli cell during the spermatogenic cycle. *Cell Tissue Res.* **257**, 93–103.

Yasuzumi, G. (1962). Spermatogenesis in animals as revealed by electron microscopy.

XII. Light and electron microscope studies on spermiogenesis of *Cipangopaludina malleata* Reeve. *J. Ultrastruct. Res.* **7,** 488–503.

Yasuzumi, G., Tanaka, H., and Tezuka, O. (1960). Spermatogenesis in animals as revealed by electron microscopy. VIII. Relation between the nutritive cells and the developing spermatids in a pond snail, *Cipangopaludina malleata* Reeve. *J. Biophys. Biochem. Cytol.* **7,** 499–504.

Yazama, F., Sawada, H., Hirosawa, K. Hayashi, Y., and Nishida, T. (1991). Visualization of the Sertoli cell (blood-testis barrier) in the boar. *Tissue Cell* **23,** 235–246.

Ye, S.-J., Ying, L., Ghosh, S., de Franca, L. R., and Russell, L. D. (1993). Sertoli cell cycle: A re-examination of the structural changes during the cycle of the seminiferous epithelium of the rat. *Anat. Rec.* **237,** 187–198.

Zong, S. D., Bardin, C. W., Phillips, D., and Cheng, C. Y. (1992). Testins are localized to the junctional complexes of rat Sertoli and epididymal cells. *Biol. Reprod.* **47,** 568–577.

Zwain, J. H., Grima, J., Stahler, M. S., Saso, L., Cailieau, J., Verhaeven, G., Bardin, C. W. and Cheng, C. Y. (1993). Regulation of Sertoli cell α_2-macroglobulin and clusterin (SGP-2) secretion by peritubular myoic cells. *Biol. Reprod.* **48,** 180–187.

Factors Controlling Growth, Motility, and Morphogenesis of Normal and Malignant Epithelial Cells

Carmen Birchmeier, Dirk Meyer, and Dieter Riethmacher
Max Delbrück Center for Molecular Medicine,
D-13122 Berlin, Germany

Factors that control epithelial growth, motility, and morphogenesis play important roles in malignancy and in normal development. Here we discuss the molecular nature and the function of two types of molecules that control the development and maintenance of epithelia: Components that regulate epithelial cell adhesion; and soluble factors and their receptors that regulate growth, motility, differentiation, and morphogenesis. In development, the establishment of epithelial cell characteristics and organization is crucially dependent on cell adhesion and the formation of functional adherens junctions. The integrity of adherens junctions is frequently disturbed late in tumor progression, and the resulting loss of epithelial characteristics correlates with the metastatic potential of carcinoma cells. Various soluble factors that induce epithelial growth, motility, or differentiation in cell culture, function via tyrosine kinase receptors. We concentrate here on receptors that are expressed exclusively or predominantly on epithelia, and on ligands that are derived from the mesenchyme. In development, these receptors and their ligands function in mesenchymal-epithelial interactions, which are known to govern growth, morphogenesis, and differentiation of epithelia. During tumor development, mutations or overexpression of the receptors are frequently observed; these alterations contribute to the development and progression of carcinomas.

KEY WORDS: E-cadherin, Catenin, Scatter factor/hepatocyte growth factor (SF/HGF), c-Met receptor, Neuregulin/Neu differentiation factor/NDF/heregulin, c-ErbB receptor, Keratinocyte growth factor (KGF).

I. Introduction

Epithelial cells are the one cell type most important in the development of human malignancies. More than 90% of all malignant tumors are carcinomas, and thus of epithelial origin. Aberrant growth and the ability to invade the underlying tissue are intrinsic properties of the fatally altered cells. Therefore, the biology of epithelia, in particular factors which control growth, motility, and invasiveness, are of considerable interest. The concepts derived from the study of such factors will have an impact on tumor diagnosis and therapy in the future. Interestingly the molecules implicated in the control of malignant properties of carcinomas turn out to play fundamental roles in the control of normal epithelial growth, differentiation, and morphogenesis during development.

During embryogenesis, epithelial cells participate in the formation of a wide array of functionally and morphologically different structures, such as the intricately branched tubular epithelium of the lung, the one-layered villi epithelium of the intestine, or the plates of hepatocytes arranged in three dimensions in the liver. These diverse cells generally develop and differentiate from the endoderm or ectoderm and thus derive from a few precursor cells which are epithelial in character. The driving forces which govern growth, differentiation, and morphogenesis of the epithelia are usually mesenchymal in origin. The importance of such paracrine signals was discovered 3–4 decades ago in organ culture experiments, but the molecular nature of the signals given by the mesenchyme and of the epithelial receptors which receive such signals is only now being elucidated.

II. Structural and Molecular Characteristics of Epithelia

Epithelia are easily recognized since they form continuous sheets of tightly adhering cells that are usually cuboidal in shape. In cell culture, epithelial cells grow in aggregates and display little cell motility compared with other cell types. Characteristic for epithelia are specialized organelles, tight junctions, adherens junctions, and desmosomes, which are responsible for the intercellular contacts between the individual cells. In addition, polarization is observed, that is, the basal, lateral, and apical surfaces of the cells can have distinct morphological features (Rodriguez-Boulan and Nelson, 1989; Schwarz *et al.*, 1990; Birchmeier and Behrens, 1994).

A consequence of the laterally located junctional complexes is the inhibition of free diffusion of proteins in the cell membrane. As a result, distinct proteins are found on the apical and basolateral surface. Polar epithelial cells have evolved special mechanisms that allow the transport of membrane proteins to either the apical or the basolateral membrane (Simons and Wandinger-Ness, 1990). Adhesion molecules or receptors for signals received by epithelia can therefore be located exclusively or predominantly on either surface.

Adherens junctions of epithelial cells are specialized structures containing the transmembrane cell-adhesion molecule, E-cadherin, which recognizes and binds E-cadherin present on the neighboring cells in a Ca^{2+}-dependent manner. The cytoplasmic portion of E-cadherin interacts with the catenins, α-, β- and γ-catenin (Takeichi, 1991; Kemler, 1993; Tsukita et al., 1993). In addition, other cytoskeletal proteins (e.g., vinculin, α-actinin, fodrin, radixin, ezrin, and moesin) are located on the cytoplasmic side of this junctional complex, which anchors the actin-containing filaments (Tsukita et al., 1992). The integrity of these junctions is critical for the maintenance of the functional characteristics and the shape of epithelia. Its disruption leads to the dissociation of epithelial sheets and thus to an increased motility of the cells (Behrens et al., 1989). A controlled loss of epithelial character concomitant with dissociation and increased motility of the cells is a prerequisite for normal morphogenic processes and occurs, for example, during mesoderm formation. A similar but uncontrolled epithelial-mesenchymal conversion is observed during development and progression of human tumors (Valles et al., 1991; Birchmeier and Behrens, 1994). Loss of epithelial differentiation and increased motility correlate with the malignancy of the carcinoma cells and with the destruction of adherens junctions by various mechanisms.

On the basal surface of epithelia, hemidesmosomes are located that correspond to junctions contacting the basement membrane (Garrod, 1993). The basement membrane separates the epithelial cell compartment from underlying mesenchymal cells and is formed by both the epithelia and the mesenchyme. The characteristic constituents of the basement membrane are laminin, collagen IV, nidogen/entactin, and basement membrane proteoglycan. The mesenchymal reticular lamina which lies below the basement membrane contains collagen types I and II as well as fibronectin (Timpl, 1989). The receptors of epithelial cells for basement membrane components are located in the basal surface of the plasma membrane. Expressed on epithelial cells are, for example, the integrins $\alpha6\beta1$ and $\alpha6\beta4$, which are receptors for laminin, and the integrin $\alpha1\beta1$, which is a receptor for collagen (A. Sonnenberg et al., 1991). The hemidesmosomes are enriched for the integrin receptor $\alpha6\beta4$. A variety of experimental evidence is accumulating that demonstrates a function for integrins not only in the adhesion of cells to the extracellular matrix, but also in the transduction of signals that can influence growth and motility (Juliano and Varner, 1993). In contrast to epithelial cells, the motility of various other cell types is strongly dependent on the substrate and can therefore be regulated by integrin signalling.

In addition, various receptors with tyrosine kinase activity exist that are expressed exclusively or predominantly on epithelial cells, such as c-Ros, c-Met, c-Neu, c-Ret or the receptor for the keratinocyte growth factor (KGF) (Kokai *et al.*, 1987; Quirke *et al.*, 1989; Press *et al.*, 1990; Mori *et al.*, 1989; E. Sonnenberg *et al.*, 1991, 1993a; Orr-Urtreger *et al.*, 1993b; Pachnis *et al.*, 1993). Since most of these receptors were identified as oncogenes in NIH3T3 fibroblasts, they can clearly transmit mitogenic signals (Aaronson, 1991). However, experiments with epithelial cells in culture have demonstrated that the receptors not only regulate growth, but can also influence motility, differentiation, and morphogenesis (Birchmeier *et al.*, 1993). Recent genetic evidence demonstrates a pivotal role for receptor tyrosine kinases in the regulation of normal epithelial physiology and development. Moreover, deregulated expression or mutation of genes encoding receptor tyrosine kinases has been found in various types of carcinomas and thus also contributes to the development of human malignancies. The receptors recognize factors which will be discussed in detail in this chapter.

III. Formation and Differentiation of Epithelial Cells in Normal Development

In the early mammalian embryo, at the morula stage, few cells with a full developmental potential exist. During compaction of the morula, these cells acquire epithelial characteristics, since they become morphologically polarized and redistribute a variety of proteins in a polar manner, for instance E-cadherin to the lateral surface (Fleming and Johnson, 1988; Kemler *et al.*, 1990; Wiley *et al.*, 1990; Fleming *et al.*, 1992).

During gastrulation, the three germ layers—ectoderm, mesoderm, and endoderm—are formed. Whereas ectoderm and endoderm remain essentially epithelial, the mesodermal cells are the first different cell types in development. The mesodermal cells display new characteristics, that is, express a set of different proteins typical for the mesenchymal program (Hay, 1990; Valles *et al.*, 1991). Mesoderm formation is thus an example of a controlled epithelial-mesenchymal transition. The signals that are responsible for mesoderm formation have been extensively studied in *Xenopus laevis,* and fibroblast growth factor (FGF), activin, and noggin have been implicated as inducers (Beddington and Smith, 1993; Harland, 1994). During further development, the different organs are formed and, in general, two distinct cell types, mesenchymal and epithelial cells, partic-ipate in this process. Whereas the mesenchymal cells are derived from either mesoderm or neural ectoderm, the epithelial cells usually arise from epithelial ectoderm or endoderm and only in a few, rare cases (for instance in the kidney) from the mesoderm after a mesenchymal-epithelial tran-sition.

An illustrative example of the morphogenesis of an organ is the lung (Fig. 1), where an outgrowth of epithelial cells from the endoderm into the splanchnic mesoderm generates the anlage of the organ. All epithelia of the differentiated lung develop by further growth, branching, and differ-entiation of the endodermal bud. In parallel, the mesenchymal cell com-partment also expands and differentiates (Spooner and Wessells, 1970). The growth of epithelia into mesenchymal tissues, and their subsequent growth, branching, and differentiation is a general theme found during development of many other organs, for example, salivary gland, pancreas, prostate, pituitary gland, kidney, or breast (Grobstein, 1953; Wessels and Cohen, 1967; Lasnitzki and Mizuno, 1980; Kusakabe et al., 1985; Saxén, 1987; Sakakura, 1991).

The anlage for the gastrointestinal tract is formed early during develop-ment. It consists then of a poorly differentiated, stratified epithelium de-rived from the endoderm that is surrounded by mesenchymal cells. In the perinatal phase, the multilayered epithelium is converted into the single-layered epithelium of the villi, and simultaneously terminal differentiation of the epithelium is initiated (Fig. 2). The first sign of this morphogenic process is the dissociation of cell-cell contacts and the formation of second-ary lumina in the multilayered epithelium. The underlying mesenchymal cells then invade the loosened epithelium. Whereas superficial epithelia are exfoliated, the epithelial cells which are in direct contact with the mesenchyme proliferate. The outcome of this rearrangement is the one-layered villi epithelium (Mathan et al., 1976). This is an example of a developmental process in which interconversions of different types of epithelial organizations are observed, a process also guided by mesenchy-

FIG. 1   Schematic representation of the developing lung. The epithelial lung bud that has grown into the splanchnic mesenchyme is indicated (left). The lung bud grows and branches repeatedly upon mesenchymal signals; the size of the mesenchyme increases in parallel (middle and right).

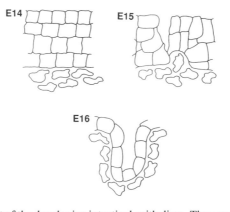

FIG. 2   Rearrangement of the developing intestinal epithelium. The appearance of the murine intestine at the indicated times in development is depicted (modified after Mathan *et al.*, 1976). The cuboidal cells and the irregularly shaped cells represent epithelia and mesenchyme, respectively. During early stages of development, a multilayered, poorly differentiated epithelium is observed; mesenchymal cells are located basally (E14). With the onset of terminal differentiation, the contacts between the epithelial cells are loosened and mesenchymal cells invade the epithelial cell layers (E15). The differentiated villi epithelium is single layered (E16).

mal epithelial interactions (Kedinger *et al.*, 1986). In addition, a regionalization of the intestinal epithelium takes place. This is characterized by the formation of morphological and functional gradients in the number and depth of the villi, in the distribution of the distinct enteric cell types, or in the expression of various proteins (Gordon, 1989; Yasugi, 1993).

The liver is a derivative of the foregut. Endodermal epithelial cells thicken to form the liver diverticulum; the cells of this liver bud proliferate and migrate into the surrounding septum transversum, which consists of loose mesenchymal cells. The epithelium subsequently differentiates into hepatocytes. Mesenchymal cells of the septum transversum also contribute to the formation of the liver and differentiate, for example, into Kupffer cells and contribute to the formation of the liver sinusoids. Again, mesenchymal-epithelial interactions are essential for the development of this organ (Le Douarin, 1975; Houssaint, 1980). The first markers for the hepatic lineage, for example albumin, transferrin, or apolipoprotein A1, are expressed at low levels in the precursor cells in the second half of embryogenesis (Meehan *et al.*, 1984). However, differentiation of the hepatocytes is a process that only ends in the perinatal period, when liver-specific genes controlled also by glucocorticoids and cAMP start to be expressed (Boshart *et al.*, 1993).

As these examples show, both epithelial and mesenchymal cells contribute to the formation of the various organs. To ensure ordered growth, differentiation, and morphogenesis, communication between the two cellular compartments has to occur. Organ culture experiments provided the first evidence for paracrine signal exchange between mesenchyme and epithelia. Such experiments demonstrate that major driving forces for the development of epithelia are mesenchymal factors.

A. Secondary Induction

Organ development can be analyzed by the removal of the respective anlagen at the appropriate time in development and subsequent culture. For example, the murine lung anlagen can be removed on day 10 of development, after the endodermal epithelia have grown into the splanchnic mesenchyme. During subsequent incubation *in vitro,* the epithelia grow and branch repeatedly. Alternatively, the epithelium can be dissociated from the mesenchyme by mild tryptic digestion. Cultured separately, the epithelial bud will not grow or branch; when recombined with the lung mesenchyme, growth and branching occur (Spooner and Wessels, 1970; Wessels, 1970).

By combining heterologous mesenchyme and epithelia, it was demonstrated that the mesenchyme does not just provide a permissive signal for the developing epithelia, such as the necessary three-dimensional matrix for the morphogenic processes. Instead, mesenchymal signals can actually be instructive in nature and can thus determine not only growth but also the exact shape of the epithelial structures and their developmental fate. Such inductive events that occur during organogenesis are commonly referred to as secondary inductions, in order to distinguish them from primary induction, which occurs early in development and results in the formation of mesoderm or neural cells.

Signalling specificity during organ development is observed, for instance, in the lung. There, the branching of the lung epithelium is induced by bronchial but not by tracheal mesenchyme (Wessels, 1970). Another example is the mammary gland epithelium, which in coculture with mammary gland mesenchyme forms long, branched tubules. When combined with the mesenchyme from the salivary gland, the mammary epithelium forms short, frequently branched structures that resemble morphologically the salivary gland epithelium (Kratochwil, 1969, 1986).

The pancreas is an extreme example of the nonspecificity of the mesenchymal contribution to epithelial morphogenesis, since the pancreatic mesenchyme can be replaced by a wide range of heterologous mesenchymes (Fell and Grobstein, 1968). In contrast, growth and branching of the ureter are only supported by the nephrogenic mesenchyme (Saxén, 1987). Therefore, some of the factors required by embryonal epithelia seem to be ubiquitously expressed in virtually every mesenchyme, whereas other epithelia require a distinct factor or a combination of factors that are only provided by one entity—the homologous mesenchyme. The complexity of the epithelial responses observed in these organ culture systems clearly indicates that more than a single mesenchymal factor controls the development of epithelial cells and that organ-specific requirements as well as an organ-specific distribution of mesenchymal factors exist.

B. E-Cadherin and the Formation of Epithelia

The importance of the cell adhesion molecule E-cadherin in the early development of epithelial-like cells was originally suggested by experiments using E-cadherin-specific antibodies, and was recently demonstrated genetically (Hyafil *et al.*, 1980; Vestweber and Kemler, 1984; Larue *et al.*, 1994; Riethmacher *et al.*, 1995). We have introduced a targeted mutation into the E-cadherin gene of the mouse genome via homologous recombination and embryonic stem cell technology (Riethmacher *et al.*, 1995). This mutation removes sequences of the E-cadherin gene that encode a Ca^{2+}-binding site essential for the adhesive function of the molecule (Ozawa *et al.*, 1990). Embryos that carry this mutation in a homozygous state undergo normal development up to and including the compaction of the morula. However, the compacted state of the morula is not sustained. The individual morula cells in the mutant embryos lose their morphological polarization; they become rounded but continue to divide (Figs. 3, 4). As a consequence, the embryos appear totally distorted at a time when wild-type or heterozygous mutant embryos form a well-organized blastocyst. Since the mutant embryos never emerge from the zona pellucida, further development, in particular implantation into the uterus, cannot take place.

A similar phenotype for an independently generated homozygous mutation of E-cadherin has also been described elsewhere (Larue *et al.*, 1994). It has previously been reported that removal of Ca^{2+} ions or treatment with anti-E-cadherin antibodies interferes with compaction of the mouse morula, indicating that even the initial polarization of the epithelial-like cells is dependent on E-cadherin (Hyafil *et al.*, 1980, 1981). Nevertheless, embryos which lack a functional E-cadherin gene can undergo normal compaction. This is apparently due to maternally derived E-cadherin protein (Sefton *et al.*, 1992) present in the mutant embryos and not to functional compensation by other Ca^{2+}-dependent adhesion molecules (Riethmacher *et al.*, 1995). The presence of maternal E-cadherin in mutant embryos at early stages was demonstrated by immunohistochemical techniques; it disappears when the embryos develop beyond the morula stage.

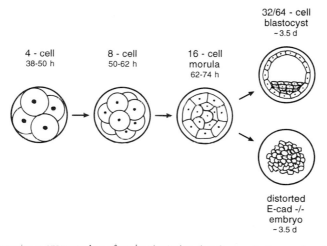

FIG. 3 Schematic representation of preimplantation development of normal and E-cadherin mutant mouse embryos. Preimplantation mouse development from the 4-cell to the blastocyst stage of normal and homozygous E cadherin embryos. The times needed to reach the depicted developmental stages are indicated.

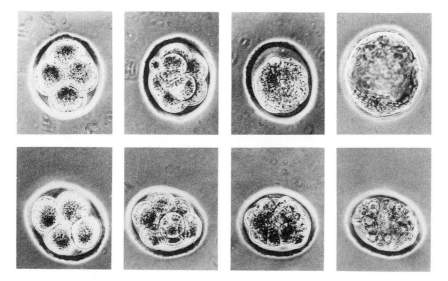

FIG. 4 Morphology of wild-type and E-cadherin mutant embryos during preimplantation development. Wild-type (top row) and homozygous E-cadherin mutant (bottom row) embryo during preimplantation development (4-cell stage, 8-cell stage, compacted morula, blastocyst, and distorted embryo, respectively). The photographs show embryos at the same stages depicted schematically in Fig. 3.

The maternal E-cadherin thus suffices for initial compaction and formation of the first epithelial-like cells, but not for development beyond the morula stage.

C. Molecular Factors That Control Epithelial Development during Organogenesis

Epithelial tyrosine kinase receptors and their specific mesenchymal ligands play an important role in the regulation of growth, differentiation, and morphogenesis of the epithelia during organogenesis. In particular, the formation of the branched, tubular structures found in many parenchymal organs can be regulated by tyrosine kinase receptors and their ligands *in vitro* and *in vivo* (Schuchardt *et al.,* 1994; Peters *et al.,* 1994; Montesano *et al.,* 1991b; Yang *et al.,* 1995). In addition, growth, survival, or differentiation can be affected by such signalling systems. Although evidence for the pivotal roles of tyrosine kinase receptors in epithelial development has recently accumulated, these are of course not the only essential molecules in epithelial development. In particular, other signalling systems such as transforming growth factor-β (TGF-β) or related molecules and the corresponding receptors, as well as cell matrix molecules or proteases have long been recognized as important components and have been reviewed elsewhere (Birchmeier and Birchmeier, 1993).

IV. Growth and Motility of Transformed Epithelial Cells

Tumors of epithelial origin are of major medical importance, since more than 90% of all fatal malignancies are carcinomas (DeVita *et al.,* 1993). Multiple genetic changes in dominant and recessive oncogenes that contribute to tumor formation and progression have been identified. Among the most frequent mutations found in carcinomas are activating mutations in the *ras* family of genes, mutations that interfere with the function of p53, and amplification or rearrangements in genes encoding tyrosine kinase receptors (Barbacid, 1987; Aaronson, 1991; Hollstein *et al.,* 1991; Vogelstein and Kinzler, 1993; Hinds and Weinberg, 1994). The mutations influence growth as well as genetic stability of the affected cells. Critical steps in the development of malignant tumors are the acquisition of unregulated growth and the ability to invade the underlying tissue. Early genetic approaches for the analysis of oncogenes have focused on the induction of transformation (i.e., unregulated growth in fibroblasts), which has led in the past decade to a wealth of data on the pathways controlling normal and aberrant growth. In comparison, there has been a lag in the under-standing of the molecular causes leading to the acquisition of the metastatic potential of carcinoma cells.

In human carcinomas, the loss of epithelial differentiation is an im-portant prognostic marker and correlates with a poor outcome of the disease (Gabbert, 1985; Gabbert *et al.*, 1985). The loss of epithelial differ-entiation is accompanied by higher motility and invasiveness of the tumor cells and by reduced intercellular adhesion (Birchmeier and Behrens, 1994). Cell biologists have been fascinated by the profound changes in cell morphology and motility associated with acquisition of metastatic ability. However, for many years the study of the characteristic morphol-ogy and motility of malignant carcinoma cells has remained essentially descriptive in nature. Only recently has an understanding in molecular terms started to emerge.

A. Adhesion Molecules in the Regulation of Motility and Metastatic Potential of Carcinoma Cells

Disruption of the adherens junctions of epithelial cells can be achieved *in vitro* by the addition of monoclonal antibodies against E-cadherin that interfere with the function of the cell adhesion molecule. The epithelial cells treated in this manner change their shape, are dissociated, and resem-ble fibroblasts in morphology. They also become more motile and able to invade a collagen matrix (Behrens *et al.*, 1989; Frixen *et al.*, 1991; Vlem-inckx *et al.*, 1991). When cocultured with heart tissue, the polar Madin-Darby canine kidney (MDCK) cells form a single-layered epithelial sheet surrounding the tissue. In the presence of anti-E-cadherin antibodies, the cells invade the tissue (Behrens *et al.*, 1989). The presence of functional adherens junctions is thus necessary for epithelial morphology; its disrup-tion increases the motility and the invasive potential of epithelial cells.

An analysis of cell lines established from different types of carcinomas demonstrated that loss of epithelial morphology and increased invasive-ness often correlates with loss or downregulation of the E-cadherin pro-tein. Carcinoma cells with a well-differentiated, epithelial morphology generally do not invade a collagen matrix *in vitro* and express E-cadherin, whereas carcinoma cells with a fibroblast-like morphology are invasive and have lost E-cadherin (Frixen *et al.*, 1991). The invasiveness of such cells could be prevented by the transfection of E-cadherin cDNA under an ectopic promotor and was again induced by functionally interfering monoclonal anti-E-cadherin antibodies. Similar results were obtained with normal murine mammary gland cells or MDCK cells transformed with v-*ras* or activated H-*ras*, respectively; these cells can display a dissociated and invasive phenotype and do then not produce E-cadherin. Expression of E-cadherin cDNA under the control of an ectopic promotor in such cells restores cell adhesion and also abolishes the ability of the cells to invade collagen matrix or heart tissue (Vleminckx *et al.*, 1991).

In many types of primary human carcinomas, for example, in carcinomas of the head and neck, in lobular breast carcinomas, or gastric and colorectal carcinomas, downregulation of E-cadherin is observed to correlate with the malignancy of the tumors (Schipper *et al.*, 1991; Moll *et al.*, 1993; Oka *et al.*, 1993; Dorudi *et al.*, 1993; Umbas *et al.*, 1994). However, tumors exist where the correlation is not apparent, such as ductal carcinomas of the breast (Moll *et al.*, 1993). Since loss of the epithelial morphology and dissociation of the cells are also manifest, different mechanisms interfering with the formation of junctional complexes might be operative.

Indeed, other molecular mechanisms that regulate the integrity of adherens junctions have been identified in normal and malignant cells. Receptor-type tyrosine kinases can phosphorylate β-catenin or plakoglobin on tyrosine residues (Shibamoto *et al.*, 1994). This phosphorylation correlates with a dissociation and increased motility of the epithelial cells. Reversible modifications, that is, transient tyrosine phosphorylation, might be a mechanism that allows short-term dissociation of epithelial cells during morphogenesis. In v-*src*-transformed epithelial cells, β-catenin or plakoglobin are phosphorylated on tyrosine (Matsuyoshi *et al.*, 1992; Behrens *et al.*, 1993; Hamaguchi *et al.*, 1993). The phosphorylation mediated by oncogenes is not temporarily regulated. In addition, mutations in the genes for E-cadherin or other cytoplasmic proteins of the junctional complex have been observed in carcinoma cell lines and primary tumors (Hirano *et al.*, 1992; Oda *et al.*, 1993, 1994; Oka *et al.*, 1993; Becker *et al.*, 1994; Risinger *et al.*, 1993). Therefore, diverse mechanisms can interfere with the formation of functional adherens junctions and result in a loss of epithelial integrity and an increase in the invasive potential of carcinoma cells.

B. Genetic Analysis of Tumor Progression in Human Carcinomas

The genetic changes associated with development and progression of various carcinomas and, in particular, of colon carcinomas have been analyzed. Frequent mutations were identified in the genes for K-ras, p53, hMSH2, DCC, and APC (Forrester *et al.*, 1987; Fearon *et al.*, 1990; Malkin *et al.*, 1990; Joslyn *et al.*, 1991; Groden *et al.*, 1991; Vogelstein and Kinzler, 1993; Su *et al.*, 1993; Fishel *et al.*, 1993). The effect of the *ras* genes on growth is well documented (McCormick, 1994). Mutations in hMSH2 lead to genetic instability (Fishel *et al.*, 1993). Although the molecular function of p53 is as yet not solved, mutations in p53 influence both growth and genetic stability, possibly through the regulation of cell cycle arrest and progression (Donehower *et al.*, 1992; Kemp *et al.*, 1993; Livingstone *et al.*, 1992). In contrast, indirect and direct evidence points toward a role for the DCC and APC genes in cell adhesion. The sequence of DCC shows homologies to N-CAM (Fearon *et al.*, 1990), whereas APC, a cytoplasmic protein, interacts with β-catenins and might thus indirectly influence the stability of the adherens junctions (Su *et al.*, 1993; Rubinfeld *et al.*, 1993; Huelsken *et al.*, 1994). Also, mutations in genes implicated in cell adhesion lead in *Drosophila* to a high incidence of carcinoma-like tumors (Gateff and Mechler, 1989; Tsukita *et al.*, 1993).

C. Tyrosine Kinase Receptors and Their Ligands in the Control of Growth and Motility of Carcinomas

A variety of tyrosine kinase receptors are expressed predominantly or exclusively on epithelia and can control the growth, differentiation, and motility of such cells (Birchmeier *et al.*, 1993). Since abnormal growth and motility are essential in tumor progression, deregulated activity of such receptors can profoundly influence carcinogenesis. Indeed, *c-neu/ HER2*, c-*met* and c-*ret* are found to be mutated or overexpressed in particular types of human carcinomas (Slamon *et al.*, 1987, 1989; Santoro *et al.*, 1990; Liu *et al.*, 1992; Di Renzo *et al.*, 1992; Kuniyasu *et al.*, 1992; Rege-Cambrin *et al.*, 1992; Hofstra *et al.*, 1994; Jucker *et al.*, 1994; Boix *et al.*, 1994). The mutations observed can often relieve the receptors from requiring the presence of the specific ligand needed for signalling under normal conditions and therefore constitutively activate the receptors (Rodrigues and Park, 1994). In addition, expression of high levels of the intact ErbB2 receptor has been reported to suffice for ligand-independent activation and the induction of transformation in fibroblasts (Di Fiore *et al.*, 1987; Hudziak *et al.*, 1987) and high levels of autophosphorylated c-Met have been observed in carcinoma cells that overexpress the intact receptor (Ponzetto *et al.*, 1991; Giordano *et al.*, 1989a). However, the presence of the specific ligands, whether produced in an autocrine or paracrine fashion, is also a variable to be evaluated and of potential importance for tumor progression.

Mainly the mitogenic effect of these receptors has been extensively studied, and growth and transformation are well-known responses to the signals mediated by the receptors in fibroblasts. Recently, the biological responses of epithelial or carcinoma cells have gained more attention. Interestingly, receptor tyrosine kinases can also control motility and invasive behavior. This is particularly well documented for c-Met and its ligand, scatter factor/hepatocyte growth factor (SF/HGF) (Stoker *et al.*, 1987; Gherardi *et al.*, 1989; Weidner *et al.*, 1990). In addition, acidic FGF (aFGF) was found to increase motility of a bladder carcinoma cell line (Jouanneau *et al.*, 1991), and, by the use of a chimeric receptor, the active Ros tyrosine kinase was demonstrated to dissociate epithelia and to increase their motility (M. Sachs, D, Riethmacher, and W. Birchmeier, unpublished observations). Such signalling specificities of tyrosine kinase receptors in epithelial or carcinoma cells might be a more common phenomenon than previously assumed. Since the motility of carcinoma cells is essential for their ability to form metastasis, receptor tyrosine kinases might contribute significantly to the development of malignancies.

V. Soluble Factors That Affect Growth, Differentiation, and Motility of Epithelia

A. Scatter Factor/Hepatocyte Growth Factor and the c-Met Receptor

Scatter factor/hepatocyte growth factor was initially characterized as a factor able to stimulate growth of primary hepatocytes (Miyazawa *et al.*, 1989; Nakamura *et al.*, 1989). A fibroblast-derived factor that induces the dissociation and motility of epithelial cells was independently isolated and found to be identical (Stoker *et al.*, 1987; Gherardi *et al.*, 1989; Weidner *et al.*, 1990). Furthermore, a growth factor derived from lung fibroblasts with mitogenic effects on epithelial or endothelial cells and on melanocytes turned out to correspond to SF/HGF (Rubin *et al.*, 1991).

The cellular responses to SF/HGF are mediated by the c-Met receptor, a transmembrane tyrosine-specific protein kinase (Fig. 5). A mutant form of c-*met* was identified in 1984 as an oncogene in a transfection/tumorigenicity assay (Blair *et al.*, 1982; Cooper *et al.*, 1984). Therefore, the gene and its protein product, which was predicted to function as a receptor, were studied extensively. When SF/HGF was found to rapidly induce tyrosine-specific phosphorylation of a protein identical in size to the β subunit of c-Met, this soon led to the identification of the specific surface receptor for SF/HGF, the c-Met receptor (Bottaro *et al.*, 1991).

1. Structure and Biochemical Properties of SF/HGF

SF/HGF is a 92-kDa glycoprotein which has unusual features, since it resembles in domain structure and sequence proteases such as plasmino-

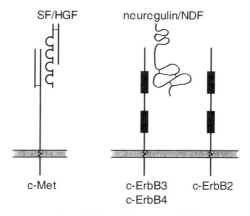

FIG. 5   Schematic representation of the c-Met and c-ErbB receptors together with their specific ligands. The c-Met receptor and its ligand SF/HGF is shown on the left. Neuregulin/ NDF interacts directly with the c-ErbB3 and c-ErbB4 receptors, and, via receptor heterodimerization, indirectly with c-ErbB2.

gen and not other known ligands for tyrosine kinase receptors. The overall sequence homology to plasminogen is 40% (Nakamura *et al.*, 1989; Miyazawa *et al.*, 1989; Weidner *et al.*, 1991). Like plasminogen, SF/HGF is synthesized as prepropeptide that is cleaved post-translationally into a heavy and a light chain; the resulting heterodimer is linked by sulfhydryl bridges. The heavy chain (60 kDa) contains the "hairpin loop" (HL) and four kringle domains (K1-K4); the light chain (30 kDa) shows extensive homologies to serine proteases. However, two of the three amino acids that form the catalytic triad of serine proteases are altered in SF/HGF. Therefore, SF/HGF appears to have no catalytic activity.

In vitro mutagenesis demonstrated the importance of the post-translational cleavage for SF/HGF activity. Mutation of Arg 494, the site of proteolytic cleavage, leads to the formation of a single-chain molecule without biological activity (Hartmann *et al.*, 1992; Lokker *et al.*, 1992). Thus, also in this respect SF/HGF resembles proteases such as plasminogen, which are produced as inactive precursors (zymogen) and activated upon proteolytic cleavage. The cleavage occurs in the extracellular environment (Naldini *et al.*, 1992; Naka *et al.*, 1992; Miyazawa *et al.*, 1993). The SF/HGF converting enzyme is a novel serine protease isolated from serum that shows extensive sequence similarities to the blood coagulation factor XII (47%) and to other serine proteases (Miyazawa *et al.*, 1993). High expression levels of the converting enzyme are found in the liver. The converting enzyme is again produced as a precursor in the liver, which is also activated by proteolysis. In addition, urokinase has been reported to cleave the SF/HGF precursor (Naldini *et al.*, 1992).

Various groups have performed extensive mutagenesis of the SF/HGF gene in order to delineate functionally important domains of the factor. These studies indicate that the N-terminal part of the heavy chain (HL plus K1) is the minimal sequence required for receptor binding (Lokker et al., 1992; Lokker and Godowski, 1993). However, additional heavy chain sequences and the light chain increase the affinity of SF/HGF to the receptor, whereas the light chain alone does not bind (Matsumoto et al., 1991; Hartmann et al., 1992; Lokker et al., 1992). Internal deletions of the various subdomains indicate that HL, K1, and K2 in particular are essential for binding (Matsumoto et al., 1991; Okigaki et al., 1992).

Human fibroblasts and tissues produce, in addition to the 90-kDa SF/HGF protein, variable amounts of a 28-kDa molecule transcribed from an alternatively spliced transcript (Chan et al., 1991; Miyazawa et al., 1991; Hartmann et al., 1992). This small SF/HGF variant contains the N-terminal part, HL, K1 and K2. The two-kringle variant (p28) can bind to the c-Met receptor with approximately threefold reduced affinity and is also able to induce tyrosine phosphorylation of c-Met. However, p28 does not evoke a mitogenic response (Chan et al., 1991; Hartmann et al., 1992). Moreover, it can compete with SF/HGF for receptor binding and specifically inhibits the mitogenic effect of SF/HGF (Chan et al., 1991). The two-kringle variant is thus a naturally occurring antagonist of SF/HGF. Furthermore, the intact heavy chain or a one-kringle variant (containing HL and K1) have similar properties (Lokker et al., 1992). Interestingly, these truncated variants are able to induce the motility effect at increased concentrations, compensating for their reduced affinity (Hartmann et al., 1992). It should be noted that the motility assay (scattering of MDCK cells) is more sensitive than the mitogenic assay (growth of hepatocytes). Whether the apparent dissociation of motogenic and mitogenic activity of such variants reflects true differences in signalling via the c-Met receptor or simply differences in the sensitivity of the assays is an unsolved point.

2. Structure and Biochemical Properties of the c-Met Receptor

The original isolate of the met oncogene was derived from an N-methyl-N-nitro-N-nitrosoguanidine-treated human osteogenic sarcoma cell line (Blair et al., 1982; Cooper et al., 1984). The oncogene is the product of a chromosomal rearrangement which fused the promotor and N-terminal coding region of tpr (translocated promotor region) on chromosome 1 to the C-terminal coding region of met on chromosome 7 (Park et al., 1986;

Dean et al., 1987).The characterization of the oncogene indicated that it encodes a tyrosine-specific protein kinase and that the oncogenic protein is located in the cytoplasm (Park et al., 1987).

The proto-oncogene encodes a glycosylated transmembrane molecule that is synthesized as a single-chain precursor and has to undergo glycosylation and disulfide bond formation before the receptor is cleaved into a 50-kDa (α) and a 145-kDa (β) subunit (Gonzatti-Haces et al., 1988; Giordano et al., 1989b). The α chain and the N-terminal part of the β chain are located on the cell surface, whereas the C-terminal part of the β chain is located cytoplasmatically and contains the tyrosine kinase domain. A minor, alternatively spliced c-met transcript exists, which encodes a variant receptor with 18 additional amino acids in the extracellular domain. The variant is transported to the cell surface but is not cleaved into two subunits (Rodrigues et al., 1991). The importance of proteolytic processing was illustrated by the identification of the LoVo colon carcinoma line deficient in the proteolysis. An otherwise mature 190-kDa receptor is produced in this cell type which is constitutively autophosphorylated. The site of cleavage of c-Met is similar in sequence to the processing site found in the insulin receptor; neither receptor is processed in LoVo cells (Mondino et al., 1991).

Because of its general structure, it was predicted that c-Met functions as a receptor that recognizes a proteinous factor on the cell surface and transmits the signal via its tyrosine kinase domain into the cytoplasm. This hypothesis was verified when the specific ligand for c-Met, SF/HGF, was identified. SF/HGF binds to c-Met with high affinity and induces rapid autophosphorylation of the receptor (Bottaro et al., 1991; Naldini et al., 1991a,b). The extracellular domain of c-Met is sufficient for ligand binding when fused to IgG sequences that also serve as a dimerization domain, and absence of the proteolytic cleavage of the extracellular c-Met sequences does not influence the affinity (Mark et al., 1992).

The active c-Met receptor transmits complex signals that evoke not only mitogenic, but also motogenic and morphogenic responses (see later discussion). The signalling pathways activated and the substrates that bind via SH2 sequences to the phosphorylated receptor are being studied intensively, in order to correlate specific molecules and pathways with particular biological responses. The phosphorylated c-Met receptor binds PI3K, a tyrosine phosphatase, the ras nucleotide exchange protein GRB-2-Sos, PLC-γ, and pp60$^{c\text{-}src}$ (Graziani et al., 1991, 1993; Villa-Moruzzi et al., 1993; Ponzetto et al., 1994; Zhu et al., 1994). It is an unusual receptor for which a multifunctional docking site has been identified, which can bind several substrates (Ponzetto et al., 1994). This multifunctional site is essential for c-Met-induced transformation and motility (Ponzetto et *al.*, 1994; Weidner *et al.*, 1995). In addition, the ras signalling pathway was demonstrated to be essential for the motility signal transmitted by c-Met (Hartmann *et al.*, 1994).

3. Biological Activities of SF/HGF

Initially, SF/HGF was found to have a mitogenic effect on primary hepato-cytes (Michalopoulos *et al.*, 1984; Nakamura *et al.*, 1984; Russel *et al.*, 1984). However, hepatocytes are not the only target cells for growth stimulation by the factor. Instead, SF/HGF acts on a wide variety of epithelial cells and, in addition, on melanocytes, endothelia, and hemato-poietic precursor cell lines (Rosen *et al.*, 1990; Rubin *et al.*, 1991; Mizuno *et al.*, 1993). SF/HGF has also been reported to act as a cytotoxic factor or to inhibit the growth of particular tumor cells (Higashio *et al.*, 1990; Shiota *et al.*, 1992).

An additional activity which led to the independent discovery of the factor is the ability to dissociate continuous epithelial sheets (scatter fac-tor) (Stoker *et al.*, 1987; Gherardi *et al.*, 1989; Weidner *et al.*, 1991). The dissociated epithelia show also increased motility and are able to invade collagen (Weidner *et al.*, 1990). Since abnormal motility is a prerequisite for metastasis of transformed epithelia, it has been suggested that SF/HGF might play a role in malignancy.

SF/HGF can also induce the formation of tubular structures from MDCK cells grown in a three-dimensional collagen matrix (Montesano *et al.*, 1991b). For this, the epithelial cells are grown for several days in collagen, where they form cysts. When SF/HGF is added, individual cells dissociate and move away from the cysts. Consequently, the cells reassociate and form continuous tubules (Weidner *et al.*, 1993). These tubules have a lumen surrounded by well-polarized epithelial cells with a smooth basal surface in contact with the collagen matrix and an apical surface rich in microvilli that faces the lumen (Montesano *et al.*, 1991a). The *in vitro*-formed structures thus resemble the tubular epithelia that are formed during development of many organs. Interestingly, a very similar activity of SF/HGF is found in organ cultures of the mammary gland. There, addition of exogenous factor to the organ in culture increases both growth and the branching of the tubular epithelia. In this way the length of tubular epithelia which are formed before further branching occurs can be increased; an increase in the number of branches formed is also ob-served (Yang *et al.*, 1995). This recent finding might shed some light on earlier organ culture studies of mammary gland development. By coculture of mammary gland epithelia with different types of heterologous mesen-chyme, striking differences in branching patterns and in the lengths of the epithelial tubules were observed (Kratochwil, 1986). Differences in the relative expression levels of SF/HGF in the various mesenchymes analyzed might account for the differences in the epithelial morphogenesis.

The diverse activities of SF/HGF on epithelia together with the fact that various truncated variants exist that show motogenic but not mitogenic activity raised the question whether all biological effects observed are mediated by c-Met. Indeed, besides the high affinity/low capacity c-Met binding sites (200–5000 per cell; K_D 0.2 nM), additional low affinity/high capacity binding sites of SF/HGF (1,000,000 per cell; K_D 2–7 nM) are detected on cell surfaces. A chimeric receptor molecule with the extracellular domain of the nerve growth factor (NGF) receptor and the tyrosine kinase domain of c-Met was produced to test the biological responses of an active c-Met tyrosine kinase on epithelia. This chimeric receptor transmits Met-specific signals, however, in response to NGF. Depending on the culture conditions of the epithelial cells expressing the chimera, NGF induces increased growth, motility, and invasiveness, as well as the formation of tubular structures (Weidner et al., 1993). Thus, all the SF/HGF effects described earlier are mediated by c-Met. The high capacity/low affinity sites might be provided by heparan sulfate proteoglycans, since SF/HGF binds strongly to heparin, a biochemical characteristic that is also used for purification of the factor (Zarnegar and Michalopoulos, 1989; Weidner et al., 1990).

SF/HGF also acts on endothelial cells and induces growth and migration of such cells in culture (Rosen et al., 1990). When matrigel, the extracellular matrix isolated from Engelbreth–Holmes–Swarm sarcoma cells, is mixed with SF/HGF and injected into mice, extensive vascularization of the matrigel plug is observed. Although at low concentrations of SF/HGF no vascularization was apparent, the matrigel plug contained significantly increased cell numbers and the majority of the cells stained with endothelial-specific markers (Grant et al., 1993). An independent assay which measures neovascularization of the eye also demonstrated the potent angiogenic activity of SF/HGF (Bussolino et al., 1992).

4. Distribution of SF/HGF and c-Met *in Vivo*

SF/HGF, like other molecules that bind strongly to heparin, can be viewed as a factor that acts locally and does not diffuse far from its sites of production. Indeed, subcutanous administration of SF/HGF demonstrates that the factor does not diffuse freely *in vivo*. Therefore, information about the sites of synthesis of the factor and its receptor can give a first indication as to potential *in vivo* function of this signalling system. Initial studies by Northern hybridization analysis demonstrated a wide organ distribution of c-*met* transcripts (Iyer et al., 1990). This was also observed for SF/HGF by tissue extraction of the factor, by immunochemistry, or by Northern hybridization analysis (Zarnegar *et al.*, 1990; Tashiro *et al.*, 1990; Wolf *et al.*, 1991a,b; Defrances *et al.*, 1992). SF/HGF expression was detected on the RNA or protein level using cultured fibroblasts derived from many organs, and using smooth muscle cells in culture (Stoker and Perryman, 1985; Rosen *et al.*, 1889; Rubin *et al.*, 1991).

In situ hybridization demonstrates a distinct distribution of transcripts in the mouse embryo: c-*met* is expressed in many different types of epithelia, whereas transcripts for SF/HGF are found in the mesenchymal cell compartment in the vicinity (Sonnenberg *et al.*, 1993a). This predominant and recurring pattern indicates that the factor and its receptor mediate a signal exchange between mesenchyme and epithelia during organogenesis. The pattern is observed in the perinatal period as well as in the adult mammary gland (Yang *et al.*, 1995). It is consistent with the observation made in cell culture, where many fibroblast cell types derived from the mesenchyme produce SF/HGF, and various epithelial cell types respond.

The organ that expresses highest levels of the factor in the embryo is the liver, which also expresses c-*met* (Sonnenberg *et al.*, 1993a). In the embryonal liver, the cell types that express the factor or its receptor have not been identified.

In the adult organ, c-*met* is expressed in hepatocytes, whereas the SF/HGF-producing cells have consistently been described as nonparenchymal (Kinoshita *et al.*, 1989). Based on *in situ* hybridization, Kupffer and sinusoidal cells have been assigned as the cells that contain the SF/HGF transcripts, whereas cell fractionation studies combined with Northern blotting analysis indicate that Ito and sinusoidal cells produce the factor (Noji *et al.*, 1990; Schirmacher *et al.*, 1992).

In addition, other transient expression sites are detected for SF/HGF and c-*met*. During gastrulation of the early chick and mouse embryo, SF/HGF and c-*met* are expressed in the organizer (Henson's node) and in the primitive streak (E. Andermacher and E. Gherardi, personal communication). This is of particular interest since SF/HGF is able to induce a second axis in the chick embryo (Stern *et al.*, 1990). Applied to extraembryonic cells of the chick, the factor also induces neuronal differentiation (Stern and Ireland, 1993). During later stages of embryogenesis, c-*met* is expressed initially in the dermomyotome and later in all myogenic cells. In the muscle lineage, the expression of the receptor is transitory (Sonnenberg *et al.*, 1993a). Thus, a potential role for SF/HGF during development of the muscle is also possible. In addition, distinct cells of the central and peripheral nervous system express the factor or its receptor (Sonnenberg *et al.*, 1993b; Jung *et al.*, 1994). Also, high levels of SF/HGF are found in the placenta, which has been used as a source for the characterization of the factor (Wolf *et al.*, 1991a; Weidner *et al.*, 1991).

5. SF/HGF and Normal Hepatic Development

The normal function of the SF/HGF gene was analyzed by the introduction of a targeted mutation into the SF/HGF gene in embryonic stem cells. The mutant ES cells were used to produce mice that carry this mutation. Whereas animals heterozygous for the SF/HGF gene are viable and fertile, a homozygous mutation is not compatible with normal development. SF/HGF $-/-$ embryos die between embryonal days 13 and 16.5 (E13-E16.5). The mutant embryos on E12 or E14 appear externally normal; however, starting on E12, their liver is considerably reduced in size. Histological examination shows damage of the embryonal liver on E14 that varies in severity and is not observed on E12. The most subtle abnormalities are a rounding and dissociation of the cells and enlarged sinusoidal spaces, as well as cell death (apoptosis) of the liver parenchyme. In extreme cases, few cytokeratin-positive cells of the parenchyme remain and basically only the empty sinusoidal linings of the liver are left (Schmidt *et al.*, 1995). In addition, the development of the placenta, particularly of the labyrinth layer, is also affected (F. Bladt and C. Birchmeier, unpublished observations). In this layer, exchange of nutrients and oxygen between maternal and fetal blood occurs.

In addition, ES cells with two mutant alleles of the c-*met* gene were produced and injected into wild-type blastocysts in order to determine the developmental fate of the mutant cells. The mutant ES cells cannot contribute to the hepatic lineage. Such cell-autonomous mutations in c-*met* demonstrated therefore that the effect observed on development of the liver in SF/HGF $-/-$ embryos is direct and proves the essential role of the SF/HGF and c-Met signalling system for the hepatic lineage (F. Bladt and C. Birchmeier, unpublished observations). It should be noted that SF/HGF is not necessary for the formation of the liver anlage in the embryo. However, in the absence of the factor, further development of the organ does not proceed. This indicates that SF/HGF has an important physiological role in the growth and survival of the cells of the hepatic lineage.

6. Roles of SF/HGF and c-Met in Pathology

SF/HGF is found in high concentrations in the serum of patients with acute liver failure, whereas in normal serum the factor is absent or is present only at very low concentrations (Nakayama *et al.*, 1985; Gohda *et al.*, 1988; Tsubouchi *et al.*, 1989). After experimentally induced liver damage, for example, by CCl_4 or partial hepatectomy, SF/HGF mRNA levels were reported to be upregulated in the liver and also in other, undamaged organs such as the lung or spleen (Kinoshita *et al.*, 1989; Zarnegar *et al.*, 1990; Gohda *et al.*, 1990). In the damaged liver, cells of the sinusoidal linings as well as infiltrating lymphocytes were reported to express the factor (Noji *et al.*, 1990; Schirmacher *et al.*, 1993). A soluble, liver-derived molecule named injurin was proposed to be responsible for upregulation of factor synthesis in the liver and in other, undamaged organs (Matsumoto *et al.*, 1992).

Since SF/HGF is the most potent growth factor known for hepatocytes, the upregulation of the factor might aid liver regeneration. This is supported by recent experiments performed in transgenic animals that carry an SF/HGF gene under the control of the albumin promotor (Shiota *et al.*, 1994). In this manner, an autocrine loop was produced in hepatocytes. After partial hepatectomy, the transgenic animals showed double the rate of liver regeneration. Intravenous administration of the factor has also been reported to aid liver regeneration (Ishiki *et al.*, 1992). Alternatively, it has been reported that administration of SF/HGF protects the liver from chemically induced damage (Roos *et al.*, 1992).

Upregulation of SF/HGF expression is also found in the psoriotic dermis of human patients (Grant *et al.*, 1993). Since c-Met is expressed in keratinocytes, the factor could act on the overlying epidermis and stimulate growth (Kamalati *et al.*, 1993). Alternatively, the high local concentration of SF/HGF in the pathological skin could cause an increase in the vascularization, which is indeed observed in such patients.

Because of the transforming potential of c-*met*, expression of the receptor has been studied in many different tumor types. There have been numerous reports on upregulation of receptor expression in carcinomas (Liu *et al.*, 1992; Di Renzo *et al.*, 1992; Kuniyasu *et al.*, 1992; Rege-Cambrin *et al.*, 1992; Jucker *et al.*, 1994). In several instances, high levels of receptor expression correlate with malignancy. In addition, in gastric carcinomas c-*met* was found to be rearranged and fused to *tpr* (Soman *et al.*, 1991); in such fusion products, Tpr provides a dimerization domain which constitutively activates the tyrosine kinase of Met (Rodrigues and Park, 1994). In addition, autocrine loops were generated by coexpressing SF/HGF and c-*met* in NIH3T3 fibroblast; such autocrine loops induce transformation and invasiveness (Rong *et al.*, 1992, 1994; Giordano *et al.*, 1993). In addition, tumors derived from such cells in nude mice are very unusual, since the tumor cells display epithelial characteristics (Tsafaty *et al.*, 1994).

B. Neuregulin/NDF and Its Receptors

Neuregulin, also called neu differentiation factor (NDF) or heregulin, was identified as a putative ligand for the c-Neu/c-ErbB2/HER2 receptor tyrosine kinase that has mitogenic and differentiation activity on epithelial cells in culture (Peles *et al.*, 1992; Wen *et al.*, 1992; Holmes *et al.*, 1992). Additional analysis demonstrated, however, that the factor interacts only indirectly with c-Neu, and that a direct interaction occurs with closely related tyrosine kinase receptor(s), c-ErbB4/HER4 and c-ErbB3/HER3 (Fig. 5) (Peles *et al.*, 1993; Carraway *et al.*, 1994; Plowman *et al.*, 1993a). Two independently identified factors that induce terminal differentiation of myotubes (acetylcholine-inducing activity, ARIA) or glial growth (glial growth factor, GGF) turned out to correspond to isoforms of neuregulin/NDF (Falls *et al.*, 1993; Marchionni *et al.*, 1993).

1. Molecular and Biochemical Characteristics of Neuregulin/NDF

Only one neuregulin/NDF gene has been characterized, from which many different mRNA transcripts are formed that produce at least ten distinct isoforms of the factor (Peles and Yarden, 1993). The basic structure of the first (Type I, Fig. 6) isoforms described (NDF, heregulin) consists of a N-terminal immunoglobulin-like domain followed by a glycodomain, an epidermal growth factor (EGF)-like domain, a hydrophobic sequence that might act as internal signal sequence and/or transmembrane domain, and finally a cytoplasmic domain (Wen *et al.*, 1992; Holmes *et al.*, 1992). Two processing sites located close to the N-terminal and immediately preceding the transmembrane domain are characteristic for these isoforms. The basic strucure of the second (Type II; Fig. 6) group of isoforms consists of an N-terminal hydrophobic signal sequence followed by a kringle domain, the immunoglobulin, and the EGF domain (Marchionni *et al.*, 1993). Of these two major structures, variants exist which differ in the absence or presence of some of these subdomains or in sequences which link them. Both major isoforms bind to heparin (Holmes *et al.*, 1992; Wen *et al.*, 1994; Marchionni *et al.*, 1993). The various neuregulin/NDF isoforms have distinct molecular weights which range from 33 to 59 kDa.

Of particular functional importance is the EGF-like domain, approximately 40 amino acids in length, since this sequence alone is biologically active, that is, it binds to the receptor (Holmes *et al.*, 1992). Two variants of the EGF domain (α and β) exist that are encoded by distinct exon combinations (Holmes *et al.*, 1992; Marchionni *et al.*, 1993; Wen *et al.*, 1994). One exon encodes the N-terminal half of the domain present in the

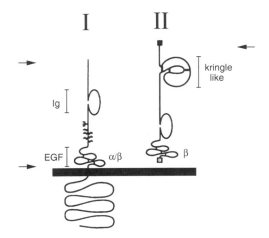

FIG. 6   Schematic representation of the structures of neuregulin/NDF isoforms. The Type I and Type II isoforms of neuregulin/NDF are shown. The different subdomains, Ig domain (Ig), the α- or β-subtype of the EGF domain (EGF), and the kringle-like domain (kringle like) are shown. The putative proteolytic processing sites are indicated by arrows.

α and the β subtypes, whereas two distinct exons encode the C-terminal half of the domain that is found in either the α or β variants, respectively (Fig. 7).

Neuregulin/NDF isoforms are expressed in a tissue-specific manner in the mouse. In general, two distinct cell types express the factor: mesenchymal and neuronal cells (Orr-Urtreger *et al.*, 1993a; Meyer and Birchmeier, 1994). Whereas cells of the central nervous system express Type I and II variants, the mesenchymal cell compartment expresses only Type I variants. Also, a preference for expression of the α subtype is notable in the mesenchymal cell compartment (Meyer and Birchmeier, 1994).

2. Molecular and Biochemical Properties of the Neuregulin/ NDF Receptors

Four members of the EGF receptor family are known: the EGF receptor also named c-ErbB1 (after the mutant variant of the EGF receptor present in the avian erythroblastosis virus) or HER1 (human EGF receptor 1); c-Neu, also named c-ErbB2 or HER2; c-ErbB3, also named HER3; and c-ErbB4, also named HER4. The four receptors have identical overall structures: a signal peptide that is followed by an extracellular domain containing two cystein-rich sequence stretches, a transmembrane domain, and a cytoplasmatic sequence that contains the tyrosine-specific protein kinase domain (Ullrich *et al.*, 1994; Bargmann *et al.*, 1986; Yamamoto *et*

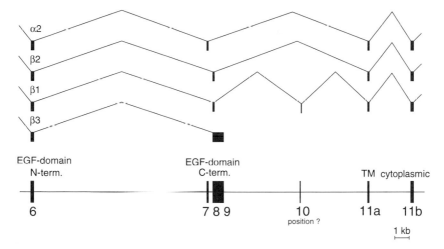

FIG. 7 Genomic structure and alternative splicing in the neuregulin/NDF gene. The genomic structure of the neuregulin/NDF gene that encodes the EGF domain and neighboring sequences are schematically depicted. The exons encode the N- or C-terminal part of the EGF domain, the putative transmembrane domain, and cytoplasmic sequences; the numbering of the exons is according to Marchionni *et al.* (1993). The exact position of exon 10, which is 24 nucleotides in length, is unknown. The exons present in the mRNAs encoding variant neuregulins/NDF (α2, β1-3) are indicated.

al., 1984; Kraus *et al.,* 1989; Plowman *et al.,* 1993b). Sequence similarities in this receptor subfamily are extensive, and the overall size of the mature, glycosylated receptors is similar.

Several specific ligands for the EGF receptor, EGF, transforming growth factor-α (TGF-α), heparin-binding EGF, and crypto- and amphiregulin have been extensively characterized. However, which specific ligand(s) interacts with the other receptors of the family is not as clear. In particular, the c-Neu ligand has received considerable attention (Peles and Yarden, 1993). Candidate ligands that interact directly with c-Neu have been identified, but the molecular structure of these factors has not been published (Dobashi *et al.,* 1991; Tarakhovsky *et al.,* 1991; Lupu *et al.,* 1992; Huang and Huang, 1992). Neuregulin/NDF, one previous candidate, does not interact directly with c-Neu but rather with c-ErbB3 and c-ErbB4 (Peles *et al.,* 1993; Culouscou *et al.,* 1993; Plowman *et al.,* 1994; Carraway *et al.,* 1994). However, extensive heterodimerization of this receptor family can take place, which seems to be responsible for the indirect activation of c-Neu by neuregulin/NDF (Carraway and Cantley, 1994). The heterodimerization results in cross-phosphorylation and thus indirect activation of the receptors. Coexpression of c-ErbB2 and c-ErbB3 has indicated that the heterodimers form high-affinity binding sites for

NDF/neuregulin (Sliwkowski *et al.*, 1994). The physiological and biological role of this "cross-talk" is unknown.

3. Biological Activities of Neuregulin/NDF *in Vitro*

The first activity of neuregulin/NDF described was the induction of growth and/or differentiation of mammary epithelial cells in culture. The activity observed depends on the exact cell line used (Holmes *et al.*, 1992; Peles *et al.*, 1992). Under normal conditions, AU-565 human mammary carcinoma cells grow as poorly differentiated epithelial cells. In the presence of neuregulin/NDF, a flattened morphology, increased nuclear size, and large cytoplasmic vesicles are observed. Milk protein (casein) typical of differentiated mammary cells is synthesized and fat droplets appear in the cytoplasm. Also, a decrease in cellular growth and an upregulation of the cell adhesion molecule ICAM-1 accompanies this differentiation (Peles *et al.*, 1992; Bacus *et al.*, 1993).

In organ cultures of the mammary gland, exogenous addition of neuregulin/NDF induces formation of lobular alveoli and terminal differentiation of the epithelial cells, which can be assessed by the expression of milk proteins. Interestingly, the factor is expressed at very high concentrations late in pregnancy, when lobulo-alveolar differentiation of the mammary gland epithelium takes place *in vivo* (Yang *et al.*, 1995).

A distinct activity that induces the terminal differentiation of myotubes led to the independent isolation of a neuregulin/NDF isoform (Falls *et al.*, 1993). Terminal differentiation of myotubes, particularly the development of the specialized neuromuscular junction, is dependent on the innervation of the muscle by motoneurons. Spinal cord or brain extracts are able to induce particular aspects of this differentiation *in vitro*, and the factor responsible turned out to correspond to a variant of neuregulin/NDF. Interestingly, neuregulin/NDF transcripts are detected in motoneurons during development and the factor is thus produced in the same cells which induce terminal differentiation *in vivo* (Falls *et al.*, 1993; Marchionni *et al.*, 1993; Orr-Urtreger *et al.*, 1993a; Meyer and Birchmeier, 1994).

A third activity of neuregulin/NDF, the induction of glial cell growth, led to its independent isolation and characterization as glial growth factor (Marchionni *et al.*, 1993). Consequently, it was found that the factor not only induces glial growth, but also determines the cellular fate of peripheral glial (Schwann) cells (Shah *et al.*, 1994). Schwann cells are one of the cell types that differentiate from the neural crest cells which migrate out of the developing neural tube. *In vitro*, neural crest cells can be isolated which differentiate into neuronal and Schwann cells. Addition of neuregulin/NDF to these precursor cells leads to the formation of only one cell type, Schwann cells, without affecting growth of the cells (Shah *et al.,* 1994). This is thus an example of a receptor tyrosine kinase signal that induces a particular cell fate.

4. Distribution of Neuregulin/NDF *in Vivo*

As described, neuregulin/NDF has many different activities *in vitro;* thus a detailed expression analysis of factor synthesis is a first step to elucidate its physiological role. A broad range of different tissues was found to express NDF when analyzed by Northern hybridization. Analysis by *in situ* hybridization demonstrates, however, that the factor is not expressed ubiquitously. Two distinct cell types, mesenchymal and neuronal cells, express NDF in the mouse embryo (Orr-Urtreger *et al.,* 1993a; Meyer and Birchmeier, 1994). NDF-expressing mesenchymal cells are found, for example, in the embryonal lung, gut, kidney, and genital ridge. Interestingly, in several organs, high transcript levels are found in mesenchymal cells located close to the epithelia (Fig. 8; see also Meyer and Birchmeier, 1994). This indicates that epithelial factors regulate the synthesis of NDF in the mesenchyme, which can then affect the epithelia. In addition, various neuronal cell types express high concentrations of NDF transcripts. Partic-

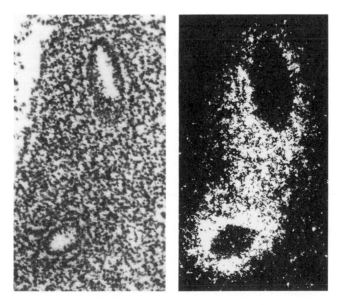

FIG. 8   Neuregulin/NDF expression in the developing lung. *In situ* hybridization of a murine lung on day 11 of development with a neuregulin/NDF probe demonstrates the presence of specific transcripts in mesenchymal cells close to the lung epithelium.

ularly, all the head ganglia and the dorsal root ganglia express the factor early in embryogenesis; motoneurons also contain high levels of the factor (Falls *et al.*, 1993; Marchionni *et al.*, 1993; Orr-Urtreger *et al.*, 1993a; Meyer and Birchmeier, 1994). This is consistent with a potential role of NDF in the determination of glial cell fate and in the innervation-dependent differentiation of muscle.

In the adult, a dynamic pattern of neuregulin/NDF expression is also found during development of the mammary gland. Whereas transcript levels are low in virgin glands or during early pregnancy, they increase 100-fold in late pregnancy and decrease sharply during lactation or involution. Again, mesenchymal cells in the gland express the factor (Yang *et al.*, 1995). Neuregulin/NDF might thus *in vivo* induce terminal differentiation of mammary epithelia.

Therefore, neuregulin/NDF might have potential functions in a variety of distinct and important developmental and physiological events. The genetic analysis of neuregulin/NDF and its receptors will contribute to further understanding of the physiological role(s) of this signalling system.

5. Roles of Neuregulin/NDF and Its Receptors in Tumor Biology

Neuregulin/NDF has been recently characterized on a molecular level, and, as a consequence, factor produced by recombinant techniques has only been available for a short time. Therefore, our knowledge of the activity of the factor, particularly on epithelial and carcinoma cells, is still fragmentary. Interestingly, neuregulin/NDF induces terminal differentiation and growth inhibition as well as expression of the adhesion molecule ICAM-1 in MDA-MB-453 and AU 565 mammary carcinoma cells, respectively (Peles *et al.*, 1992; Bacus *et al.*, 1993). Experiments in organ culture also indicate a role for neuregulin/NDF in differentiation of normal mammary epithelia (Yang *et al.*, 1995). If this differentiation activity can also be observed on a broad panel of carcinoma cell lines *in vitro* and *in vivo*, it might be of interest for the treatment of mammary tumors. However, the observation that neuregulin/NDF can induce the growth of other mammary carcinoma cell lines indicates that the factor might play a dual role in the control of growth and differentiation. Therefore, a necessary prerequisite for therapy might be to predetermine the response to the factor on cells obtained by tumor biopsy. Since more than one receptor can transmit the neuregulin/NDF signal, the different responses observed could reflect distinct expression patterns or expression levels of the receptors in the tumor cells.

Only few mammary carcinoma cell lines express neuregulin/NDF (Wen *et al.*, 1992). Thus, the factor seems to act generally in a paracrine modus on carcinoma cells. During embryogenesis, neuregulin/NDF expression is confined to the mesenchymal and neuronal cell compartment. Expression in the mesenchymal compartment seems to be under the control of epithelial factors, because neuregulin/NDF transcript levels are found to be high in the direct vicinity of growing epithelia (Fig. 8). Such an epithelial control of the expression of mesenchymal factors could explain the reciprocal interaction observed between the two cell types in development (cf. Birchmeier and Birchmeier, 1993). Interactions between stroma (surrounding mesenchyme) and carcinoma cells have also been postulated to affect tumor progression (DeVita *et al.*, 1993). It might thus be of interest to determine whether mesenchymal expression of neuregulin/NDF is similarly controlled by the growing carcinoma cells in developing tumors.

Both receptors that interact directly with NDF/neuregulin, c-ErbB3 and c-ErbB4, are overexpressed in some carcinoma cells. Expression and/ or overexpression of c-ErbB3 in carcinoma cells or primary tumors of various origins (mammary tumors, pancreatic tumors, tumors of the head and neck, adenocarcinomas) have been reported (Kraus *et al.*, 1989, 1993; Plowman *et al.*, 1993b; Skinner and Hurst, 1993; Issing *et al.*, 1993; Gasparini *et al.*, 1994; Mandai *et al.*, 1994). Overexpression might be a frequent event in tumor progression, but amplification of the c-*erbB3* gene, previously observed frequently for c-*neu*/erbB2, has not been observed. The mechanism implicated in overexpression is an altered transcriptional regulation in tumor cells (Skinner and Hurst, 1993).

The c-Neu receptor can be indirectly phosphorylated and activated by NDF/neuregulin. Fast phosphorylation and receptor activation induced by neuregulin/NDF has been observed in various mammary carcinoma cell lines that express high levels of both c-Neu and c-ErbB4. This indirect activation is the result of receptor "cross-talk" (Carraway and Cantley, 1994). To what extent such a cross-talk is important for neuregulin/NDF signalling in normal or transformed cells remains to be elucidated.

C. Keratinocyte Growth Factor and Other Members of the Fibroblast Growth Factor Family That Affect Epithelial Growth, Differentiation, and Morphogenesis

Nine genes that encode fibroblast growth factors are known in mammals. Some of the genes also encode different isoforms which are produced from alternatively spliced mRNAs or by the use of alternative initiation codons. These factors interact with a family of tyrosine kinase receptors, which are encoded by four genes. However, transcripts of some of the genes are alternatively spliced and produce therefore more than one receptor isoform. The different receptors and receptor isoforms have distinct affinities for the different types of FGFs (Burgess and Maciag, 1989; Johnson and Williams, 1993; Mason et al., 1994).

Keratinocyte growth factor (KGF, FGF-7), a member of the family, has been implicated in mesenchymal-epithelial interactions. KGF was discovered as a fibroblast-derived factor that induces growth of keratinocytes (Rubin et al., 1989; Finch et al., 1989). The biological activity of KGF is distinct from other members of the family because it is not mitogenic for fibroblasts or endothelial cells, but is specific for epithelia. Targeted expression of KGF to keratinocytes in transgenic mice elicits a striking change in epithelial differentiation of the skin (Guo et al., 1993). KGF is expressed in the mesenchymal compartment in organs of the embryo and the adult, as has been demonstrated by the analysis of cell lines derived from such sources and by in situ hybridization analysis. In addition, the factor is produced in myogenic and neuronal cells (Mason et al., 1994). During wound healing, expression of KGF mRNA is upregulated 100-fold in the mesenchyme of the wound (Werner et al., 1992).

KGF binds with high affinity a particular splice variant of the FGF-2 receptor, FGF2-IIIb; the same isoform also binds acidic FGF. The receptor was identified by an elegant approach which relied on the formation of an autocrine loop in NIH3T3 cells. NIH3T3 cells express KGF, but not the corresponding receptor. By transfection of a cDNA library synthesized from keratinocyte mRNA and consequent selection for foci, a cDNA that encodes the high-affinity KGF receptor (FGF2-IIIb receptor) was identified (Miki et al., 1991). In vivo, transcripts for the FGF-2 receptor are predominantly found in embryonic epithelia (Orr-Urtreger et al., 1993b).

Transdominant FGF receptor mutants which lack the tyrosine kinase domain have been used for functional studies in vivo. Since the FGF receptors, like other receptor tyrosine kinases, are activated by ligand-dependent dimerization and ensuing cross-phosphorylation of the receptors in the dimer, such truncated receptors interfere with the activation of the intact receptors when expressed in the same cells at high concentration (Amaya et al., 1991). The blocking response of FGF receptors depends on the endogenous and truncated receptor binding the same ligand (Ueno et al., 1992).

Transdominant variants of the FGF2-IIIb receptors were expressed in transgenic animals, and, by the use of the appropriate promotors, targeted to the skin and the lung epithelium. Expressed under the control of the keratin 6 promotor, which directs the transdominant receptor to the suprabasal layer of the skin, epidermal thickening and disrupted organization of epidermal keratinocytes was observed (Werner et al., 1993). Expressed under the control of the keratin 14 promotor, which directs the transdominant receptor to undifferentiated keratinocytes in the basal level of the skin, epidermal atrophy was induced (Werner et al., 1994). Upon skin

injury, a severe reduction of the proliferation rate of epidermal keratinocytes at the wound edge was observed in the transgenic animals, which
resulted in a marked delay of reepithelialization of the wound (Werner *et
al.*, 1994).

In addition, KGF/FGF-7 plays also an important role in the development, differentiation, and morphogenesis of epithelia during development
of internal organs. When expression of the transdominant receptor was
directed to the lung epithelium, it interfered in a striking manner with
development of the lung (Peters *et al.*, 1994). Branching morphogenesis
and differentiation of the lung epithelium was completely abolished. Instead, two undifferentiated tubes that extended from the bifurcation of
the trachea down to the diaphragm were formed. Thus, KGF plays a
role in skin development and wound healing and is also essential for
the branching morphogenesis and differentiation of the developing lung
epithelium.

Gene amplification of the FGF2-IIIb receptor has been observed in
human tumor cell lines and led to the isolation of this receptor isoform,
then named K-sam (Hattori *et al.*, 1990). In addition, the transforming
activity of FGFs in NIH3T3 cells or in mouse mammary tumor virus
(MMTV)-induced mammary carcinomas of the mouse have been well
documented (Burgess and Maciag, 1989). aFGF has been also reported
to stimulate motility of carcinoma cells, which is a prerequisite for the
invasiveness of the tumor cells (Valles *et al.*, 1991). In addition, basic
FGF (bFGF) is a potent angiogenesis factor, that is, it induces growth of
endothelial cells and vascularization in various *in vivo* assays. Neovascularization of tumors is an absolutely essential step in tumor progression
(Folkman and Shing, 1992). Therefore, this family of growth factors can
influence multiple and equally crucial steps in the development of malignant carcinomas.

VI. Summary and Future Perspectives

During the past few years, factors that control epithelial growth, motility,
and morphogenesis have been characterized on a molecular level, and
found to play important roles in development as well as in tumor progression. The Ca^{2+}-dependent cell adhesion molecule, E-cadherin, is essential
for the generation and maintenance of epithelial cell morphology and
polarization. Homozygous mutation of the E-cadherin gene results in a
total disorganization of the early mammalian embryo. In carcinoma cells,
the integrity of the adherens junction is frequently destroyed by various
molecular mechanisms, the most frequent being loss or downregulation

of E-cadherin expression. Dysfunction of the adherens junction correlates with an increase in motility of the carcinoma cells and with their ability to become invasive and to form metastasis.

In addition, several mesenchymal factors, SF/HGF, neuregulin/NDF, or members of the FGF family can induce motility, growth, and morphogenesis of epithelial cells in culture. These factors function as specific ligands for epithelial receptor-type tyrosine kinases. They were originally identified as growth factors, or the corresponding receptors were found because of their oncogenic potential when mutated. In carcinoma cells, overexpression and/or mutation of the receptors can be observed, and they have therefore been implicated in tumor development. Mesenchymal factors have long been known to drive epithelial development as assessed by organ culture experiments. Recent genetic analysis in the mouse has corroborated the essential roles of these factors and their receptors in the control of growth, morphogenesis, and differentiation of epithelia during organ formation.

Although a pivotal role of tyrosine kinase receptors and their ligands in organ development, particularly in the control of epithelial growth, differentiation, and morphogenesis is emerging, our knowledge today is in many respects still fragmentary. During development, these factors and their receptors seem to function in an organ-specific manner. However, expression of receptors and ligands is not necessarily restricted to particular organs. Furthermore, not all mesenchymal factors essential for the control of epithelial development and morphogenesis have been identified yet. In addition, substrates of tyrosine kinase receptors that regulate growth have been intensely studied, but little is as yet known about the molecules that regulate motility or morphogenesis. The initial evidence indicates that there could be a partial overlap between the signalling cascades that regulate growth and motility or morphogenesis.

Acknowledgments

We thank Walter Birchmeier and Eva Riethmacher-Sonnenberg for critically reading the manuscript, Friedhelm Bladt for help with the bibliography, Silvia Deis for help with preparing the manuscript, and Udo Ringeisen for preparing the figures.

References

Aaronson, S. (1991). Growth factors and cancer. *Science* **254,** 1146–1153.
Amaya, E., Musci, J., and Kirschner, M. W. (1991). Expression of a dominant negative mutant of the FGF receptor disrupts mesoderm formation in Xenopus embryos. *Cell (Cambridge, Mass.)* **66,** 257–270.

Bacus, S. S., Gudkov, A. V., Zelnick, C. R., Chin, D., Stern, R., Stancovski, I., Peles, E., Ben-Baruch, N., Farbstein, H., Lupu, R., Wen, D., Sela, M., and Yarden, Y. (1993). Neu differentiation factor (Heregulin) induces expression of intercellular adhesion molecule 1: Implications for mammary tumors. *Cancer Res.* **53,** 5251–5261.

Barbacid, M. (1987). Ras genes. *Annu. Rev. Biochem.* **56,** 779–827.

Bargmann, C. I., Hung, M. C., and Weinberg, R. A. (1986). The neu oncogene encodes an epidermal growth factor receptor-related protein. *Nature (London)* **319,** 226–230.

Becker, K.-F., Atkinson, M. J., Reich, U., Becker, I., Nekarda, H., Siewert, J. R., and Hoefler, H. (1994). E-Cadherin gene mutations provide clues to diffuse type gastric carcinomas. *Cancer Res.* **54,** 3845–3852.

Beddington, R., and Smith, J. C. (1993). Control of vertebrate gastrulation: Inducing signals and responding genes. *Curr. Opin. Genet. Dev.* **3,** 655–661.

Behrens, J., Mareel, M., Van Roy, F., and Birchmeier, W. (1989). Dissecting tumor cell invasion: Epithelial cells acquire invasive properties after the loss of uvomorulin-mediated cell-cell adhesion. *J. Cell Biol.* **108,** 2435–2447.

Behrens, J., Vakaet, L., Friis, R., Winterhager, E., Van Roy, F., Mareel, M. M., and Birchmeier, W. (1993). Loss of epithelial differentiation and gain of invasiveness correlates with tyrosine phosphorylation of the E-cadherin/beta-catenin complex in cells transformed with a temperature-sensitive v-src gene. *J. Cell Biol.* **120,** 757–766.

Birchmeier, C., and Birchmeier, W. (1993). Molecular aspects of mesenchymal-epithelial interactions. *Annu. Rev. Cell Biol.* **9,** 511–540.

Birchmeier, C., Sonnenberg, E., Weidner, K. M., and Walter, B. (1993). Tyrosine kinase receptors in the control of epithelial growth and morphogenesis during development. *BioEssays* **15,** 185–190.

Birchmeier, W., and Behrens, J. (1994). Cadherin expression in carcinomas: Role in the formation of cell junctions and the prevention of invasiveness. *Biochim. Biophys. Acta* **1198,** 11–26.

Blair, D. G., Cooper, C. S., Oskarsson, M. K., Eader, L. A., and Vande Woude, G. F. (1982). New method for detecting cellular transforming genes. *Science* **218,** 1122–1125.

Boix, L., Rosa, J. L., Ventura, F., Castells, A., Bruix, J., Rodes, J., and Bartrons, R. (1994). C-met mRNA overexpression in human hepatocellular carcinoma. *Hepatology* **19,** 88–91.

Boshart, M., Nitsch, D., and Schuetz, G. (1993). Extinction of gene expression in somatic cell hybrids—a reflection of important regulatory mechanisms? *Trends Genet.* **9,** 240–245.

Bottaro, D. P., Rubin, J. S., Faletto, D. L., Chan, A. M., Kmiecik, T. E., Vande Woude, G. F., and Aaronson, S. A. (1991). Identification of the hepatocyte growth factor receptor as the c-met proto-oncogene product. *Science* **251,** 802–804.

Burgess, W. H., and Maciag, T. (1989). The heparin-binding (fibroblast) growth factor family of proteins. *Annu. Rev. Biochem.* **58,** 575–606.

Bussolino, F., Di Renzo, M. F., Ziche, M., Bocchietto, E., Olivero, M., Naldini, L., Gaudino, G., Tamagnone, L., Coffer, A., and Comoglio, P. M. (1992). Hepatocyte growth factor is a potent angiogenic factor which stimulates endothelial cell motility and growth. *J. Cell Biol.* **119,** 629–641.

Carraway, K. L., and Cantley, L. C. (1994). A neu acquaintance for ErbB3 and ErbB4: A role for receptor heterodimerization in growth signalling. *Cell (Cambridge, Mass.)* **78,** 5–8.

Carraway, K. L., Sliwkowski, M. X., Akita, R., Platko, J. V., Guy, P. M., Nuijens, A., Diamonti, A. J., Vandlen, R. L., Cantley, L. C., and Cerione, R. A. (1994). The erbB3 gene product is a receptor for heregulin. *J. Biol. Chem.* **269,** 14303–14306.

Chan, A. M. L., Rubin, J. S., Bottaro, D. P., Hirschfield, D. W., Chédid, M., and Aaronson, S. A. (1991). Identification of a competitive HGF antagonist encoded by an alternative transcript. *Science* **254,** 1382–1385.

Cooper, C. S., Park, M., Blair, D. G., Tainsky, M. A., Huebner, K., Croce, C. M., and Vande Woude, G. F. (1984). Molecular cloning of a new transforming gene from a chemically transformed human cell line. *Nature (London)* **311**, 29–34.

Culouscou, J. M., Plowman, G. D., Carlton, G. W., Green, J. M., and Shoyab, M. (1993). Characterization of a breast cancer cell differentiation factor that specifically activates the HER4/p180^{evbB4} receptor. *J. Biol. Chem.* **268**, 18407–18410.

Dean, M., Park, M., and Vande Woude, G. F. (1987). Characterization of the rearranged tpr-met oncogene breakpoint. *Mol. Cell. Biol.* **7**, 921–924.

Defrances, M. C., Wolf, H. K., Michalopoulos, G. K., and Zarnegar, R. (1992). The presence of hepatocyte growth factor in the developing rat. *Development (Cambridge, UK)* **116**, 387–395.

DeVita, V. T., Hellman, S., and Rosenberg, S. A. (1993). "Cancer: Principles and Practice of Oncology," 4th ed. Lippincott, Philadelphia.

Di Fiore, P. P., Pierce, J. H., Kraus, M. H., Segatto, O., King, C. R., and Aaronson, S. A. (1987). *erb*B-2 is a potent oncogene when overexpressed in NIH/3T3 cells. *Science* **237**, 178–182.

Di Renzo, M. F., Olivero, M., Ferro, S., Prat, M., Bongarzone, I., Pilotti, S., Belfiore, A., Costantino, A., Vigneri, R., Pierotti, M., and Comoglio, P. (1992). Overexpression of the c-MET/HGF receptor gene in human thyroid carcinomas. *Oncogene* **7**, 2549–2553.

Dobashi, K., Davis, J. G., Mikami, Y., Freeman, J. K., Hamuro, J., and Greene, M. (1991). Characterization of a neu/c-erbB-2 protein-specific activating factor. *Proc. Natl. Acad. Sci. U.S.A.* **88**, 8582–8586.

Donehower, L. A., Harvey, M., Slagle, B. L., McArthur, M. J., Montgomery, C. A., Jr., Butel, J. S., and Bradley, A. (1992). Mice deficient for p53 are developmentally normal but susceptible to spontaneous tumours. *Nature (London)* **356**, 215–221.

Dorudi, S., Sheffield, J. P., Poulsom, R., Northover, J. M., and Hart, I. R. (1993). E-cadherin expression in colorectal cancer, An immunocytochemical and in situ hybridization study. *Am. J. Pathol.* **142**, 981–986.

Falls, D., Rosen, K., Corfas, G., Lane, W., and Fischbach, G. (1993). ARIA, a protein that stimulates acetylcholine receptor synthesis, is a member of the neu ligand family. *Cell* **72**, 801–815.

Fearon, E. R., Cho, K. R., Nigro, J. M., Kern, S. E., Simons, J. W., Ruppert, J. M., Hamilton, S. R., Preisinger, A., Thomas, G., Kinzler, K., and Vogelstein, B. (1990). Identification of a chromosome 18q gene that is altered in colorectal cancers. *Science* **247**, 49–56.

Fell, P. E., and Grobstein, C. (1968). The influence of extra-epithelial factors on the growth of embryonic mouse pancreatic epithelium. *Exp. Cell Res.* **53**, 301–304.

Finch, P. W., Rubin, J. S., Miki, T., Ron, D., and Aaronson, S. A. (1989). Human KGF is FGF-related with properties of a paracrine effector of epithelial cell growth. *Science* **245**, 752–755.

Fishel, R., Lescoe, M. K., Rao, M. R., Copeland, N. G., Jenkins, N. A., Garber, J., Kane, M., and Kolodner, R. (1993). The human mutator gene homolog MSH2 and its association with hereditary nonpolyposis colon cancer. *Cell (Cambridge, Mass.)* **75**, 1027–1038.

Fleming, T. P., and Johnson, M. H. (1988). From egg to epithelium. *Annu. Rev. Cell Biol.* **4**, 459–485.

Fleming, T. P., Javed, Q., and Hay, M. (1992). Epithelial differentiation and intercellular junction formation in the mouse early embryo. *Development (Cambridge, UK)*, *Suppl.*, pp. 105–113.

Folkman, J., and Shing, Y. (1992). Angiogenesis. *J. Biol. Chem.* **267**, 10931–10934.

Forrester, K., Almoguera, C., Han, K., Grizzle, W. E., and Perucho, M. (1987). Detection of high incidence of K-ras oncogenes during human colon tumorigenesis. *Nature (London)* **327**, 298–303.

Frixen, U., Behrens, J., Sachs, M., Eberle, G., Voss, B., Warda, A., Loechner, D., and Birchmeier, W. (1991). E-cadherin mediated cell-cell adhesion prevents invasiveness of human carcinoma cell lines. *J. Cell Biol.* **111,** 173–185.

Gabbert, H. (1985). Mechanism of tumor invasion: Evidence from in vivo observations. *Cancer Metastasis Rev.* **4,** 293–309.

Gabbert, H., Wagner, R., Moll, R., and Gerharz, C. D. (1985). Tumor dedifferentiation: An important step in tumor invasion. *Clin. Exp. Metastasis* **3,** 257–279.

Garrod, D. R. (1993). Desmosomes and hemidesmosomes. *Curr. Opin. Cell Biol.* **5,** 30–40.

Gasparini, G., Gullick, W. J., Maluta, S., Palma, P. D., Caffo, O., Leonardi, E., Boracchi, P., Pozza, F., Lemonoine, N. R., and Bevilacqua, P. (1994). c-erbB-3 and c-erbB-2 protein expression in node-negative breast carcinoma—an immunocytochemical study. *Eur. J. Cancer* **30,** 16–22.

Gateff, E. A., and Mechler, B. M. (1989). Tumor-suppressor genes of Drosophila melanogaster. *CRC Crit. Rev. Oncogene* **1,** 221–245.

Gherardi, E., Gray, J., Stoker, M., Perryman, M., and Furlong, R. (1989). Purification of scatter factor, a fibroblast-derived basic protein that modulates epithelial interactions and movement. *Proc. Natl. Acad. Sci. U.S.A.* **86,** 5844–5848.

Giordano, S., Ponzetto, C., Di Renzo, M. F., Cooper, C. S., and Comoglio, P. M. (1989a). Tyrosine kinase receptor indistinguishable from the c-met protein. *Nature (London)* **339,** 155–156.

Giordano, S., Di Renzo, M. F., Narsimhan, R. P., Cooper, C. S., Rosa, C., and Comoglio, P. M. (1989b). Biosynthesis of the protein encoded by the c-met proto-oncogene. *Oncogene* **4,** 1383–1388.

Giordano, S., Zhen, Z., Medico, E., Gaudino, G., Galimi, F., and Comoglio, P. M. (1993). Transfer of motogenic and invasive response to scatter factor/hepatocyte growth factor by transfection of human MET protooncogene. *Proc. Natl. Acad. Sci. U.S.A.* **90,** 649–653.

Gohda, E., Tsubouchi, H., Nakayama, H., Hirono, S., Sakiyama, O., Takahashi, K., Miyazaki, H., Hashimoto, S., and Daikuhara, Y. (1988). Purification and partial characterization of hepatocyte growth factor from plasma of a patient with hepatic failure. *J. Clin. Invest.* **81,** 414–419.

Gohda, E., Hayashi, Y., Kawaida, A., Tsubouchi, H., and Yamamoto, I. (1990). Hepatotrophic growth factor in blood of mice treated with carbon tetrachloride. *Life Sci.* **46,** 1801–1808.

Gonzatti-Haces, M., Seth, A., Park, M., Copeland, T., Oroszlan, S., and Vande Woude, G. F. (1988). Characterization of the tpr-met oncogene p65 and the met proto-oncogene cell surface p140 tyrosine kinases. *Proc. Natl. Acad. Sci. U.S.A.* **85,** 21–25.

Gordon, J. I. (1989). Intestinal epithelial differentiation: New insights from chimeric and transgenic mice. *J. Cell Biol.* **108,** 1187–1194.

Grant, D. S., Kleinman, H. K., Goldberg, I. D., Bhargava, M. M., Nickoloff, B. J., Kinsella, J. L., Polverini, P., and Rosen, E. M. (1993). Scatter factor induces blood vessel formation in vivo. *Proc. Natl. Acad. Sci. U.S.A.* **90,** 1937–1941.

Graziani, A., Gramaglia, D., Cantley, L. C., and Comoglio, P. M. (1991). The tyrosine phosphorylated hepatocyte growth factor/scatter factor receptor associates with phosphatidylinositol 3-kinase. *J. Biol. Chem.* **266,** 22087–22090.

Graziani, A., Gramaglia, D., dalla Zonca, P., and Comoglio, P. M. (1993). Hepatocyte growth factor/scatter factor stimulates the Ras guanine nucleotide exchanger. *J. Biol. Chem.* **268,** 9165–9168.

Grobstein, C. (1953). Morphogenetic interaction between embryonic mouse tissues separated by a membrane filter. *Nature (London)* **172,** 869–871.

Groden, J., Thliveris, A., Samowitz, W., Carlson, M., Gelbert, L., Albertsen, H., Joslyn, G., Stevens, J., Spirio, L., Robertson, M., Sargeant, L., Krapcho, K., Wolff, E., Burt, R., Hughes, J. P., Warrington, J., McPherson, J., Wasmuth, J., Le Paslier, D.,

Abderrahim, H., Cohen, D., Leppert, M., and White, R. (1991). Identification and characterization of the familial adenomatous polyposis coli gene. *Cell (Cambridge, Mass.)* **66,** 589–600.

Guo, L., Yu, Q. C., and Fuchs, E. (1993). Targeting expression of keratinocyte growth factor to keratinocytes elicits striking changes in epithelial differentiation in transgenic mice. *EMBO J.* **12,** 973–986.

Hamaguchi, M., Matsuyoshi, N., Ohnishi, Y., Gotoh, B., Takeichi, M., and Nagai, Y. (1993). p60^{v-src} causes tyrosine phosphorylation and inactivation of the N-cadherin-catenin cell adhesion system. *EMBO J.* **12,** 307–314.

Harland, R. M. (1994). Neural induction in *Xenopus. Curr. Opin. Genet. Dev.* **4,** 543–549.

Hartmann, G., Naldini, L., Weidner, K. M., Sachs, M., Vigna, E., Comoglio, P. M., and Birchmeier, W. (1992). A functional domain in the heavy chain of scatter factor/hepatocyte growth factor binds the *c-Met* receptor and induces cell dissociation but not mitogenesis. *Proc. Natl. Acad. Sci. U.S.A.* **89,** 11574–11578.

Hartmann, G., Weidner, M., Schwarz, H., and Birchmeier, W. (1994). The motility signal of scatter factor/hepatocyte growth factor mediated through the receptor tyrosine kinase met requires intracellular action of ras. *J. Biol. Chem.* **269,** 21936–21939.

Hattori, Y., Odagiri, H., Nakatani, H., Miyagawa, K., Naito, K., Sakamoto, H., Katoh, O., Yoshida, T., Sugimura, T., and Terada, M. (1990). K-sam, an amplified gene in stomach cancer, is a member of the heparin-binding growth factor receptor genes. *Proc. Natl. Acad. Sci. U.S.A.* **87,** 5983–5987.

Hay, E. D. (1990). Role of cell-matrix contacts in cell migration and epithelial-mesenychamal transformation. *Cell. Differ. Dev.* **32,** 367–375.

Higashio, K., Shima, N., Goto, M., Itagaki, Y., Nagao, M., Yasuda, H., and Morinaga, T. (1990). Identity of a tumor cytotoxic factor from human fibroblasts and hepatocyte growth factor. *Biochem. Biophys. Res. Commun.* **170,** 397–404.

Hinds, P. W., and Weinberg, R. A. (1994). Tumor suppressor genes. *Curr. Opin. Genet. Dev.* **4,** 135–141.

Hirano, S., Kimoto, N., Shimoyama, Y. H., Hirohashi, S., and Takeichi, M. (1992). Identification of a neural alpha-catenin as a key regulator of cadherin function and multicellular organization. *Cell (Cambridge, Mass.)* **70,** 293–301.

Hofstra, R. M. W., Landsvater, R. M., Ceccherini, I., Stulp, R. P., Stelwagen, T., Luo, Y., Pasini, B., Hoeppener, J. W. M., van Amstel, H. K., Romeo, G., Lips, C. J. M., and Buys, C. H. C. M. (1994). A mutation in the RET proto-oncogene associated with multiple endocrine neoplasia type 2B and sporadic medullary thyroid carcinoma. *Nature (London)* **367,** 375–376.

Hollstein, M., Sidransky, D., Vogelstein, B., and Harris, C. C. (1991). p53 mutations in human cancers. *Science* **253,** 49–53.

Holmes, W. E., Sliwkowski, M. X., Akita, R. W., Henzel, W. J., Lee, J., Park, J. W., Yansura, D., Abadi, N., Raab, H., Lewis, G. D., Shepard, H. M., Kuang, W. J., Wood, W. I., Goeddel, D. V., and Vandlen, R. L. (1992). Identification of Heregulin, a specific activator of p185^{erbB2}. *Science* **256,** 1205–1210.

Houssaint, E. (1980). Differentiation of the mouse hepatic primordian. I. An analysis of tissue interactions in hepatocyte differentiation. *Cell Differ.* **9,** 269–279.

Huang, S. S., and Huang, J. S. (1992). Purification and characterization of the neu/erb B2 ligand-growth factor from bovine kidney. *J. Biol. Chem.* **267,** 11508–11512.

Hudziak, R. M., Schlessinger, J., and Ullrich, A. (1987). Increased expression of the putative growth factor receptor p185^{HER2} causes transformation and tumorigenesis of NIH 3T3 cells. *Proc. Natl. Acad. Sci. U.S.A.* **84,** 7159–7163.

Huelsken, J., Birchmeier, W., and Behrens, J. (1994). E-Cadherin and APC compete for the interaction with -catenin and the cytoskeleton. *J. Cell Biol.* **127,** 2061–2069.

Hyafil, F., Morello, D., Babinet, C., and Jacob, F. (1980). A cell surface glycoprotein involved in the compaction of embryonal carcinoma cells and cleavage stage embryos. *Cell (Cambridge, Mass.)* **21**, 927–934.

Hyafil, F., Babinet, C., and Jacob, F. (1981). Cell-cell interactions in early embryogenesis: A molecular approach to the role of calcium. *Cell (Cambridge, Mass.)* **26**, 447–454.

Ishiki, Y., Ohnishi, H., Muto, Y., Matsumoto, K., and Nakamura, T. (1992). Direct evidence that hepatocyte growth factor is a hepatotrophic factor for liver regeneration and for potent antihepatitis action in vivo. *Hepatology* **16**, 1227–1235.

Issing, W. J., Heppt, W. J., and Kastenbauer, E. R (1993). erbB-3, a third member of the erbB/epidermal growth factor receptor gene family: Its expression in head and neck cancer cell lines. *Eur. Arch. Oto-Rhino-Laryngol.* **250**, 392–395.

Iyer, A., Kmiecik, T. E., Park, M., Daar, I., Blair, D., Dunn, K. J., Sutrave, P., Ihle, J. N., Bodescot, M., and Vande Woude, G. F. (1990). Structure, tissue-specific expression and transforming activity of the mouse *met* protooncogene. *Cell Growth Differ.* **1**, 87–95.

Johnson, D. E., and Williams, L. T. (1993). Structural and functional diversity in the FGF receptor multigene family. *Adv. Cancer Res.* **60**, 1–41.

Joslyn, G., Carlson, M., Thliveris, A., Albertsen, H., Gelbert, L., Samowitz, W., Groden, J., Stevens, J., Spirio, L., Robertson, M., Sargeant, L., Krapcho, K., Wolff, E., Burt, R., Hughes, J. P., Warrington, J., McPherson, J., Wasmuth, J., Le Paslier, D., Abderrahim, H., Cohen, D., Leppert, M., and White, R. (1991). Identification of deletion mutations and three new genes at the familial polyposis locus. *Cell (Cambridge, Mass.)* **66**, 601–613.

Jouanneau, J., Gavrilovic, J., Caruelle, D., Jaye, M., Moens, G., Caruelle, J. P., and Thiery, J. P. (1991). Secreted or nonsecreted forms of acidic fibroblast growth factor produced by transfected epithelial cells influence cell morphology, motility, and invasive potential. *Proc. Natl. Acad. Sci. U.S.A.* **88**, 2893–2897.

Jucker, M., Gunther, A., Gradl, G., Fonatsch, C., Krueger, G., Diehl, V., and Tesch, H. (1994). The Met/hepatocyte growth factor receptor (HGFR) gene is overexpressed in some cases of human leukemia and lymphoma. *Leuk. Res.* **18**, 7–16.

Juliano, R. L., and Varner, J. A. (1993). Adhesion molecules in cancer: The role of integrins. *Curr. Opin. Cell Biol.* **5**, 812–818.

Jung, W., Castren, E., Odenthal, M., Vande Woude, G. F., Ishii, T., Dienes, H. P., Lindholm, D., and Schirmacher, P. (1994). Expression and functional interaction of hepatocyte growth factor-scatter factor and its receptor c-met in mammalian brain. *J. Cell Biol.* **126**, 485–494.

Kamalati, T., Brooks, R. F., Holder, N., and Buluwela, L. (1993). *In vitro* regulation of HGF-SF expression by epithelial-mesenchymal interactions. *In* "Hepatocyte Growth Factor-Scatter Factor and the c-Met Receptor" (I. D. Goldberg and E. M. Rosen, eds.), pp. 201–224. Birkhaeuser, Basel.

Kedinger, M., Simon-Assmann, P. M., Lacroix, B., Marxer, A., Hauri, H. P., and Haffen, K. (1986). Fetal gut mesenchyme induces differentiation of cultured intestinal endodermal and crypt cells. *Dev. Biol.* **113**, 474–483.

Kemler, R. (1993). From cadherins to catenins: Cytoplasmic protein interactions and regulation of cell adhesion. *Trends Genet.* **9**, 317–321.

Kemler, R., Gossler, A., Mansouri, A., and Vestweber, D. (1990). The cell adhesion molecule uvomorulin. *In* "Morphoregulatory Molecules" (G. M. Edelmann, B. A. Cunningham, and J. P. Thiery, eds.), pp. 41–56. Wiley, New York.

Kemp, C. J., Donehower, L. A., Bradley, A., and Balmain, A. (1993). Reduction of p53 gene dosage does not increase initiation or promotion but enhances malignant progression of chemically induced skin tumors. *Cell (Cambridge, Mass.)* **74**, 813–822.

Kinoshita, T., Tashiro, K., and Nakamura, T. (1989). Marked increase of HGF mRNA in non-parenchymal liver cells of rats treated with hepatotoxins. *Biochem. Biophys. Res. Commun.* **165,** 1229–1234.

Kokai, Y., Cohen, J. A., Drebin, J. A., and Greene, M. I. (1987). Stage- and tissue-specific expression of the *neu* oncogene in rat development. *Proc. Natl. Acad. Sci. U.S.A.* **84,** 8498–8501.

Kratochwil, K. (1969). Organ specificity in mesenchymal induction demonstrated in the embryonic development of the mammary gland of the mouse. *Dev. Biol.* **20,** 46–71.

Kratochwil, K. (1986). Tissue combination and organ culture studies in the development of the embryonic mammary gland. *In* "Developmental Biology" (R. B. L. Gwatkin, ed.), Vol. 4, pp. 315–333. Plenum, New York.

Kraus, M. H., Issing, W., Miki, T., Popescu, N., and Aaronson, S. A. (1989). Isolation and characterization of ERBB3, a third member of the ERBB/epidermal growth factor receptor family: Evidence for overexpression in a subset of human mammary tumors. *Proc. Natl. Acad. Sci. U.S.A.* **86,** 9193–9197.

Kraus, M. H., Fedi, P., Starks, V., Muraro, R., and Aaronson, S. A. (1993). Demonstration of ligand-dependent signaling by the *erb*B-3 tyrosine kinase and its constitutive activation in human breast tumor cells. *Proc. Natl. Acad. Sci. U.S.A.* **90,** 2900–2904.

Kuniyasu, H., Yasui, W., Kitadai, Y., Yokozaki, H., Ito, H., and Tahara, E. (1992). Frequent amplification of the c-met gene in scirrhous type stomach cancer. *Biochem. Biophys. Res. Commun.* **189,** 227–232.

Kusakabe, M., Sakakura, T., Sano, M., and Nishizuka, Y. (1985). A pituitary-salivary mixed gland induced by tissue recombination of embryonic pituitary epithelium and embryonic submandibular gland mesenchyme in mice. *Dev. Biol.* **110,** 382–391.

Larue, L., Ohsugi, M., Hirchenhain, J., and Kemler, R. (1994). E-cadherin null mutant embryos fail to form a trophoectoderm epithelium. *Proc. Natl. Acad. Sci. U.S.A.* **91,** 8263–8267.

Lasnitzki, I., and Mizuno, T. (1980). Prostatic induction: Interaction of epithelium and mesenchyme from normal wild-type mice and androgen-insensitive mice with testicular feminization. *J. Endocrinol.* **85,** 423–428.

Le Douarin, N. M. (1975). An experimental analysis of liver development. *Med. Biol.* **53,** 427–455.

Liu, C., Park, M., and Tsao, M. S. (1992). Overexpression of c-met proto-oncogene but not epidermal growth factor receptor or c-erbB-2 in primary human colorectal carcinomas. *Oncogene* **7,** 181–185.

Livingstone, L. R., White, A., Sprousse, J., Livanos, E., Jacks, T., and Tisty, T. D. (1992). Altered cell cycle arrest and gene amplification potential accompany loss of wild-type p53. *Cell (Cambridge, Mass.)* **70,** 923–935.

Lokker, N. A., and Godowski, P. J. (1993). Generation and characterization of a competitive antagonist of human hepatocyte growth factor, HGF/NK1. *J. Biol. Chem.* **268,** 17145–17150.

Lokker, N. A., Mark, M. R., Luis, E. A., Bennett, G. L., Robbins, K. A., Baker, J. B., and Godowski, P. J. (1992). Structure-function analysis of hepatocyte growth factor: Identification of variants that lack mitogenic activity yet retain high affinity receptor binding. *EMBO J.* **11,** 2503–2510.

Lupu, R., Colomer, R., Kannan, B., and Lippman, M. (1992). Characterization of a growth factor that binds exclusively to the erbB2 receptor and induces cellular responses. *Proc. Natl. Acad. Sci. U.S.A.* **89,** 2287–2291.

Malkin, D., Li, F., Strong, L., Fraumeni, J., Nelson, C., Kim, D., Kassel, J., Gryka, M., Bischoff, F., Tainsky, M., and Friend, S. (1990). Germ line p53 mutations in a familial syndrome of breast cancer, sarcomas and other neoplasms. *Science* **250,** 1233–1238.

Mandai, M., Konishi, I., Koshiyama, M., Mori, T., Arao, S., Tashiro, H., Okamura, H.,

Nomura, H., Hiai, H., and Fukumoto, M. (1994). Expression of metastasis related nm23-H1 and nm23-H2 genes in ovarian carcinomas: Correlation with clinicopathology, EGFR, c-erbB2 and c-erbB3 genes and sex steroid receptor expression. *Cancer Res.* **54,** 1825–1830.

Marchionni, M., Goodearl, A., Chen, M., Bermingham-McDonogh, O., Kirk, C., Hendricks, M., Danehy, F., Misumi, D., Sudhalter, J., Kobayashi, K., Wroblewski, D., Lynch, C., Baldassare, M., Hiles, I., Davis, J., Hsuan, J., Totty, N., Otsu, M., McBurney, R., Waterfield, M., Stroobant, P., and Gwynne, D. (1993). Glial growth factors are alternatively spliced erbB2 ligands expressed in the nervous system. *Nature (London)* **362,** 312–318.

Mark, M., Lokker, N., Zioncheck, T., Luis, E., and Godowski, P. (1992). Expression and characterization of hepatocyte growth factor receptor-IgG fusion proteins. *J. Biol. Chem.* **267,** 26166–26171.

Mason, I. (1994). The ins and outs of fibroblast growth factors. *Cell (Cambridge, Mass.)* **78,** 547–552.

Mason, I., Fuller-Pace, F., Smith, R., and Dickson, C. (1994). FGF-7 (keratinocyte growth factor) expression during mouse development suggests roles in myogenesis, forebrain regionalization and epithelial-mesenchymal interactions. *Mech. Dev.* **45,** 15–30.

Mathan, M., Moxey, P., and Trier, J. (1976). Morphogenesis of fetal rat duodenal villi. *Am. J. Anat.* **146,** 73–92.

Matsumoto, K., Tekahara, T., Inoue, H., Hagiya, M., Shimizu, S., and Nakamura, T. (1991). Deletion of kringle domains or the N-terminal hairpin structure in hepatocyte growth factor results in marked decrease in related biological activities. *Biochem. Biophys. Res. Commun.* **181,** 691–699.

Matsumoto, K., Tajima, H., Hamanoue, M., Kohno, S., Kinoshita, T., and Nakamura, T. (1992). Identification and characterization of "injurin," an inducer of the gene expression of hepatocyte growth factor. *Proc. Natl. Acad. Sci. U.S.A.* **89,** 3800–3804.

Matsuyoshi, N., Hamaguchi, M., Taniguchi, S., Nagafuchi, A., Tsukita, S., and Takeichi, M. (1992). Cadherin-mediated cell-cell adhesion is perturbed by v-src tyrosine phosphorylation in metastatic fibroblasts. *J. Cell Biol.* **118,** 703–714.

McCormick, F. (1994). Activators and effectors of ras p21 proteins. *Curr. Opin. Genet. Dev.* **4,** 71–76.

Meehan, R., Barlow, D., Hill, R., Hogan, B., and Hastie, N. (1984). Pattern of serum protein gene expression in mouse visceral yolk sac and foetal liver. *EMBO J.* **3,** 1881–1885.

Meyer, D., and Birchmeier, C. (1994). Distinct isoforms of neuregulin are expressed in mesenchymal and neuronal cells during mouse development. *Proc. Natl. Acad. Sci. U.S.A.* **91,** 1064–1068.

Michalopoulos, G., Houck, K., Dolan, M., and Luetteke, C. (1984). Control of hepatocyte replication by two serum factors. *Cancer Res.* **44,** 4414–4419.

Miki, T., Fleming, T., Bottaro, D., Rubin, J., Ron, D., and Aaronson, S. (1991). Expression cDNA cloning of the KGF receptor by creation of a transforming autocrine loop. *Science* **251,** 72–75.

Miyazawa, K., Tsubouchi, H., Naka, D., Takahashi, K., Okigaki, M., Arakaki, N., Nakayama, H., Hirono, S., Sakiyama, O., Takashi, K., Gohda, E., Daikuhara, Y., and Kitamura, N. (1989). Molecular cloning and sequence analysis of cDNA for human hepatocyte growth factor. *Biochem. Biophys. Res. Commun.* **163,** 967–973.

Miyazawa, K., Kitamura, A., Naka, D., and Kitamura, M. (1991). An alternatively processed mRNA generated from human hepatocyte growth factor gene. *Eur. J. Biochem.* **197,** 15–22.

Miyazawa, K., Shimomura, T., Kitamura, A., Kondo, J., Morimoto, Y., and Kitamura, N. (1993). Molecular cloning and sequence analysis of the cDNA for a human serine protease responsible for activation of hepatocyte growth factor. *J. Biol. Chem.* **268,** 10024–10028.

Mizuno, K., Higuchi, O., Ihle, J., and Nakamura, T. (1993). Hepatocyte growth factor stimulates growth of hematopoietic progenitor cells. *Biochem. Biophys. Res. Commun.* **194,** 178–186.

Moll, R., Mitze, M., Frixen, U., and Birchmeier, W. (1993). Differentiated loss of E-cadherin expression on infiltrating ductal and lobular breast carcinomas. *Am. J. Pathol.* **143,** 1731–1742.

Mondino, A., Giordano, S., and Comoglio, P. (1991). Defective posttranslational processing activates the tyrosine kinase encoded by the met protooncogene. *Mol. Cell. Biol.* **11,** 6084–6092.

Montesano, R., Schaller, G., and Orci, L. (1991a). Induction of epithelial tubular morphogenesis in vitro by fibroblast-derived soluble factors. *Cell (Cambridge, Mass.)* **66,** 697–711.

Montesano, R., Matsumoto, K., Nakamura, T., and Orci, L. (1991b). Identification of a fibroblast-derived epithelial morphogen as hepatocyte growth factor. *Cell (Cambridge, Mass.)* **67,** 901–908.

Mori, S., Akiyama, T., Yamada, Y., Morishita, Y., Sugawara, I., Toyoshima, K., and Yamamoto, T. (1989). C-erbB2 gene product, a membrane protein commonly expressed on human fetal epithelial cells. *Lab. Invest.* **61,** 93–97.

Naka, D., Ishii, T., Yoshiyama, Y., Miyazawa, K., Hara, H., Hishida, T., and Kitamura, N. (1992). Activation of hepatocyte growth factor by proteolytic conversion of a single chain form to a heterodimer. *J. Biol. Chem.* **267,** 20114–20119.

Nakamura, T., Nawa, K., and Ichihara, A. (1984). Partial purification and characterization of hepatocyte growth factor from serum of hepatectomized rats. *Biochem. Biophys. Res. Commun.* **122,** 1450–1459.

Nakamura, T., Nishizawa, T., Hagiya, M., Seki, T., Shimonishi, M., Sugimura, A., Tashiro, K., and Shimizu, S. (1989). Molecular cloning and expression of human hepatocyte growth factor. *Nature (London)* **342,** 440–443.

Nakayama, H., Tsubouchi, H., Gohda, E., Koura, M., Nagahama, J., Yoshida, H., Daikuhara, Y., and Hashimoto, S. (1985). Stimulation of DNA synthesis in adult rat hepatocytes in primary culture by sera from patients with fulminant hepatic failure. *Biomed. Res.* **6,** 231–237.

Naldini, L., Weidner, K. M., Vigna, E., Gaudino, G., Bardelli, A., Ponzetto, C., Narsimhan, R., Hartmann, G., Zarnegar, R., Michalopoulos, G. K., Birchmeier, W., and Comoglio, P. M. (1991a). Scatter factor and hepatocyte growth factor are indistinguishable ligands for the met receptor. *EMBO J.* **10,** 2867–2878.

Naldini, L., Vigan, E., Nashiman, R., Gaudino, G., Zarnegar, R., Michalopoulos, G. K., and Comoglio, P. M. (1991b). Hepatocyte growth factor (HGF) stimulates the tyrosine kinase activity of the receptor encoded by the protooncogene c-met. *Oncogene* **6,** 501–504.

Naldini, L., Tamagnone, L., Vigna, E., Sachs, M., Hartmann, G., Birchmeier, W., Daikuhara, Y., Tsubouchi, H., Blasi, F., and Comoglio, P. M. (1992). Extracellular proteolytic cleavage by urokinase is required for activation of hepatocyte growth factor. *EMBO J.* **11,** 4825–4833.

Noji, S., Tashiro, K., Koyama, E., Nohno, T., Ohyama, K., Tanigushi, S., and Nakamura, T. (1990). Expression of hepatocyte growth factor gene in endothelial and Kupffer cells of damaged rat livers, as revealed by in situ hybridisation. *Biochem. Biophys. Res. Commun.* **173,** 42–47.

Oda, T., Kanai, Y., Shimoyama, Y., Nagafuchi, A., Tsukita, S., and Hirohashi, S. (1993). Cloning of the human alpha-catenin cDNA and its aberrant mRNA in a human cancer cell line. *Biochem. Biophys. Res. Commun.* **193,** 897–904.

Oda, T., Kanai, T., Oyama, K., Yoshiura, Y., Shimoyama, Y., Birchmeier, W., Sugimura,

T., and Hirohashi, S. (1994). E-cadherin gene mutations in human gastric carcinoma cell lines. *Proc. Natl. Acad. Sci. U.S.A.* **91**, 1858–1862.

Oka, H., Shiozaki, H., Kobayashi, K., Inoue, M., Tahara, H., Kobayashi, T., Takatsuka, Y., Matsuyoshi, N., Hirano, S., Takeichi, M., and Mori, T. (1993). Expression of E-cadherin cell adhesion molecules in human breast cancer tissues and its relationship to metastasis. *Cancer Res.* **53**, 1696–1701.

Okigaki, M., Komada, M., Uehara, Y., Miyazawa, K., and Kitamura, N. (1992). Functional characterization of human hepatocyte growth factor mutants obtained by deletion of structural domains. *Biochemistry* **31**, 9555–9561.

Orr-Urtreger, A., Trakhtenbrot, L., Ben-Levy, R., Wen, D., Rechavi, G., Lonai, P., and Yarden, Y. (1993a). Neural expression and chromosomal mapping of neu differentiation factor to 8p12-p21. *Proc. Natl. Acad. Sci. U.S.A.* **90**, 1867–1871.

Orr-Urtreger, A., Bedford, M., Burakova, T., Arman, E., Zimmer, Y., Yayon, A., Givol, D., and Lonai, P. (1993b). Developmental localization of the splicing alternatives of fibroblast growth factor receptor-2 (FGFR2). *Dev. Biol.* **158**, 475–486.

Ozawa, M., Engel, R., and Kemler, R. (1990). Single amino acid substitutions in one Ca^{2+}-binding site of uvomorulin abolish the adhesive function. *Cell (Cambridge, Mass.)* **63**, 1033–1038.

Pachnis, V., Mankoo, B., and Constantini, F. (1993). Expression of the c-ret protooncogene during mouse embryogenesis. *Development (Cambridge, UK)* **119**, 1005–1017.

Park, M., Dean, M., Cooper, C., Schmidt, M., O'Brian, S., Blair, D., and Vande Woude, G. F. (1986). Mechanism of met oncogene activation. *Cell (Cambridge, Mass.)* **25**, 895–904.

Park, M., Dean, M., Kaul, K., Braun, M., Gonda, M., and Vande Woude, G. F. (1987). Sequence of MET protooncogene cDNA has features characteristic of the tyrosine kinase family of growth factor receptors. *Proc. Natl. Acad. Sci. U.S.A.* **84**, 6379–6383.

Peles, E., and Yarden, Y. (1993). Neu and its ligands: From an oncogene to neural factors. *BioEssays* **15**, 815–824.

Peles, E., Bacus, S., Koski, R., Lu, H., Wen, D., Ogden, R., Ben-Levy, R., and Yarden, Y. (1992). Isolation of the Neu/HER-2 stimulatory ligand: A 44 kd glycoprotein that induces differentiation of mammary tumor cells. *Cell (Cambridge, Mass.)* **69**, 205–216.

Peles, E., Ben-Levy, R., Tzahar, E., Liu, N., Wen, D., and Yarden, Y. (1993). Cell-type specific interaction of neu differentiation factor (NDF/heregulin) with Neu/Her2 suggests complex ligand-receptor relationships. *EMBO J.* **12**, 961–971.

Peters, K., Werner, S., Liao, X., Wert, S., Whitsett, J., and Williams, L. (1994). Targeted expression of a dominant negative FGF receptor blocks branching morphogenesis and epithelial differentiation of the mouse lung. *EMBO J.* **13**, 3296–3301.

Plowman, G., Green, J., Culouscou, J.-M., Carlton, G., Rothwell, V., and Buckley, S. (1993a). Heregulin induces tyrosine phosphorylation of HER4/p180erbB4. *Nature (London)* **366**, 473–475.

Plowman, G., Culouscou, J.-M., Whitney, G., Green, J., Carlton, G., Foy, L., Neubauer, M., and Shoyab, M. (1993b). Ligand-specific activation of HER4/p180, a fourth member of the epidermal growth factor receptor family. *Proc. Natl. Acad. Sci. U.S.A.* **90**, 1746–1750.

Plowman, G., Green, J., Culouscou, J., Carlton, G., Rothwell, V., and Buckley, S. (1993). Heregulin induces tyrosine phosphorylation of HER4. *Nature (London)* **366**, 473–475.

Ponzetto, C., Giordano, S., Peverali, F., Della Valle, G., Abate, M., Vaula, G., and Comoglio, P. M. (1991). c-met is amplified but not mutated in a cell line with an activated met tyrosine-kinase. *Oncogene* **6**, 553–559.

Ponzetto, C., Bardelli, A., Zhen, Z., Maina, F., dalla Zonca, C., Giordano, S., Graziani, A., Panayotou, and Comoglio, P. M. (1994). A multifunctional docking site mediates signalling and transformation by the hepatocyte growth factor/scatter factor receptor family. *Cell (Cambridge, Mass.)* **77**, 261–271.

Press, M., Cordon-Cardo, C., and Slamon, D. (1990). Expression of the Her2/neu protoonco-gene in normal human and fetal tissues. *Oncogene* **5**, 953–962.

Quirke, P., Pickles, A., Tuzi, N., Mohamdee, O., and Gullick, W. (1989). Pattern of expression of c-erbB2 oncoprotein in human fetuses. *Br. J. Cancer* **60**, 64–69.

Rege-Cambrin, G., Scaravaglio, P., Carozzi, F., Giordano, S., Ponzetto, C., Comoglio, P. M., and Saglio, G. (1992). Karyotype analysis of gastric carcinoma cell lines carrying an amplified c-met oncogene. *Cancer Genet. Cytogenet.* **64**, 170–173.

Riethmacher, D., Brinkmann, V., and Birchmeier, C. (1995). A targeted mutation in the mouse E-cadherin gene results in defective preimplantation development. *Proc. Natl. Acad. Sci. U.S.A.* **92**, 885–889.

Risinger, J., Berchuck, A., Kohler, M., and Boyd, J. (1993). Mutations of the E-cadherin gene in human gynecologic cancers. *Nat. Genet.* **7**, 98–102.

Rodrigues, G., and Park, M. (1994). Oncogenic activation of tyrosine kinases. *Curr. Opin. Genet. Dev.* **4**, 15–24.

Rodrigues, G., Naujokas, M., and Park, M. (1991). Alternative splicing generates isoforms of the met receptor tyrosine kinase which undergo differential processing. *Mol. Cell. Biol.* **11**, 2962–2970.

Rodriguez-Boulan, E., and Nelson, W. (1989). Morphogenesis of the polarized epithelial cell phenotype. *Science* **245**, 718–725.

Rong, S., Bodescot, M., Blair, D., Dunn, J., Nakamura, T., Mizuno, K., Park, M., Chan, A., Aaronson, S., and Vande Woude, G. F. (1992). Tumorigenicity of the met protoonco-gene and the gene for hepatocyte growth factor. *Mol. Cell. Biol.* **12**, 5152–5158.

Rong, S., Segal, S., Anver, M., Resau, J., and Vande Woude, G. F. (1994). Invasiveness and metastasis of NIH3T3 cells induced by Met-hepatocyte growth factor/scatter factor autocrine stimulation. *Proc. Natl. Acad. Sci. U.S.A.* **91**, 4731–4735.

Roos, F., Terell, T., Godowski, P., Chamow, S., and Schwall, R. (1992). Reduction of alpha-naphtylisothiocyanate-induced hepatoxicity by recombinant human hepatocyte growth factor. *Endocrinology* **131**, 2540–2544.

Rosen, E., Goldberg, I., Kacinski, B., Buckholz, T., and Vinter, D. (1989). Smooth muscle releases an epithelial cell scatter factor which binds to heparin. *In Vitro Cell Dev. Biol.* **25**, 163–173.

Rosen, E., Meromsky, L., Setter, E., Vinter, D., and Goldberg, I. (1990). Purified scatter factor stimulates epithelial and vascular endothelial cell migration. *Proc. Soc. Exp. Biol. Med.* **195**, 34–43.

Rubin, J., Osada, H., Finch, P., Taylor, W., Rudikoff, S., and Aaronson, S. (1989). Purification and characterization of a newly identified growth factor specific for epithelial cells. *Proc. Natl. Acad. Sci. U.S.A.* **86**, 802–806.

Rubin, J., Chan, A., Bkottaro, D., Burgess, W., Taylor, W., Cech, A., Hirschfield, D., Wong, J., Miki, T., Finch, P., and Aaronson, S. (1991). A broad-spectrum human lung fibroblast-derived mitogen is a variant of hepatocyte growth factor. *Proc. Natl. Acad. Sci. U.S.A.* **88**, 415–419.

Rubinfeld, B., Souza, B., Albert, I., Mueller, O., Chamberlain, S., Masiarz, F., Munemitsu, S., and Polakis, P. (1993). Association of the APC gene product with beta-catenin. *Science* **262**, 1731–1741.

Russel, W., McGowan, J., and Bucher, N. (1984). Partial characterization of a hepatocyte growth factor from rat platelets. *J. Cell. Physiol.* **119**, 183–192.

Sakakura, T. (1991). New aspects of stroma-parenchyma relations in mammary gland differentiation. *Int. Rev. Cytol.* **125**, 165–202.

Santoro, M., Rosati, M., Grieco, M., Berlingieri, M., D'Amato, V., de Franciscis, V., and Fusco, A. (1990). The ret protooncogene is expressed in human pheochromocytomas and thyroid medullary carcinomas. *Oncogene* **5**, 1595–1598.

Saxen, L. (1987). "Organogenesis of the Kidney." Cambridge Univ. Press, Cambridge, UK.

Schipper, J., Frixen, U., Behrens, J., Unger, A., Jahnke, K., and Birchmeier, W. (1991). E-cadherin expression in squamous cell carcinomas of head and neck: Invasive correlation with tumor dedifferentiation and lymph node metastasis. *Cancer Res.* **51**, 6328–6337.

Schirmacher, P., Geerts, A., Pictrangelo, A., Dienes, H., and Rogler, C. (1992). Hepatocyte growth factor/Hepatopoietin A is expressed in fat-storing cells from rat liver but not myofibroblast-like cells derived from fat-storing cells. *Hepatology* **15**, 5–11.

Schirmacher, P., Geerts, A., Jung, W., Pietrangelo, A., Rogler, C., and Dienes, H. (1993). The role of Ito cells in the biosynthesis of HGF-SF in the liver. In "Hepatocyte Growth Factor-Scatter Factor and the c-Met Receptor" (I. D. Goldberg and E. M. Rosen, eds.), pp. 285–299. Birkhaeuser, Basel.

Schmidt, C., Bladt, F., Goedecke, S., Brinkmann, V., Zschiesche, W., Sharpe, M., Gheradi, E., and Birchmeier, C. (1995). Scatter factor/hepatocyte growth factor is essential for liver development. *Nature (London)* **373**, 699–702.

Schuchardt, A., D'Agati, V., Larsson-Blomberg, L., Constantini, F., and Pachnis, V. (1994). Defects in the kidney and enteric system of mice lacking the tyrosine kinase receptor ret. *Nature (London)* **267**, 380–383.

Schwarz, M., Owaribe, K., Kartenbeck, J., and Franke, W. (1990). Desmosomes and hemidesmosomes: Constitutive molecular components *Annu. Rev. Cell Biol.* **6**, 461–491.

Sefton, M., Johnson, M., and Clayton, L. (1992). Synthesis and phoshorylation of uvomorulin during mouse early development. *Development (Cambridge, UK)* **115**, 313–318.

Shah, N., Marchionni, M., Isaacs, I., Stroobant, P., and Anderson, D. (1994). Glial growth factor restricts mammalian neural crest stem cells to a glial fate. *Cell (Cambridge, Mass.)* **77**, 349–360.

Shibamoto, S., Hayakawa, M., Takeuchi, K., Hori, T., Oku, N., Miyazawa, K., Kitamura, N., Takeichi, M., and Ito, F. (1994). Tyrosine phosphorylation of beta-catenin and plakoglobin enhanced by hepatocyte growth factor and epidermal growth factor in human carcinoma cells. *Cell Adhes. Commun.* **4**, 295–305.

Shiota, G., Rhoads, D., Wang, T., Nakamura, T., and Schmidt, E. (1992). Hepatocyte growth factor inhibits growth of hepatocellular carcinoma cells. *Proc. Natl. Acad. Sci. U.S.A.* **89**, 373–377.

Shiota, G., Wang, T., Nakamura, T., and Schmidt, E. (1994). Hepatocyte growth factor in transgenic mice: Effects on hepatocyte growth, liver regeneration and gene expression. *Hepatology* **19**, 913–919.

Simons, K., and Wandinger-Ness, A. (1990). Polarized sorting in epithelia. *Cell (Cambridge, Mass.)* **62**, 207–210.

Skinner, A., and Hurst, H. (1993). Transcriptional regulation of the erbB-3 gene in human breast carcinoma cell lines. *Oncogene* **8**, 3393–3401.

Slamon, D., Clark, G., Wong, S., Levin, W., Ullrich, A., and McGuire, W. (1987). Human breast cancer: Correlation of relapse and survival with amplification of the HER2/neu oncogene. *Science* **235**, 177–182.

Slamon, D., Godolphin, W., Jones, L., Holt, J., Wong, S., Keith, D., Levin, W., Stuart, S., Udove, J., Ullrich, A., and Press, M. (1989). Studies of the HER2/neu protooncogene in human breast and ovarian cancer. *Science* **244**, 707–712.

Sliwkowski, M., Schaefer, G., Akita, R., Lofgren, J., Fitzpatrick, V., Nuijens, A., Fendly, B., Cerione, R., Vandlen, R., and Carraway, K. (1994). Coexpression of erbB2 and erbB3 proteins reconstitutes a high affinity receptor for heregulin. *J. Biol. Chem.* **269**, 14661–14665.

Sonnenberg, A., Calafat, J., Janssen, H., Daams, H., van der Raaij-Helmer, L., Falcioni, R., Kennel, S., Aplin, J., Baker, J., Loizidou, M., and Garrod, D. (1991). Integrin alpha6/

beta 4 complex is located in hemidesmosomes, suggesting a major role in epidermal cell-basement membrane adhesion. *J. Cell Biol.* **113,** 907–917.

Sonnenberg, E., Goedecke, A., Walter, B., Bladt, F., and Birchmeier, C. (1991). Transient and locally restricted expression of the ros1 protooncogene during mouse development. *EMBO J.* **10,** 3693–3702.

Sonnenberg, E., Meyer, D., Weidner, K., and Birchmeier, C. (1993a). Scatter factor/hepatocyte growth factor and its receptor, the c-met tyrosine kinase, can mediate a signal exchange between mesenchyme and epithelia during mouse development. *J. Cell Biol.* **123,** 223–235.

Sonnenberg, E., Weidner, K., and Birchmeier, C. (1993b). Expression of the met-receptor and its ligand SF/HGF during mouse embryogenesis. *In* "Hepatocyte Growth Factor-Scatter Factor and the c-Met Receptor" (I. D. Goldberg and E. M. Rosen, eds.), pp. 381–395. Birkhaeuser, Basel.

Spooner, B., and Wessells, N. (1970). Mammalian lung development: Interactions in primordium formation and bronchial morphogenesis. *J. Exp. Zool.* **175,** 445–454.

Stern, C., and Ireland, G. (1993). HGF-SF: A neural inducing molecule in vertebrate embryos? *In* "Hepatocyte Growth Factor-Scatter Factor and the c-Met Receptor" (I. D. Goldberg and E. M. Rosen, eds.), pp. 369–381. Birkhaeuser, Basel.

Stern, C., Ireland, G., Herrick, S., Gherardi, E., Gray, J., Perryman, M., and Stoker, M. (1990). Epithelial scatter factor and development of the chick embryonal axis. *Development (Cambridge, UK)* **110,** 1271–1284.

Stoker, M., and Perryman, M. (1985). An epithelial scatter factor released by embryo fibroblasts. *J. Cell Sci.* **77,** 209–223.

Stoker, M., Gherardi, E., Perryman, M., and Gray, J. (1987). Scatter factor is a fibroblast-derived modulator of epithelial cell motility. *Nature (London)* **327,** 239–242.

Su, L., Vogelstein, B., and Kinzler, K. (1993). Association of the APC tumor suppressor protein with catenins. *Science* **262,** 1734–1737.

Takeichi, M. (1991). Cadherin cell adhesion receptors as a morphogenic regulator. *Science* **252,** 1451–1455.

Tarakhovsky, A., Zaichuk, T., Prassolov, V., and Butenko, Z. (1991). A 25 kDa polypeptide is the ligand for p185neu and is secreted by activated macrophages. *Oncogene* **6,** 2187–2196.

Tashiro, K., Hagiya, M., Nishizawa, T., Seki, T., Shimonishi, M., Shimizu, S., and Nakamura, T. (1990). Deduced primary structure of rat hepatocyte growth factor and expression of the mRNA in rat tissues. *Proc. Natl. Acad. Sci. U.S.A.* **87,** 3200–3204.

Timpl, R. (1989). Structure and biological activity of basement membrane proteins. *Eur. J. Biochem.* **180,** 487–502.

Tsafaty, I., Resau, J. H., Rulong, S., Keydar, I., Faletto, D. L., and Vande Woude, G. F. (1992). The met protooncogene receptor and lumen formation. *Science* **257,** 1258–1261.

Tsubouchi, H., Hirono, S., Gohda, E., Nakayama, H., Takahashi, K., Sakiyama, O., Miyazaki, H., Sugihara, J., Tomita, E., Muto, Y., Daikuhara, Y., and Hashimoto, S. (1989). Clinical significance of human hepatocyte growth factor in blood from patients with fulminant hepatic failure. *Hepatology* **9,** 875–881.

Tsukita, S., Tsukita, S., Nagafuchi, A., and Yonemura, S. (1992). Molecular linkage between cadherins and actin filaments in cell-cell adherens junctions. *Curr. Opin. Cell Biol.* **4,** 834–839.

Tsukita, S., Itoh, M., Nagafuchi, A., Yonemura, S., and Tsukita, S. (1993). Submembranous junctional plaque proteins include potential tumor suppressor molecules. *J. Cell Biol.* **123,** 1049–1053.

Ueno, H., Gunn, M., Dell, K., Tseng, A., and Williams, L. (1992). A truncated form of fibroblast growth factor receptor 1 inhibits signal transduction by multiple types of fibroblast growth factor receptor. *J. Biol. Chem.* **267,** 1470–1476.

Ullrich, A., Coussens, L., Hayflick, J. S., Dull, T. J., Gray, A., Tam, A. W., Lee, J., Yarden, Y., Libermann, T. A., Schlessinger, J., Downward, J., Mayes, E. L., Whittle, N., Waterfield, M. D., and Seeburg, P. H. (1984). Human epidermal growth factor receptor cDNA sequence and aberrant expression of the amplified gene in A341 epidermoid carcinoma cells. *Nature (London)* **309,** 418–425.

Umbas, R., Isaacs, W. B., Bringuier, P. P., Schaafsma, H. E., Karthaus, H. F., Oosterhof, G. O., Debruyne, F. M., and Schalken, J. A. (1994). Decreased E-cadherin expression is associated with poor prognosis in patients with prostate cancer. *Cancer Res.* **54,** 3929–3933.

Valles, A. M., Boyer, B., and Thiery, J. P. (1991). Adhesion systems in embryonic epithelial to mesenchymal transformation and in cancer invasion and metastasis. *In* "Cell Motility Factors" (I. D. Goldberg, ed.), pp. 17–34, Birkhaeuser, Basel.

Vestweber, D., and Kemler, R. (1984). Rabbit antiserum against a purified surface glycoprotein decompacts mouse preimplantation embryos and reacts with specific adult tissues. *Exp. Cell Res.* **152,** 169–178.

Villa-Moruzzi, E., Lapi, S., Prat, M., Gaudino, G., and Comoglio, P. M. (1993). A protein tyrosine phosphatase activity associated with the hepatocyte growth factor/scatter factor receptor. *J. Biol. Chem.* **268,** 18176–18180.

Vleminckx, K., Vakaet, L., Mareel, M., Fier, W., and Van Roy, F. (1991). Genetic manipulation of E-cadherin expression by epithelial tumor cells reveals an invasion suppressor role. *Cell (Cambridge, Mass.)* **66,** 107–119.

Vogelstein, B., and Kinzler, K. W. (1993). The multistep nature of cancer. *Trends Genet.* **9,** 138–141.

Weidner, K. M., Behrens, J., Vandekerckhove, J., and Birchmeier, W. (1990). Scatter factor: Molecular characteristics and effect on the invasiveness of epithelial cells. *J. Cell Biol.* **111,** 2097–2108.

Weidner, K. M., Arakaki, N., Hartmann, G., Vandekerckhove, J., Weingart, S., Rieder, H., Fonatsch, C., Tsubushi, H., Hishida, T., Daikuhara, Y., and Birchmeier, W. (1991). Evidence for the identity of human scatter factor and human hepatocyte growth factor. *Proc. Natl. Acad. Sci. U.S.A.* **88,** 7001–7005.

Weidner, M., Sachs, M., and Birchmeier, W. (1993). The Met receptor tyrosine kinase transduces motility, proliferation, and morphogenic signals of scatter factor/hepatocyte growth factor in epithelial cells. *J. Cell Biol.* **121,** 145–154.

Weidner, M., Sachs, M., Riethmacher, D., and Birchmeier, W. (1995). Mutation of the juxtamembrane tyrosine residue 1001 suppresses loss-of-function mutations of the met receptor in epithelial cells. *Proc. Natl. Acad. Sci. U.S.A.* **92,** in the press.

Wen, D., Peles, E., Cupples, R., Suggs, S., Bacus, S., Luo, Y., Trail, G., Hu, S., Silbinger, S., Levy, R., Koski, R., Lu, H., and Yarden, Y. (1992). Neu differentiation factor: A transmembrane glycoprotein containing an EGF domain and an immunoglobulin homology unit. *Cell (Cambridge, Mass.)* **69,** 559–572.

Wen, D., Suggs, S., Karunagaran, D., Liu, N., Cupples, R., Luo, Y., Janssen, A. M., Ben-Baruch, N., Trollinger, D., Jacobson, V. L., Meng, S. Y., Lu, H. S., Hu, S., Chang, D., Yang, W., Yanigahara, D., Koski, R. A., and Yarden, Y. (1994). Structural and functional aspects of the multiplicity of neu differentiation factor. *Mol. Cell. Biol.* **14,** 1909–1919.

Werner, S., Peters, K., Longaker, M., Fuller-Pace, F., Banda, M., and Williams, L. (1992). Large induction of keratinocyte growth factor expression in the dermis during wound healing. *Proc. Natl. Acad. Sci. U.S.A.* **89,** 6896–6900.

Werner, S., Weinberg, W., Liao, X., Peters, K., Blessing, M., Yuspa, S., Weiner, R., and Williams, L. (1993). Targeted expression of a dominant negative FGF receptor mutant in the epidermis of transgenic mice reveals a role of FGF in keratinocyte organization and differentiation. *EMBO J.* **12,** 2635–2643.

Werner, S., Smola, H., Liao, X., Longaker, M., Krieg, T., Hofschneider, P., and Williams,

L. (1994). The function of KGF in morphogenesis of epithelium and reepithelization of wounds. *Science* **266,** 819–822.

Wessells, N. (1970). Mammalian lung development: Interactions in formulation and morphogenesis of tracheal buds. *J. Exp. Zool.* **175,** 455–466.

Wessells, N., and Cohen, J. (1967). Early pancreatic organogenesis: Morphogenesis, tissue interactions and mass effects. *Dev. Biol.* **15,** 237–270.

Wiley, L., Kidder, G., and Watson, A. (1990). Cell polarity and development of the first epithelium. *BioEssays* **12,** 67–73.

Wolf, H., Zarnegar, R., Oliver, L., and Michalopoulos, G. K. (1991a). Hepatocyte growth factor in human placenta and trophoblastic disease. *Am. J. Pathol.* **138,** 1035–1043.

Wolf, M., Zarnegar, R., and Michalopoulos, G. K. (1991b). Localisation of hepatocyte growth factor in human and rat tissues: An immunohistochemical study. *Hepatology* **14,** 488–494.

Yamamoto, T., Ikawa, S., Akiyama, T., Sembra, K., Nomura, N., Miyajima, N., Saito, T., and Toyoshima, K. (1986). Similarity of protein encoded by the human c-erbB2 gene to epidermal growth factor receptor. *Nature (London)* **319,** 230–234.

Yang, J., Spitzer, E., Meyer, D., Sachs, M., Birchmeier, C., and Birchmeier, W. (1995). Sequential requirement of scatter factor/hepatocyte growth factor and neu differentiation factor/heregulin in the morphogenesis and differentiation of the mammary gland. Submitted for publication.

Yasugi, S. (1993). Role of epithelial-mesenchymal interactions in differentiation of epithelium of vertebrate digestive organs. *Dev. Growth Differ.* **35,** 1–9.

Zarnegar, R., and Michalopoulos, G. K. (1989). Purification and biological characterization of human hepatopoietin A, a polypeptide growth factor for hepatocytes. *Cancer Res.* **49,** 3314–3320.

Zarnegar, R., Muga, S., Rahdija, R., and Michalopoulos, G. K. (1990). Tissue distribution of hematopoietin A: A heparin binding polypeptide growth factor for hepatocytes. *Proc. Natl. Acad. Sci. U.S.A.* **87,** 1252–1256.

Zhu, H., Naujokas, M., Fixman, E., Torossian, K., and Park, M. (1994). Tyrosine 1356 in the carboxy terminal tail of the HGF/SF receptor is essential for the transduction of signals for cell motility and morphogenesis. *J. Biol. Chem.* **269,** 29946–29948.

Dynamics of the Seminal Vesicle Epithelium

Lucinda R. Mata
Department of Cell Biology, Gulbenkian Institute of Science, 2781 Oeiras Codex, Portugal

In newborn rodents, seminal vesicle epithelium (SVEP) cells display a poorly developed rough endoplasmic reticulum (RER) and Golgi complex, and they show no signs of secretion. From puberty onward, secretory material starts to appear, and the RER and Golgi complex progressively develop and reorganize until the adult ultrastructure is established around 40–60 days of age. Multivesicular bodies and lysosomes follow in this development but lysosomes evolve to lipofucsin granules with aging. The duration of the secretory cycle in SVEP cells is shorter than in other exocrine cells and the secretory protein pattern depends on the animal species, androgen status, and sexual activity.

SVEP cells are also involved in endocytosis, which is coupled to exocytosis, and their endocytic pathway intersects the exocytic pathway in Golgi cisterns. The structure and function of SVEP cells depends mainly on testosterone, but other hormones and factors, such as the neuropeptide VIP, also influence their activity. Castration leads to programmed death and regression of SVEP cells to an extent that depends on the animal species. In addition, castration induces changes in the secretory protein pattern and delays its intracellular transport. Endocytic kinetics is also delayed following castration. Primary cultures of SVEP cells in a bicameral system are proposed as a model to investigate the activities of SVEP cells further.

KEY WORDS: Seminal vesicle epithelium, Secretion, Endocytosis.

I. Introduction

The seminal vesicles (SV), together with the prostate and Cowper's gland, constitute the male sex accessory glands. These glands are present in most families of placental mammals (Rumpler, 1985) where their secretions can reach more than 99% of the seminal plasma (Luke and Coffey, 1994).

Among male accessory glands, the prostate, which is prone to tumors seriously afflicting humans, has been extensively studied, whereas seminal vesicles, which seldom display such pathology (Mazzucchelli *et al.*, 1992), have not been paid as much attention. In addition, because seminal vesicle secretions do not appear to be an absolute requirement for *in vitro* fertilization (Williams-Ashman, 1983; Curry and Atherton, 1990), their significance in reproduction was until recently a matter of debate (Williams-Ashman, 1983).

However, removal of the seminal vesicles can severely reduce, or even impair, *in vivo* fertility (Chow *et al.*, 1986; Sofikitis *et al.*, 1992). Evidence accumulated during the past decade has led to recognition that seminal vesicle secretions are of great importance in ensuring *in vivo* fertility (Adjiman, 1985; Curry and Atherton, 1990).

The relevance of seminal vesicles for fertility is related to the influence of their secretions on sperm motility, metabolism, and surface components (Clavert *et al.*, 1985; Curry and Atherton, 1990) and their role in the immunology of reproduction (Clavert *et al.*, 1985; Emoto *et al.*, 1990; Thaler *et al.*, 1990; Porta *et al.*, 1993). The characterization of seminal vesicle secretions and the understanding of the regulation of their secretory activity has thus become a matter of utmost importance.

In addition to secretion, the seminal vesicle epithelial (SVEP) cells are also involved in endocytosis of proteins from the gland lumen content. The study of endocytosis in the hamster seminal vesicle provided the first evidence supporting reutilization of internalized membrane in the exocytic pathway of exocrine cells (Mata and David-Ferreira, 1973) and made seminal vesicles eligible as a model for investigating endocytosis and membrane traffic (Mata and Christensen, 1990a,b). Although the physiological role of endocytosis in these glands remains to be fully clarified, it has been suggested that, besides its connection with cell membrane reutilization, endocytosis might be involved in the regulation of secretion and in mucosal immune response (Mata, 1994).

The structure and function of seminal vesicles, which depend mainly on androgens, has been extensively studied in small mammals in the laboratory. In this chapter, the dynamics of the seminal vesicle epithelium of laboratory rodents will be discussed mainly in connection with cell ultrastructure, secretion, and endocytosis.

II. Ultrastructural Dynamics of Seminal Vesicle Epithelium during Postnatal Development and Aging

The SVEP cells are subjected to morphological changes during two main periods of animal life. The first period extends from birth to the young adult and is related to cell differentiation for secretion. The second period starts during adulthood and continues with increasing age. The ultrastructural changes related to postnatal differentiation and secretion are conspicuous and rapid whereas those involved in aging are less obvious and develop slowly.

Seminal vesicles develop as paired diverticula from the caudal end of Wolffian ducts following the onset of fetal androgens but only after birth, when the testosterone rate increases, does the gland epithelium differentiate as a secretory one (Aumuller, 1979).

From an ultrastructural study of seminal vesicles during postnatal development (Aumuller, 1979), a general pattern emerges for the structural differentiation of the SV secretory epithelium which also agrees with observations described later in the guinea pig (Barham et al., 1980) and hamster (Mata, 1981).

In the newborn hamster, the seminal vesicles appear as compact diverticula at the proximal end of the ductus deferens; they are smaller than 1 mm and virtually without a lumen, whereas in the adult animal they are membranous tubuloalveolar bags with translucent walls, about 20 mm long with a large lumen fully loaded with hyaline secretion. While this transformation is occurring, the epithelium lining the lumen is evolving from a pseudostratified to a tall cell columnar epithelium and differentiating for secretion. This differentiation was followed at the ultrastructural level in hamsters up to 62 days old (Mata, 1981) and, based on the development and organization of the rough endoplasmic reticulum (RER) and Golgi complex (GC), the presence of secretory material, and the relative volume of cellular components, four phases were established. (The main features of such phases are summarized in Table I.)

The most obvious changes occur from phase 1 to phase 2 (Figs. 1 and 2). They include an increase and reorganization in the RER and GC, and the appearance of secretory vacuoles observed at the earliest at 12 days after birth. In phase 2, apical secretory vacuoles either appear full of dense secretion or display a halo of poorly condensed material (Fig. 2b). Apical secretory granules are concentric with their vacuoles. From phase 2 to phase 3, the relative volumes of RER and GC increase about twofold and the intracellular secretory material is increased by about sixfold. In addition, the cisternae of the RER become narrower and clearly oriented along the major cell axis; the GC is enriched in secretory vacuoles associ-

TABLE I

Postnatal Development of Hamster Seminal Vesicle Secretory Cells

Phase 1	Age < 12 days
Lumen	Narrow, usually empty, occasionally containing cytoplasmic debris
Apical membrane	Rare and small microvilli and occasional cilia
Cytoplasmic matrix	Fibrillar, displaying scattered microfilaments as well as microtubules and numerous ribosomes occasionally in polysomes of 5–7
Lysosomes	Rare; $V_v,\% = 0.2$[a]
RER	Scarce, with few membrane-associated ribosomes; small cisterns dilated with a fibrillar content; $V_v,\% = 4.3$
Golgi complex	Scattered or associated stacks of small and flat cisterns (3–6/stack) without any electronic dense content; $V_v,\% = 2.3$
Intracellular secretory material	Absent

Phase 2	Age: 12–23 days
Lumen	Larger than in phase 1 and containing clumps of condensed secretion occasionally included in, and continuous with, noncondensed fibrillar secretory material; cell debris is rare
Apical membrane	Long and numerous microvilli
Cytoplasmic matrix	Similar to phase 1 but including bundles of microfilaments
Lysosomes	Common; $V_v,\% = 1.2$
RER	Cisterns occasionally grouped, dilated and oriented toward the cell apex; membranes presenting ribosome-free areas; $V_v,\% = 6.6$
Golgi complex	Stacks of mostly clear cisterns, occasionally associated as a crescent with vacuoles associated with the trans side and displaying a dense content; $V_v,\% = 3.8$
Intracellular secretory material	Present in the Golgi complex and in apical vacuoles as concentric granules occasionally presenting a halo with poorly condensed secretion; $V_v,\% = 0.4$

Phase 3	Age: 24–47 days
Lumen	Well developed, presenting crypts and loaded with dense secretion
Apical membrane	Numerous and well-developed microvilli
Cytoplasmic matrix	Similar to phase 2 but reduced to the narrow space left between well-developed cell organelles

(continued)

TABLE I *(continued)*

Phase 3	Age: 24–47 days
Lysosomes	Common; $V_v,\% = 1.2$
RER	Cisterns are long, narrow, and well packed, mostly oriented toward the cell apex; membranes studded with numerous ribosomes; $V_v,\% = 12.4$
Golgi complex	Conspicuous cup-shaped stacks of cisterns with dilated rims and displaying a dense content in small discrete areas; large secretory granules in vacuoles associated with the trans side; $V_v,\% = 7.2$
Intracellular secretory material	Large eccentric granules accumulated in Golgi and apical vacuoles presenting a large clear halo; $V_v,\% = 2.3$

Phase 4	Age: 48–62 days
Lumen	Enlarged and deep crypts; fully loaded with dense, compact secretions
Apical membrane	Coated pits and vesicles are often associated with the apical membrane, which displays numerous and long microvilli
Cytoplasmic matrix	Similar to phase 3
Lysosomes	Conspicuous, dense and pleomorphic; $V_v,\% = 12.0$
RER	Similar to phase 3; $V_v,\% = 11.6$
Golgi complex	Enlarged cisterns and accumulation of secretory vacuoles in the trans side; $V_v,\% = 9.8$
Intracellular secretory material	Abundant; increased number of secretory granules in the Golgi complex and in the cell apex; $V_v,\% = 4.5$

^{a} V_v = relative volume.

ated with Golgi stacks (Fig. 3a); and the secretory granules are fully condensed and eccentric with their vacuoles.

From phase 3 to phase 4, the most striking change is an increase of about twofold in the intracellular secretory material and an increase in lysosomes (Fig. 3b) whose relative volume is by then 16 times larger than in phase 1. The Golgi complex is increased by only 30% and no changes are noticed in the RER from phase 3 to phase 4. The contact between the secretory cells is interdigitated in the basolateral region (Fig. 4a) and presents junctional complexes in the apical zone (Fig. 4b).

The main features of the fine structure of the seminal vesicle secretory cells as described in adult hamsters (Mifune *et al.*, 1986) are established during phase 4 and conform to the general pattern previously proposed for accessory male glands in adult rats (Brandes, 1974a).

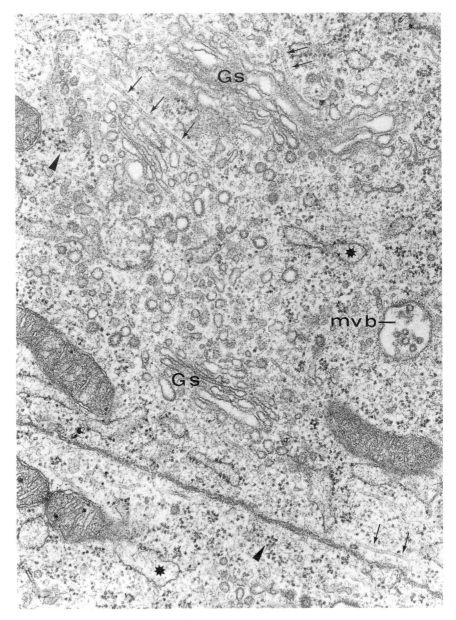

FIG. 1   Electron micrograph of a seminal vesicle epithelial cell of a 6-day-old hamster to illustrate phase 1 in postnatal differentiation. The RER (asterisk) presents dilated and scarce cisterns with a few membrane-associated ribosomes. The Golgi apparatus appears as scattered small stacks of flat and clear cisterns. Free ribosomes or polysomes (arrowheads) and microtubules (arrows) can be seen in the ground cytoplasm. Multivesicular bodies (mvb) are occasionally observed. ×36,000.

As noticed in the rat (Dahnke, 1970) and mouse (Deane and Wurtzel-mann, 1965), the differentiation of seminal vesicle secretory cells of the hamster is not synchronous and secretory vacuoles do not appear in all cells at the same time. In 12-day-old hamsters, only very few cells present secretory vacuoles and their detection in ultrathin sections is thus a rare event. In fact, the presence of intracellular secretory material at such an early age could be easily missed if it wasn't for the systematic sampling designed for ultrastructural stereological analysis. This is probably the reason why the beginning of secretion in hamster seminal vesicles was previously set at 21 days after birth (Cavazos and Belt, 1965).

The earliest age at which secretory granules are observed in hamster seminal vesicles agrees with what was reported in mice, where secretory granules were first noticed at 11–14 days after birth (Dean and Wurtzel-mann, 1965), whereas in the rat, the onset of secretion, based on the presence of secretory vacuoles, was first dated at 16 days of age (Dahnke, 1970). Care should, nevertheless, be taken to avoid confusing the earliest time of detection of secretory granules by electron microscopy with the onset of secretory protein synthesis.

The ultrastructural observation of secretory granules requires enough protein to be synthesized, transported, condensed, and accumulated in secretory vacuoles to become visible as granules at the electron micro-scope level. A lag time might therefore be expected between the onset of secretory protein synthesis and the ultrastructural detection of secretory granules. In fact, immunological (Kistler et al., 1981) and immunocyto-chemical (Aumuller and Seitz, 1986; Fawell and Higgins, 1986) techniques, which are more sensitive than morphology by itself, have allowed the detection of secretory proteins in the rat seminal vesicle earlier than the 16 days of age previously set by ultrastructural observations (Dahnke, 1970).

In addition to dynamic changes involved in differentiation for secretion, ultrastructural changes of the seminal vesicle epithelium, although less dramatic, are also observed with increasing age. The accumulation of lipid droplets and secondary lysosomes, including lipofucsin pigment and myelinic figures, as well as a general tendency for a decrease in epithelium height, are features common to the aging process (Mainwaring and Bran-des, 1974; Cavazos, 1975; Aumuller, 1979). The accumulation of amyloid in human seminal vesicles, which is also related to aging (Elbadawi, 1979; Coyne and Kealy, 1993), was recently shown to originate from seminal vesicle epithelial cells (Cornwell et al., 1992).

Changes of the seminal vesicle epithelium toward atrophy and inactivity have, in addition, been reported in connection with exposure to short daylight conditions in animals sensitive to photoperiod (Schindelmeiser et al., 1988).

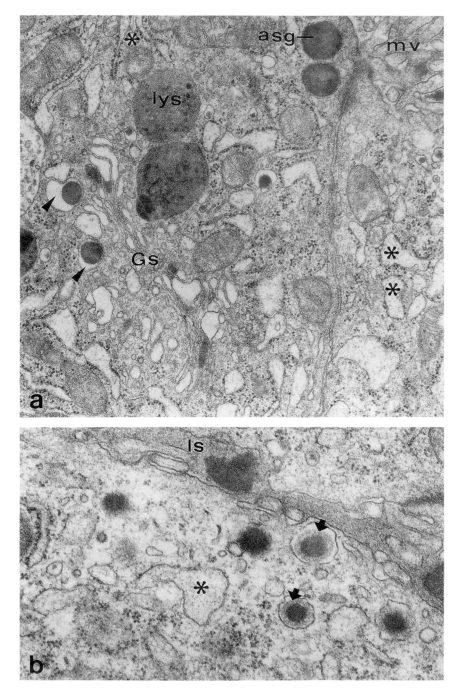

III. Dynamics of Protein Secretion

Seminal vesicle secretions constitute the major fraction of the seminal plasma, reaching 60–70% of the total volume of ejaculate in humans (Spring-Mills and Hafez, 1979), thus influencing greatly the chemical makeup of the seminal fluid. The contribution of seminal vesicles to the seminal plasma includes ions (e.g., Na^+, K^+, Zn^+), sugars (e.g., fructose, glucose), peptides (e.g., glutathione) and, above all, highly concentrated proteins. Protein concentrations in seminal vesicle secretions can reach 200 mg/ml in the hamster (Mata, 1981) and well above that in the rat (Ostrowsky et al., 1979) and in humans (Williams-Ashman, 1983).

Seminal vesicle secretory proteins play a relevant role in mammalian reproduction owing to their involvement in important biological processes, such as (1) calcium transport to spermatozoa (Coronel et al., 1993), (2) sperm-egg recognition (Jonakova et al., 1992; Sanz et al., 1992), (3) capacitation and acrosome reaction (Kwok et al., 1993; Manjunath et al., 1994) and (4) suppression of the local immune response in the female reproductive tract (Galdiero et al., 1989; Metafora et al., 1989a, b; Porta et al., 1993). The total number of the major proteins identified in seminal vesicle secretions by SDS-PAGE electrophoresis depends on the animal species, ranging from three in the hamster (Fawell et al., 1987; Pinho and Mata, 1992) to seven in the mouse (Chen et al., 1987).

The pattern of those proteins depends, in each species, on the animal's androgen status (Norvitch et al., 1989; Hagstrom et al., 1992; Pinho and Mata, 1992) and sexual activity (Seitz and Aumuller, 1991). Seminal vesicle secretory proteins are immunologically related in the same species and in different species as well (Fawell et al., 1987; Carballada and Esponda, 1991). Such similarity speaks of a common biochemical origin in the same species (Norvitch et al., 1988; Hagstrom et al., 1992) and of the same evolutionary origin in different species (Fawell et al., 1987).

In laboratory rodents, secretory proteins are synthesized in parallel in every seminal vesicle epithelial cell (Fawell and Higgins, 1986; Aumuller and Seitz, 1986) and can appear subcellularly compartmentalized (Aumuller and Seitz, 1986, 1990). The dynamics of the secretory cycle of

---

FIG. 2   Electron micrographs of seminal vesicle epithelial cells of (a) 22- and (b) 18-day-old hamsters to illustrate phase 2 in postnatal differentiation. The RER includes numerous cisterns, occasionally grouped and dilated (asterisks), whose membranes present ribosome-free areas. Golgi stacks (Gs) are organized as a crescent presenting associated secretory vacuoles (arrowheads). Apical secretory granules (asg) are concentric with their vacuoles and occasionally display a halo of poorly condensed secretion (arrows). mv, apical microvilli; ls, lumenal secretion. (a), ×28,000; (b), ×36,000.

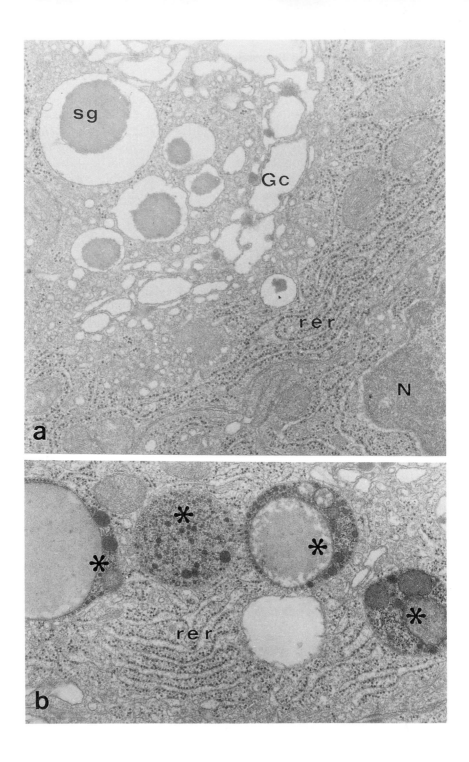

seminal vesicle proteins was studied by electron microscope autoradiography (EMAR) in the rat (Flickinger, 1974a), hamster (Mata, 1979), and mouse (Samuel and Flickinger, 1987). The time sequence and pathway of seminal vesicle secretory protein as depicted by electron microscopy conform to the general pattern of other exocrine cells (Case, 1978), as first described in the exocrine pancreas (Palade, 1975).

In the hamster, [³H]leucine was used *in vitro* in pulse-labeling experiments, whereas in the rat and mouse *in vivo,* labeling was performed using either [³H]leucine or [³H]threonine, respectively. The results obtained are substantially similar despite the differences in experimental protocols.

The incorporation of radioactive amino acids as studied by EMAR in seminal vesicle secretory cells shows a wave of label moving through the RER, GC, and apical secretory granules which reaches the luminal content within 30 min. The maximum labeling in the RER is observed within 5 min from onset of labeling whereas in the Golgi complex labeling peaks after 15 min. Apical secretory granules appear heavily labeled after 25 min of onset.

The time required for maximal labeling of RER in seminal vesicle secretory cells is in the same range as that found in other secretory tissues (Case, 1978). In the Golgi complex, labeling peaks between 11 and 37 min after onset in other secretory cells, which includes what was observed in seminal vesicles. However, labeled secretory proteins start to decline earlier and faster in the Golgi complex of the seminal vesicle than in other cells. In addition, the release of apical secretory granules to the gland lumen is also faster than in other tissues. As a consequence, the time of intracellular storage of secretory protein in seminal vesicle epithelial cells is reduced, leading to a much shorter secretory cycle (30 min) than in the prostate (Flickinger, 1974b), coagulating gland (Samuel and Flickinger, 1986, 1987), and other mammalian tissues, where it ranges from 1 to 6 hr (Case, 1978).

The high efficiency of such a short secretory cycle and the fact that seminal vesicle secretory proteins are released continuously to the gland lumen (Aumuller, 1979), where they accumulate to be expelled at ejaculation, justify the high concentration of protein found in seminal vesicle secretions.

FIG. 3 Electron micrographs of seminal vesicle epithelial cells of (a) 32- and (b) 62-day-old hamsters to illustrate phases 3 and 4, respectively, in postnatal differentiation. The rough endoplasmic reticulum (rer) presents narrow and well-packed cisterns whose membranes are studded with numerous ribosomes. The Golgi complex (Gc) displays conspicuous and cup-shaped organized cisterns and includes numerous and large secretory granules (sg) associated with the trans side. At phase 4 (b) the lysosomes (asterisks) are conspicuous and pleomorphic. N, nucleus. ×28,000.

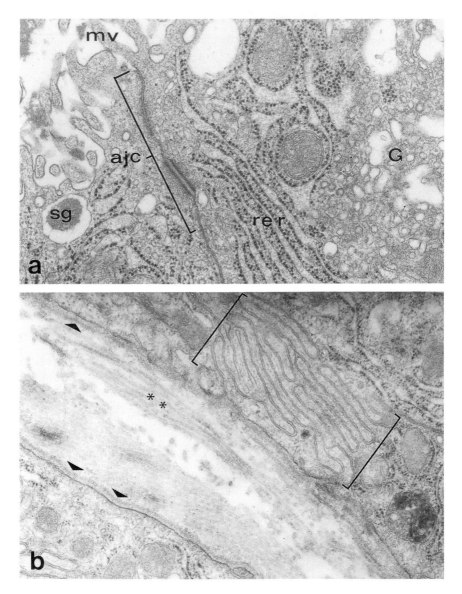

FIG. 4   Electron micrographs of seminal vesicle epithelial cells of (a) 42- and (b) 62-day-old hamsters corresponding to phase 4 in postnatal differentiation. Notice the presence of well-packed cisterns of rough endoplasmic reticulum (rer) oriented toward the cell apex and an apical vacuole containing an eccentric secretory granule (sg) surrounded by a clear halo. The contact between secretory cells is interdigitated in the basolateral region (in brackets) and presents junctional complexes in the apical zone (ajc). G, Golgi region; mv, apical microvilli; asterisks, subepithelial collagen fibers; arrowheads, epithelial basal lamina. (a), ×36,000; (b), ×30,000.

IV. Endocytic Activity and Membrane Recycling

In addition to the production of exportable secretory proteins, which is their most obvious function, seminal vesicle epithelial cells are also involved in taking up ions, water, and macromolecules from the gland lumen.

Seminal vesicle epithelial cells were shown to take up water and electrolytes (Levine and Kelly, 1980) and, in agreement with that finding, the presence of the CHIP 28 water channel was recently demonstrated in the apical and basolateral membranes of the seminal vesicle secretory cells (Brown *et al.*, 1993).

As to the absorption of macromolecules, ultrastructural features suggestive of endocytic activity were first noticed during the postnatal differentiation of the mouse seminal vesicle (Deane and Wurtzelmann, 1965) and later confirmed in the hamster's gland (Mata and David-Ferreira, 1973).

In the hamster, apical coated vesicles, known to be involved in protein endocytosis (Roth and Porter, 1964; van Deurs *et al.*, 1989), are observed during postnatal differentiation of the seminal vesicle epithelial cells (Mata, 1981) as soon as secretory material starts to be released to the lumen (Fig. 5a) and are frequently present in the secretory cells of adult animals (Fig. 5b). In addition, the presence of complex interdigitations between the cells (Fig. 4b) is consistent with their possible involvement in the absorption and transcellular transport of materials stored in the gland lumen. Endocytosis was, in fact, demonstrated in the hamster seminal vesicle secretory cells using horseradish peroxidase (HRP) as an intraluminal tracer (Mata and David-Ferreira, 1973; Mata, 1976) and later confirmed using instead radioiodinated secretory proteins (Mata and Maunsbach, 1982).

HRP is endocytosed from the gland lumen by coated vesicles which detach from the apical membrane into the cytoplasm and transfer the tracer to other compartments through membrane fusion. In addition to endocytic vesicles, the endocytic pathway includes multivesicular bodies, secondary lysosomes, secretory vacuoles, and Golgi cisterns from which HRP is cleared, probably along with secretions. The tracer is also transferred to the intercellular spaces below tight junctions and to the subepithelial space and blood vessels. The distribution pattern of HRP through those compartments depends on time and displays a cycle of 30 min. In this respect, the most striking observation was the detection of two peaks in the number of labeled apical vesicles, at 15 and 45 min after labeling onset and coinciding with the labeling of Golgi cisterns. This finding was viewed as indicating a possible correlation between the dynamics of endocytosis and protein secretion in the epithelial cells of the hamster seminal

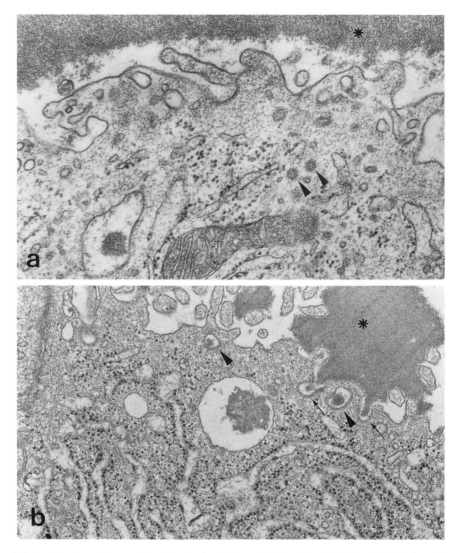

FIG. 5 Electron micrographs of seminal vesicle epithelial cells of (a) 16- and (b) 48-day-old
hamsters corresponding to phases 2 and 4, respectively, in postnatal differentiation. Notice
in (a) the presence of apical coated vesicular profiles (arrowheads) displaying a dense content
similar to lumenal secretion (asterisks) at phase 2 when secretory material starts to be
released to the lumen. In young adult animals (b), coated pits (arrows) and apical vesicular
profiles (arrowheads) containing dense material similar to lumenal secretion (asterisk) are
often observed. ×36,000.

vesicle (Mata, 1976), which was later confirmed (Mata and David-Ferreira, 1990).

The results obtained with HRP were substantially confirmed by EMAR when radioiodinated secretory proteins of the hamster seminal vesicle were used instead (Mata and Maunsbach, 1982). This confirmation rules out the possible argument of artifactual endocytosis induced by an exogenous tracer such as HRP. Endocytosis as shown in hamster seminal vesicle secretory cells, including the labeling of Golgi cisterns, has also been observed in the rat and guinea pig seminal vesicles (Mata, 1994). The inclusion of Golgi cisterns in the endocytic pathway, although shown first in seminal vesicle secretory cells, is now recognized in other cells (van Deurs et al., 1989).

The physiological role of endocytosis in the seminal vesicle epithelium has been discussed in connection with (1) a possible feedback regulation of protein secretion (Mata and Maunsbach, 1982), (2) the development of mucosal immune response (Mata, 1994) and (3) membrane recovery in secretory cells (Mata and David-Ferreira, 1973; Mata, 1976).

The possible involvement of endocytosis in the regulation of secretory activity in seminal vesicle epithelial cells deserves some consideration. As mentioned above, seminal vesicle epithelial cells are able to take up and transcytose secretory proteins accumulated in the gland lumen. Following endocytosis, seminal vesicle secretory proteins might eventually bind to androgens, either intra- or extracellularly, and in some way modulate their action on the secretory cells. Although speculative, this idea would fit with pieces and hints which can be found in the literature concerning other secretory cells, such as salivary, prostate, and Sertoli cells.

Salivary (Dlouhy et al., 1986) and prostate (Traish et al., 1981) secretory cells were shown to secrete androgen-binding protein (ABP). Considering that endocytosis of secretory proteins has also been reported in salivary gland cells (Hand et al., 1987), it seems reasonable to assume that their secreted ABP is not excluded from the endocytic process. Although endocytosis in the epithelial cells of the prostate has not been shown directly, the detection of one of their secretory proteins in the subepithelial space supports that possibility (Carmo-Fonseca and Vaz, 1989).

ABP is also secreted by Sertoli cells to the lumen of seminiferous tubules, from where it is most likely taken up and transcytosed. In fact, Sertoli cells were shown to secrete ABP (Hermo et al., 1994) apically and basally in the proportion of 80% and 20% (Hadley et al., 1987), respectively, and, in addition, to take up and transcytose protein from the lumen of the seminiferous tubules (Soares-Pessoa and David-Ferreira, 1980; Hermo et al., 1994). It is therefore conceivable that some of the ABP secreted apically might then be endocytosed and released basally by Sertoli cells. ABP released into the lumen of seminiferous tubules can also reach the epididymis to be endocytosed by principal cells (Gerard *et al.*, 1988; Hermo *et al.*, 1994).

Endocytosis as discussed earlier in seminal vesicle and prostate secretory cells; and in Sertoli, salivary duct, and epididymal principal cells as well, raises some intriguing questions: (1) Why should seminal vesicle and prostate epithelial cells, or even Sertoli cells, take up and transcytose proteins from their own secretion? (2) Why would epithelial cells, such as salivary duct and epididymal principal cells, endocytose ABP secreted by other cells? (3) What do these cells have in common that might be correlated with such behavior?

Sertoli cells, epidydimal epithelial cells, and epithelial cells of the prostate, salivary gland, and seminal vesicles, have in common the fact that they are target cells for androgens. In addition, salivary and prostate secretory cells, as well as Sertoli cells, secrete ABP which they might eventually reabsorb, or/and provide to be endocytosed by other androgendependent cells. As to the seminal vesicle epithelial cells, which can take up their own secretory proteins, the presence of ABP in their secretions, although not systematically investigated, has been suggested by a few immunocytochemical data (Aumuller and Heyns, 1981).

Although the physiological role of ABPs is still a mystery, they have been proposed to be involved in influencing uptake, accumulation, and storage of androgens, hence modulating androgen action on target cells (Traish *et al.*, 1981). In this context, it could be hypothesized that endocytosis as observed in those androgen-dependent cells might be involved in making ABP available to interact with androgens, either intra- or extracellularly, thus modulating their action on those target cells.

As mentioned earlier, endocytosis in the seminal vesicle secretory cells has also been interpreted as possibly being related to mucosal immune response (Mata, 1994). It is well established that epithelial cells lining the digestive, respiratory, and reproductive tracts play key roles in immune defense. Epithelial cells lining the lumen ducts of such systems are able to endocytose antigens which they present to intraepithelial lymphocytes, so triggering the mucosal immune response that leads to the delivery of antibodies into secretions (Kraehenbuhl *et al.*, 1992).

The presence of intraepithelial lymphocytes has already been demonstrated in human male accessory glands, including the seminal vesicle (El-Demiry and James, 1988). In laboratory rodents, although specific markers have not been used to demonstrate intraepithelial lymphocytes in the male tract, the morphological detection of lymphocyte-like cells, sometimes referred as "halo cells" or "migratory cells," has been reported (Dahl and Tveter, 1973; Hoffer *et al.*, 1973; Dym and Romrell, 1975) in the seminal vesicles (Allison and Cearley, 1972), and their possible immunological significance has been discussed (Allison and Cearley, 1972; Hoffer *et al.*, 1973). In addition, the presence of IgA and secretory component have recently been demonstrated in seminal vesicles (Stern *et al.*, 1992). It is in this context that transcytosis in the seminal vesicle secretory cells has been interpreted as possibly involved in the mucosal immune response of these glands.

The best-substantiated interpretation of the role of endocytosis in the seminal vesicle secretory cells is, nevertheless, its connection with apical cell membrane recovery. Membrane recovery was first suggested by Palade and associates who realized, during their studies on exocrine cells, that a process must exist to remove the excess membrane added to the apical surface of those cells during exocytosis in order to maintain a constant cell volume (Jamieson and Palade, 1971). Whether the membrane to be removed following exocytosis would be destroyed or reutilized became then a matter of debate (Meldolesi *et al.*, 1978) although the prevailing idea was that it would more likely be degraded in lysosomes (Holtzman *et al.*, 1977). Biochemical evidence showing that the turnover of membrane proteins in secretory cells is slower than that required for the amount of membrane to be used in secretion (Meldolesi, 1974; Morré *et al.*, 1979; Steinman *et al.*, 1983), together with the demonstration that the endocytic pathway includes the Golgi cisterns and secretory vacuoles of exocrine cells (Mata and David-Ferreira, 1973; Farquhar, 1983), strengthened the idea of apical membrane retrieval by endocytosis followed by reutilization in the exocytic pathway.

The extensive information supplied on the possible reutilization of apical membrane in the exocytic pathway of exocrine cells was, until the 1990s, based on tracers noncovalently bound to the cell membrane and hence considered as indirect evidence of such a process. The covalent labeling of apical membrane glycoproteins with [^{3}H]galactose, allowing the pathway of membrane internalized during endocytosis to be followed by EMAR, demonstrated membrane reutilization and recycling in seminal vesicle secretory cells, thus contributing to the clarification of the matter (Mata and Christensen, 1990b).

It was shown that internalized membrane is transferred to several components, including Golgi cisterns, secretory vacuoles, and the basolateral domain of the cell membrane, and then recycled back to the apical domain of the membrane of seminal vesicle secretory cells. This study, demonstrating that internalized membrane is reutilized and recycled through the exocytic pathway, together with evidence on endocytosis coupled to exocytosis (Mata and David-Ferreira, 1990), substantiates the early interpretation that endocytosis is involved in membrane recovery in seminal vesicle secretory cells (Mata and David-Ferreira, 1973). The recycling of membrane internalized during endocytosis has also been suggested in cells other than exocrine cells (van Deurs *et al.,* 1989).

V. Effect of Hormones and Other Factors on Seminal Vesicle Epithelium

A. Secretory Activity

Testosterone is the main hormonal regulator of the structure and function of seminal vesicle epithelial cells. Fetal morphogenesis, growth, and functional differentiation occurring at puberty, as well as the maintenance of epithelial ultrastructure and secretory activity in adulthood, are strictly controlled by testosterone. The information available on the effects of testosterone withdrawal has recently been extensively reviewed (Aumuller and Seitz, 1990).

Testosterone depletion, either physiological or experimental, leads to a decline in DNA synthesis and RNA content of seminal vesicles. As a consequence, a (1) severe reduction in the size and capacity of secretion of the whole glands, and (2) programmed cell death leading to (3) a dramatically reduced number of epithelial cells are observed following castration.

However, seminal vesicle epithelial cells surviving after castration are still capable of producing secretory proteins (Veneziale *et al.,* 1977; Higgins and Fuller, 1981; Ostrowsky *et al.,* 1982; Mata and David-Ferreira, 1982, 1985; Pinho and Mata, 1992) whose extent of accumulation in secretory granules depends on the animal species, as discussed later. In addition, the qualitative pattern (Higgins and Fuller, 1981; Norvitch *et al.,* 1989; Pinho and Mata, 1992) of seminal vesicle secretory proteins also depends on testosterone.

The ultrastructural alterations induced by testosterone depletion in the nucleus and cytoplasm of seminal vesicle epithelial cells of laboratory rodents correlate well with the biochemical findings mentioned earlier.

In fact, castration leads to a decrease in nuclear and nucleolar volume of seminal vesicle secretory cells and induces qualitative and quantitative changes in their nuclear and nucleolar compartments (Arnold *et al.,* 1983). These observations were interpreted in connection with a decrease in RNA transcription and rRNA synthesis following castration. Nuclear pore density is also decreased following testosterone withdrawal (Ortiz and Cavicchia, 1990), which might be correlated with a decrease in migration of RNA through the nuclear envelope, as shown in the prostate of castrated rats (Carmo-Fonseca, 1986).

The dynamics of postcastration involution of seminal vesicle epithelium as shown in the rat (Dahl and Tveter, 1973) includes a decrease in epithelial cell size and number, a reduction in the endoplasmic reticulum and Golgi complex, and a pronounced decrease in secretory granule size and number, leading to the disappearance of secretory material. This general pattern of involution is also observed in other species (Toner and Baillie, 1966; Tse and Wong, 1980) although with variations in the chronology and the extent of regressive events. For instance, following orchidectomy, secretory granules disappear in the rat (Szirmai and Van Der Linde, 1965; Dahl and Tveter, 1973; Cavazos, 1975); are reduced in the mouse (Toner and Baillie, 1966) and guinea pig (Tse and Wong, 1980); and are increased in size and number on a per cell basis in the hamster (Mata and David-Ferreira, 1985).

The effect of testosterone withdrawal on the secretory activity of the hamster seminal vesicle was studied at the ultrastructural level by autoradiography (Mata and David-Ferreira, 1982) and morphometric analysis (Mata and David-Ferreira, 1985). The stereological approach showed a reduction in epithelial cell size and number to about 50–60% and a significant decrease in the rough endoplasmic reticulum and Golgi complex. Unexpectedly, and apparently in contradiction to such results, the number and size of apical secretory granules as well as the amount of intracellular secretory material were increased following castration. These findings were interpreted as indicating that testosterone withdrawal does not abolish the synthesis of secretory proteins but instead delays their intracellular transport and exocytic rate in epithelial cells lasting after orchidectomy. In fact, the study of the secretory cycle by electron microscope autoradiography showed that in castrated animals a relatively higher amount of newly synthesized secretory proteins is retained in the RER and Golgi complex, whereas a relatively smaller amount is transported to apical secretory vacuoles and released to the lumen (Mata and David-Ferreira, 1982).

In the hamster, secretory granules persist and seminal vesicle epithelial cells keep their secretory activity up to at least 6 months after castration (Mata and David-Ferreira, 1985). Biochemical results (Mata and David-Ferreira, 1990) showing that the synthesis of secretory proteins is maintained and exocytosis reduced (Fig. 6) in the seminal vesicle epithelial cells of castrated animals are in agreement with the findings obtained by stereologic and autoradiographic approaches. Overall, these studies allow the conclusion that testosterone interferes with the dynamics of intracellular transport and exocytosis of secretory proteins in the seminal vesicle epithelium. The castration-resistant secretion detected in the hamster seminal vesicle was a fortunate situation that allowed the discovery of these effects of testosterone, which are otherwise impossible to perceive.

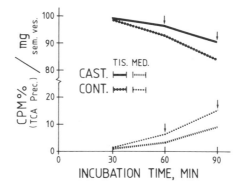

FIG. 6 Proportion of trichloroacetic acid (TCA)-precipitable radioactivity released to the incubation medium (MED.) and kept in seminal vesicle tissue slices (TIS.) of castrated (CAST.) and control (CONT.) hamsters. The proportion of TCA-precipitable radioactivity in the incubation medium of control gland slices is significantly higher (arrows, $P < 5\%$) than in that of gland slices from castrated animals, showing that exocytosis is reduced following testosterone withdrawal. (Reproduced by permission from Mata and David-Ferreira, 1990.)

The maintenance of secretory activity in the seminal vesicle epithelium of castrated hamsters, which was also observed in the guinea pig (Tse and Wong, 1980; Tam *et al.*, 1992; Tam and Wong, 1993) and, to a certain extent, in the mouse (Toner and Baillie, 1966), was interpreted as indicating the possible existence of an extratesticular source of androgens, most likely the adrenals. This hypothesis was tested in guinea pig (Tam *et al.*, 1985) and hamster seminal vesicles (Pinho and Mata, 1992).

Adrenalectomy or cyproterone acetate (CPA) treatment of castrated hamsters did not impair the secretory activity of seminal vesicle epithelial cells that remained after castration (Pinho and Mata, 1992). Following the treatment of castrates with CPA, which is an established antiandrogen, total protein synthesis remained unchanged and a secretagogue effect of the drug was detected by stereologic and biochemical analysis (Fig. 7).

In addition, the pattern of newly synthesized secretory proteins, as analyzed by SDS-PAGE and fluorography, was similar in CPA-treated or untreated castrates. Adrenalectomy neither affected secretory cell activity nor induced any further ultrastructural changes in the seminal vesicle epithelium of castrated hamsters. Therefore, in the hamster seminal vesicle, apparently unlike the guinea pig (Tam *et al.*, 1985), the secretory activity of epithelial cells remaining after testosterone withdrawal does not depend on nontesticular androgens.

The maintenance of secretion in castrated hamsters might be due to either an irreversible imprinting by neonatal androgens, as shown in male

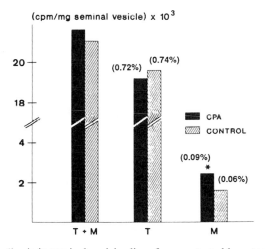

FIG. 7 Protein synthesis in seminal vesicle slices from castrated hamsters either treated or nontreated with CPA. After 90 min of incubation with [³H]leucine, the proportion of TCA-precipitable radioactivity is significantly higher in the medium (M) containing gland slices from CPA-treated than in that from nontreated castrates (asterisk, $P < 5\%$), thus showing that CPA exerts a secretagogue effect on secretory cells. Total TCA-precipitable material (T + M) is similar in both situations. (Reproduced from Fig. 4 of Pinho and Mata, 1992 by permission of Blackwell Scientific Publications Ltd, Oxford.)

sex tissues of other laboratory rodents (Desjardins and Jones, 1970; Jean-Faucher et al., 1985, 1986, 1989; Yamane et al., 1987), or to the involvement of other hormones and nonhormonal factors. In any case, the regulation of this castration-resistant secretion remains still to be unraveled.

In addition to testosterone, other hormones, such as estrogens and prolactin, and factors such as cyclic AMP, neurotransmitters, and neuropeptides, might also influence the dynamics of seminal vesicle epithelial cells.

The synergistic action of prolactin with androgens on the size and secretory activity of the seminal vesicle is well accepted (Williams-Ashman, 1983) but the effect of prolactin by itself is less clear.

Specific binding sites for prolactin have been demonstrated in seminal vesicle membranes (Barkey et al., 1979; Amit et al., 1983) and prolactin potentiation of the effect of testosterone on the growth of seminal vesicles (Negro-Vilar et al., 1973, 1977), namely by increasing the height of epithelial cells, has been suggested (Antliff et al., 1960; Spring-Mills et al., 1982). In addition, prolactin alone was suggested to exert a direct stimulatory effect on the growth of the whole seminal vesicle in mice genetically deficient in that hormone (Gonzalez et al., 1991), but the contribution of the seminal vesicle epithelium to that increase in growth cannot be determined from that study. As to seminal vesicle secretory activity, the results on the action of prolactin are somewhat conflicting. Evidence obtained in the rat (Aumuller, 1979) and mouse (Spring-Mills *et al.*, 1982) suggests that prolactin is necessary to maintain seminal vesicle secretion, whereas in guinea pig, prolactin has been suggested to exert some inhibitory effects on seminal vesicle secretory activity in castrated animals (Tam *et al.*, 1992). In the hamster, no further ultrastructural effects were observed on seminal vesicle secretory cells when bromocryptine, a potent inhibitor of prolactin release, was given to castrated animals (unpublished observations from our laboratory). The effects of prolactin on seminal vesicle secretory activity thus appear to depend on the animal species and hormonal status but, considering that the details of experimental design are varied in these studies, further investigation is still required to clarify this matter.

Estrogen treatment of adult males leads to involution of accessory glands, including seminal vesicles, as revealed biochemically (Neubauer *et al.*, 1981, 1989) and morphologically (Cavazos, 1975; Davies and Danzo, 1981; Gaytan *et al.*, 1986). This general effect was proposed to be mediated through a decrease in circulating androgens (Mawhinney and Neubauer, 1979). In addition, although they are unable to stimulate secretory activity, estrogens can increase the DNA content in the seminal vesicle of castrated rats (Mariotti *et al.*, 1981) and stimulate epithelial cell proliferation in the mouse gland (Yamane *et al.*, 1986; Okamoto *et al.*, 1982). In guinea pig seminal vesicle, estrogens might interfere with the process of glycosilation in secretory cells (Tam and Wong, 1993), but biochemical studies on the separated seminal vesicle epithelium of guinea pig showed that estrogens do not stimulate epithelial cell growth (Neubauer and Mawhinney, 1981; Mariotti *et al.*, 1983). In addition, binding studies showed that estrogen binding sites are mostly localized in seminal vesicle muscles of guinea pig (Blume and Mawhinney, 1978; Weinberger, 1984) and mouse (Schleicher *et al.*, 1985). Therefore, the information available on the effects of estrogens on the seminal vesicle suggests, overall, that estrogens might not interfere directly with the dynamics of the gland epithelium. Hence, any possible contribution of such an effect to the changes induced by estrogens in the seminal vesicle as a whole is, most likely, very limited.

Although the effects of other extragonadal hormones have not been intensively investigated, a few results favor the influence of growth hormone, insulin, and thyroid hormones on the maintenance of functions of seminal vesicle epithelium (Dadoune, 1985). In agreement with that is the impairment of secretory glycoproteins by hypothyroidism recently reported in the seminal vesicle (Senthilkumaran *et al.*, 1993).

In addition to testosterone and extragonadal hormones as discussed earlier, other factors, such as cyclic AMP, neurotransmitters, and neuropeptides might also interfere with the activity of seminal vesicle epithelial cells.

Cyclic AMP was shown to mimic testosterone stimulation in the seminal vesicle epithelium (Brandes, 1974b) and to increase exocytosis of seminal vesicle secretory proteins (Koenig *et al.*, 1976). Neurotransmitters were demonstrated to exert a permissive effect on androgen-regulated synthesis of secretory proteins specific to the seminal vesicle (Kinghorn *et al.*, 1987), and neuropeptides, which are present in the seminal vesicle of several species (Gonzales, 1989), have been suggested as possibly involved in the regulation of seminal vesicle secretory activity (Aumuller and Seitz, 1990). In fact, recent work from our laboratory showed that vasoactive intestinal peptide (VIP) interferes with the secretory activity of seminal vesicle epithelial cells and is involved as well in the modulation of muscarinic function (Pinho *et al.*, 1994).

Briefly, in the hamster seminal vesicle, VIP-immunoreactive varicose nerve fibers form a loose network in the mucosal layer just beneath the epithelium where VIP colocalizes with acetylcholinesterase activity (Figs. 8a and 8b). VIP binding studies followed by autoradiography showed specific binding sites for VIP (Figs. 8c and 8d) to be significantly associated with the basal region of the epithelium. In addition, the incorporation of [^{3}H]leucine as studied using a system of gland slices revealed that VIP significantly increases exocytosis. Specific binding sites for VIP were also localized in the muscle layer and VIP was shown to significantly inhibit the increase in muscle tension induced by carbachol. These results allowed the conclusion that VIP exerts secretagogue and relaxant effects on the hamster seminal vesicle which might lead to accumulation of secretion in the gland lumen and help to avoid fluid expulsion from the organ in between ejaculations. This study on the distribution, binding sites, and possible functions of VIP illustrates how important neuropeptides might be for the dynamics of the seminal vesicle epithelium and of the whole gland as well.

B. Endocytic Activity

The dynamics of the endocytic activity displayed by seminal vesicle epithelial cells is also affected by testosterone depletion (Mata and David-Ferreira, 1990). The endocytic pathway in castrated hamsters involves the same compartments as in intact animals but endocytosis is decreased and the kinetics of compartment labeling is delayed following testosterone withdrawal. The labeling of Golgi cisterns by HRP, which correlates with a significant peak in the number of endocytic vesicles, occurs 20 min later in castrates than in intact animals. The increase in exocytosis due to pilocarpine treatment of castrated hamsters leads to an increase in endocy-

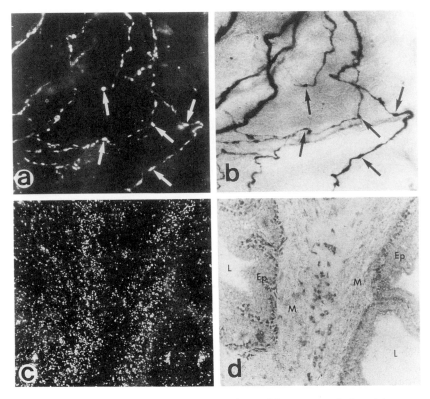

FIG. 8 The same field in whole-mount preparations of hamster seminal vesicle mucosal layer is shown in (a) and (b) to illustrate an extensive network of varicose fibers containing both VIP immunoreactivity (a, white arrows) and acetylcholinesterase activity (b, black arrows). In (c) and (d), the same area of an autoradiograph is shown in dark field and bright field, respectively, to illustrate ^{125}I-VIP binding sites mostly associated with the muscular coat. (a) and (b), ×426; (c) and (d), ×106. (Reprinted from Figs. 2 and 4 of Pinho *et al.*, 1994 by kind permission of Elsevier Science Ltd., The Boulevard, Langford Lane, Kidlington, OX5 1GB, UK.)

tosis and advances the kinetics of Golgi labeling by 20 min (Fig. 9), hence reverting the endocytic dynamics to that of intact animals. This study demonstrates that (1) endocytosis is coupled to exocytosis and (2) testosterone interferes with the kinetics of endocytosis in the seminal vesicle epithelial cells.

VI. Seminal Vesicle Epithelial Cells in Culture

The information reviewed here shows that seminal vesicle epithelial cells are a good model for studying secretion, endocytosis, and membrane

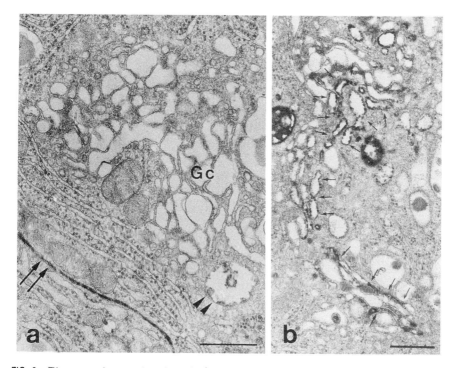

FIG. 9 Electron micrographs of seminal vesicle epithelial cells from castrated hamsters, either (a) untreated or (b) treated with pilocarpine, obtained after 40 and 20 min of HRP endocytosis, respectively. HRP reaction is seen in (a) in multivesicular bodies (arrowheads) and intercellular spaces (long arrows) of secretory cells whose Golgi cisterns (GC) are clear from the tracer, whereas in (b) Golgi stacks are heavily labeled with HRP (small arrows). These findings show that the increase in exocytosis due to pilocarpine treatment of castrated hamsters (b) leads to an increase in endocytosis, thus advancing the kinetics of labeling of Golgi cisternae by 20 min. (a), ×30,000; (b), ×24,000. (Reproduced by copyright permission from Mata and David-Ferreira, 1990.)

traffic in exocrine cells. In fact, seminal vesicle epithelial cells are regulated secretory cells which display a short secretory cycle, are very active in secretion as well as in endocytosis and, in addition, those activities depend on testosterone, which allows their experimental manipulation. However, a further insight on the effect of hormones and other factors, namely neurotransmitters and regulatory peptides, in the dynamics of seminal vesicle epithelial cells would require their cultivation in a chemically defined medium, to and from which those components can be rapidly added or removed in precise concentrations.

Several attempts to establish seminal vesicle epithelial cells in culture were described during the 1980s. In most cases 5–20% of serum was required for cell growth (Lieber *et al.*, 1980; Kierszenbaum *et al.*, 1983; Tajana *et al.*, 1984; Kinghorn *et al.*, 1987), which is a serious drawback for studying hormone action. When cultivated in serum-free medium, the growth of seminal vesicle epithelial cells was reported either to require a three-dimensional collagen matrix (Tomooka *et al.*, 1985) or to be very limited (Mata *et al.*, 1986).

Guinea pig seminal vesicle epithelial cells cultivated in chemically defined medium in collagen-coated substrates were shown to grow in distinct monolayer islets displaying sustained growth for 8–10 days but, although a few islets were able to merge, full confluence was not reached. SVEPs were polarized, displaying apical microvilli and desmosome-like junctions, but were rather flat in shape and no signs of secretory activity were observed. In addition, and in contrast to the *in vivo* behavior, the endocytic pathway did not include the Golgi cisterns (Mata *et al.*, 1986).

Primary cultures of seminal vesicle epithelial cells of a much better quality are at present grown in our laboratory in serum-free medium as previously described (Mata *et al.*, 1986), but using instead a two-chamber system provided with a permeable collagen membrane insert (Transwell-col, Costar).

The cells grown in Transwell inserts are the same height as *in vivo* (10 μm tall), are morphological and functionally well polarized, and are active in secretion as well as in endocytosis.

Confluent cells are organized as a polygonal lattice, which is characteristic of epithelial cells, and are positive for immunocytokeratin reaction. Polarization is expressed by the presence of apical microvilli, junctional complexes, and high electrical transepithelial resistance (>1000 ohms). Cultivated cells are able to produce secretory proteins as shown by the (1) presence of apical secretory vesicles (Fig. 10a), (2) linear accumulation of protein in the medium with increasing time, and (3) inhibition of protein release by monensin (Fig. 10b). In addition, epithelial cells grown in such inserts give an immunopositive reaction with antibodies raised against the hamster seminal vesicle proteins (Fig. 10c). The secretory activity of the cultured HSVEP cells is nevertheless lower than *in vivo*, as illustrated by (1) the size of secretory vesicles, which is about one-third of the *in vivo* secretory vacuoles, (2) the duration of the secretory cycle, which is twice that observed *in vivo*, and (3) the amount of secreted protein, which is about 5% instead of 10% of total protein as in seminal vesicle slices.

Seminal vesicle epithelial cells in these primary cultures are also able to take up HRP and cationized ferritin apically, although at a slower rate than *in vivo*, and the endocytic pathway is the same as *in vivo* including, namely, the Golgi cisterns (Mata, 1994).

In conclusion, SVEP-cultured cells as described here, although different from, are close enough to the cells *in vivo*, both morphological and func-

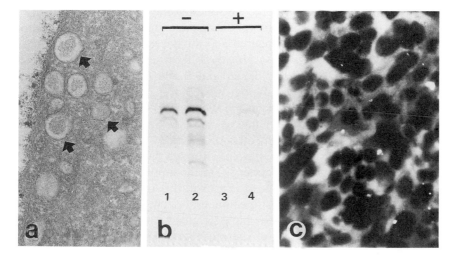

FIG. 10 Illustration of the secretory activity of cultivated seminal vesicle epithelial cells. (a) Electron micrograph demonstrating the presence of apical secretory vesicles (arrows) in cultivated seminal vesicle epithelial cells. (b) Cultivated epithelial cells were pulse labeled for 30 min with [^{35}S]Met and chased for 60 min either in the absence (−) or in presence (+) of monensin. The media from the inner and outer chambers were run in SDS-PAGE followed by fluorography. Lanes 1 and 3, inner chambers; lanes 2 and 4, outer chambers. Monensin completely inhibits apical secretion (inner chambers) and nearly completely inhibits basal secretion (outer chambers). (c) Micrograph illustrating the positive reaction of cultivated epithelial cells with antibodies against a 79-KDa protein from seminal vesicle secretion as obtained by immunofluorescence. Nuclei are ovoid shaped as seen in black (negative reaction) and surrounded by the cytoplasm as revealed in white due to bright immunofluorescence. (a), ×42,000; (c), ×412.

tionally, to be used to answer specific questions on the dynamics of seminal vesicle epithelial cells, particularly in relation to the effects of androgens, neurotransmitters, and neuropeptides. In addition, considering that much of our present knowledge on endocytosis, membrane traffic, and sorting in polarized epithelial cells comes from transformed cultured cells, and should therefore be validated in nontransformed cells, these primary cultures of SVEP cells might become useful as a model for approaching such matters.

VII. Concluding Remarks

In my reading in preparing this chapter, some subjects emerged which I felt exciting enough to deserve a final comment. The effects of testosterone withdrawal on seminal vesicle epithelial cells is one of them.

The general idea which could be found some 20–15 years ago in the literature of ultrastructural morphology was that, following castration, the secretory granules disappeared and the secretory activity of seminal vesicle epithelial cells was lost. However, a few old morphological observations suggest (Ortiz, 1953; Toner and Baillie, 1966), and some more recent ones confirm (Tse and Wong, 1980; Mata and David-Ferreira, 1982, 1985; Tam et al., 1985, 1992; Tam and Wong, 1993) that it can be otherwise.

From the literature accumulated to data (see Section V), it is now clear that seminal vesicle epithelial cells of castrated laboratory rodents are still able to produce secretory proteins although the extent of their accumulation in secretory granules depends on the animal species. The rat and hamster are extreme and opposite cases in this respect.

Castration leads to the disappearance of secretory granules in the rat seminal vesicle, whereas in the hamster gland it leads, in contrast, to an increase in size and number of secretory granules per average epithelial cell. This situation raises some interesting questions.

Why is it that secretory proteins synthesized by the seminal vesicle epithelial cells of castrated rats do not accumulate in secretory granules? Are they degraded intracellularly and therefore not numerous enough to accumulate in granules? Or are they changed in such a way that their intracellular targeting excludes their possible accumulation in secretory vacuoles? Or is it that the condensing machinery involved in their accumulation in granules is affected by testosterone withdrawal?

Why is the amount of intracellular secretion increased in the hamster seminal vesicle following castration? The data are consistent with an increase in intracellular secretory material resulting from a decrease in the intracellular transport and exocytic rate of secretory proteins, thus allowing the conclusion that testosterone interferes with secretory protein transport and exocytosis. If so, where and how is such interference exerted?

In exocrine cells, the efficiency of transport and exocytosis of secretory proteins was shown to depend on ATP (Jamieson and Palade, 1968, 1971). Can it be that testosterone withdrawal primarily decreases ATP availability in seminal vesicle epithelial cells and through that decreases their efficiency in transport and exocytosis of secretory proteins? Or is testosterone acting through other kinds of mediators such as, for instance, neuropeptides? Vasoactive intestinal peptide (VIP) was shown to increase exocytosis in seminal vesicle secretory cells (Pinho et al., 1994). Does testosterone withdrawal decrease VIP release and thus decrease exocytosis? Or does testosterone instead influence the number of VIP receptor sites available for binding to that neuropeptide? or both?

If asking questions takes some time, searching for the right answers takes, undoubtedly, much longer. These few questions, together with those I hope will come up through the reading of this chapter, will keep us busy for many years to come.

Acknowledgments

I would like to thank my colleagues Sérgio Gulbenkian, Mário Pinho, and Gabriela Rodrigues for their determined and enthusiastic collaboration in some of the work included in this review. Thanks are also due to Ms. M. Alpiarça, A. Homem, and R. M. Santos for excellent technical assistance. Part of this work was supported by a grant from Junta Nacional de Investigação Científica e Tecnológica (PBIC/C/CEN/1033/92).

References

Adjiman, M. (1985). Disfunction of the seminal vesicle and male infertility. In "Seminal Vesicles and Fertility. Biology and Pathology" (C. Bollack and A. Clavert, eds.), pp. 158–161. Karger, Basel.

Allison, V. F., and Cearley, G. W., Jr. (1972). Electron microscopic study of cells within the epithelium of the seminal vesicle of aging rats. Anat. Rec. 172, 262.

Amit, T., Barkey, R. J., and Youdim, M. B. H. (1983). Effect of prolactin, testosterone and estrogen on prolactin binding in the rat testis, prostate, seminal vesicle and liver. Mol. Cell. Endocrinol. 30, 179–188.

Antliff, H. R., Prasad, M. R. N., and Meyer, R. K. (1960). Action of prolactin on seminal vesicles of guinea pig. Proc. Soc. Exp. Biol. Med. 103, 77–80.

Arnold, C., Gulbenkian, S., Carmo-Fonseca, M., and David-Ferreira, J. F. (1983). Androgen-dependent changes in nuclear ultrastructure. A stereological study on rat ventral prostate and seminal vesicle. Biol. Cell. 47, 161–170.

Aumuller, G. (1979). Prostate gland and seminal vesicles. In "Handbuch der mikroskopischen Anatomie des Menschen" (A. Oksche and L. Vollrath, eds.), Vol. 7, Part 6, p. 380. Springer-Verlag, Berlin and New York.

Aumuller, G., and Heyns, W. (1981). Immunocytochemistry of prostatic binding protein in the rat ventral prostate. In "The Prostatic Cell: Structure and Function. Part A. Morphologic, Secretory and Biochemical Aspects" (G. P. Murphy, A. A. Sandberg, and J. P. Karr, eds.), pp. 409–415. Alan R. Liss, New York.

Aumuller, G., and Seitz, J. (1986). Immunoelectron microscopic evidence for different compartments in the secretory vacuoles of the rat seminal vesicles. Histochem. J. 18, 15–23.

Aumuller, G., and Seitz, J. (1990). Protein secretion and secretory processes in male accessory sex glands. Int. Rev. Cytol. 121, 127–231.

Barham, S. S., Lieber, M. M., and Veneziale, C. M. (1980). Ultrastructural and biochemical studies of guinea pig seminal vesicle during postnatal development and after castration. Invest. Urol. 18, 13–20.

Barkey, R. J., Shani, J., Amit, J., and Barzilai, D. (1979). Characterization of the specific binding of prolactin to binding sites in the seminal vesicle of the rat. J. Endocrinol. 80, 181–189.

Blume, C. D., and Mawhinney, M. G. (1978). Estrophilic molecules in the male guinea pig. J. Steroid Biochem. 9, 515–525.

Brandes, D. (1974a). Fine structure and cytochemistry of male sex accessory organs. *In* "Male Accessory Sex Organs. Structure and Function in Mammals" (D. Brandes, ed.), pp. 17–113. Academic Press, London.

Brandes, D. (1974b). Hormonal regulation of fine structure. *In* "Male Accessory Sex Organs. Structure and Function in Mammals" (D. Brandes, ed.), pp. 184–222. Academic Press, London.

Brown, D., Verbavatz, J.-M., Valenti, G., Lui, B., and Sabolic, I. (1993). Localization of the CHIP28 water channel in reabsorptive segments of the rat male reproductive tract. *Eur. J. Cell Biol.* **61**, 264–273.

Carballada, R., and Espanda, P. (1991). Electroforetic pattern of rodent seminal vesicle proteins as revealed by silver staining. *Int. J. Androl.* **14**, 52–57.

Carmo-Fonseca, M. (1986). Quantitative ultrastructural autoradiographic study of RNA transport in rat ventral prostate. *J. Ultrastruct. Mol. Struct. Res.* **94**, 63–76.

Carmo-Fonseca, M., and Vaz, Y. (1989). Immunocytochemical localization and lectin-binding properties of the 22kDa secretory protein from rat ventral prostate. *Biol. Reprod.* **40**, 153–164.

Case, R. M. (1978). Synthesis, intracellular transport and discharge of exportable proteins in the pancreatic acinar cell and other cells. *Biol. Rev. Cambridge Philos. Soc.* **53**, 211–354.

Cavazos, L. F. (1975). Fine structure and functional correlates of male accessory sex glands of rodents. *In* "Handbook of Physiology" (R. O. Greep, E. B. Astwood, D. W. Hamilton, and S. R. Geiger, eds.), Sect. 7, Vol. V, pp. 353–381. Am. Physiol. Soc., Washington, DC.

Cavazos, L. F., and Belt, W. D. (1965). The fine structure of the seminal vesicle during development. *Anat. Rec.* **151**, 333.

Chen, Y. H., Pentecost, B. T., McLachlan, J. A., and Teng, C. T. (1987). The androgen-dependent mouse seminal vesicle secretory protein. IV. Characterization and complementary deoxyribonucleic acid cloning. *Mol. Endocrinol.* **1**, 707–716.

Chow, P. H., Pang, S. F., Ng, K. W., and Wong, T. M. (1986). Fertility, fecundity, sex ratio and the accessory sex glands in male golden hamsters. *Int. J. Androl.* **9**, 312–320.

Clavert, A., Gabriel-Robez, O., and Montagnon, D. (1985). Physiological role of the seminal vesicles. *In* "Seminal Vesicles and Fertility. Biology and Pathology" (C. Bollack and A. Clavert, eds.), pp. 80–94. Karger, Basel.

Cornwell, G. G., III, Westermark, G. T., Pitkanen, P., and Westermark, P. (1992). Seminal vesicle amyloid: The first example of exocrine cell origin of an amyloid fibril precursor. *J. Pathol.* **167**, 297–303.

Coronel, C. E., Novella, M. L., Winnica, D. E., and Lardy, H. A. (1993). Isolation and characterization of a 54-kilodalton precursor of caltrin, the calcium transport inhibitor protein from seminal vesicles of the rat. *Biol. Reprod.* **48**, 1326–1333.

Coyne, J. D., and Kealy, W. F. (1993). Seminal vesicle amyloidosis: Morphological, histochemical and immunohistochemical observations. *Histopathology* **22**, 173–176.

Curry, P. T., and Atherton, R. W. (1990). Seminal vesicles: Development, secretory products, and fertility. *Arch. Androl.* **25**, 107–113.

Dadoune, J. P. (1985). Functional morphology of the seminal vesicle epithelium. *In* "Seminal Vesicles and Fertility. Biology and Pathology" (C. Bollack and A. Clavert, eds.), pp. 18–35. Karger, Basel.

Dahl, E., and Tveter, K. J. (1973). The ultrastructure of the accessory sex organs of the male rat. III. The post-castration involution of the coagulating gland and the seminal vesicle. *Z. Zellforsch. Mikrosk. Anat.* **144**, 179–189.

Dahnke, H.-G. (1970). Die postnatale Entwickling des Vesikulardrusenepithels von Ratten. *Cytobiologie* **2**, 445–456.

Davies, J., and Danzo, B. J. (1981). Hormonally responsive areas of the reproductive system of the male guinea pig. II. Effects of estrogens. *Biol. Reprod.* **25**, 1149–1158.

Deane, H. W., and Wurtzelmann, S. (1965). Electron microscope observations on the postnatal differentiation of the seminal vesicle epithelium of the laboratory mouse. *Am. J. Anat.* **117**, 91–134.

Desjardins, C., and Jones, R. A. (1970). Differential sensitivity of rat accessory-sex-tissues to androgen following neonatal castration or androgen treatment. *Anat. Rec.* **166**, 299.

Dlouhy, S. R., Nichols, W. C., and Karn, R. C. (1986). Production of an antibody to mouse salivary androgen binding protein (ABP) and its use in identifying a prostate protein produced by a gene distinct from Abp. *Biochem. Genet.* **24**, 743–763.

Dym, M., and Romrell, L. J. (1975). Intraepithelial lymphocytes in the male reproductive tract of rats and Rhesus monkeys. *J. Reprod. Fertil.* **42**, 1–7.

Elbadawi, A. (1979). Pathology. Part I: Nonproliferative diseases. *In* "Accessory Glands of the Male Reproductive Tract" (E. S. E. Hafez and E. Spring-Mills, eds.), pp. 175–224. Ann Arbor Sci., Ann Arbor, MI.

El-Demiry, M., and James, K. (1988). Lymphocyte subsets and macrophages in the male genital tract in health and disease. *Eur. Urol.* **14**, 226–235.

Emoto, M., Kita, E., Nishikawa, F., Katsui, N., Hamuro, A., Oku, D., and Kashiba, S. (1990). Biological functions of mouse seminal vesicle fluid. II. Role of water-soluble fraction of seminal vesicle fluid as a nonspecific immunomodulator. *Arch. Androl.* **25**, 75–84.

Farquhar, M. G. (1983). Multiple pathways of exocytosis, endocytosis, and membrane recycling: Validation of a Golgi route. *Fed. Proc., Fed. Am. Soc. Exp. Biol.* **42**, 2407–2413.

Fawell, S. E., and Higgins, S. J. (1986). Tissue distribution, developmental profile and hormonal regulation of androgen-responsive secretory proteins of rat seminal vesicles studied by immunocytochemistry. *Mol. Cell. Endocrinol.* **48**, 39–49.

Fawell, S. E., McDonald, C. J., and Higgins, S. J. (1987). Comparison of seminal vesicle secretory proteins of rodents using antibody and nucleotide probes. *Mol. Cell. Endocrinol.* **50**, 107–114.

Flickinger, C. J. (1974a). Synthesis, intracellular transport, and release of secretory protein in the seminal vesicle of the rat, as studied by electron microscope radioautography. *Anat. Rec.* **180**, 407–427.

Flickinger, C. J. (1974b). Protein secretion in the rat ventral prostate and the relation of Golgi vesicles, cisternae and vacuoles, as studied by electron microscope radioautography. *Anat. Rec.* **180**, 427–449.

Galdiero, F., Tufano, M. A., De Martino, L., Capasso, C., Porta, R., Ravagnan, G., Peluso, G., and Metafora, S. (1989). Inhibition of macrophage phagocytic activity by SV-IV, a major protein secreted from the rat seminal vesicle epithelium. *J. Reprod. Immunol.* **16**, 269–284.

Gaytan, F., Bellido, C., Aguilar, R., and Lucena, M. C. (1986). Morphometric analysis of the rat ventral prostate and seminal vesicles during prepubertal development: Effects of neonatal treatment with estrogen. *Biol. Reprod.* **35**, 219–226.

Gérard, A., Khanfri, J., Guéant, J. L., Frémont, S., Nicolas, J. P., Grignon, G., and Gérard, H. (1988). Electron microscope radioautographic evidence of in vivo androgen-binding protein internalization in the rat epididymis principal cells. *Endocrinology (Baltimore)* **122**, 1297–1307.

Gonzales, G. F. (1989). Functional structure and ultrastructure of seminal vesicles. *Arch. Androl.* **22**, 1–13.

Gonzalez, S. I., Chandrashekar, V., Shire, J. G. M., Luthy, I. A., Bartke, A., and Calandra, R. S. (1991). Effects of hyperprolactinemia on ornithine decarboxilase activity and poly-amine levels in seminal vesicles of genetically prolactin-deficient adult dwarf mice. *Biol. Reprod.* **44**, 321–326.

Hadley, M. A., Djakiew, D., Byers, S. W., and Dym, M. (1987). Polarized secretion of androgen-binding protein and transferrin by Sertoli cells grown in a bicameral culture system. *Endocrinology (Baltimore)* **120,** 1097–1103.

Hagström, J., Harvey, S., and Wieban, E. (1992). Androgens are necessary for the establishment of secretory protein expression in the guinea pig seminal vesicle epithelium. *Biol. Rep.* **47,** 768–775.

Hand, A. R., Coleman, R., Mazariegos, M. R., Lustmann, J., and Lotti, L. V. (1987). Endocytosis of secretory proteins by salivary gland duct cells. *J. Dent. Res.* **66,** 412–419.

Hermo, L., Oko, R., and Morales, C. R. (1994). Secretion and endocytosis in the male reproductive tract: A role in sperm maturation. *Int. Rev. Cytol.* **154,** 105–189.

Higgins, S. J., and Fuller, F. M. (1981). Effects of testosterone on protein synthesis in rat seminal vesicles analysed by two-dimensional gel electroforesis. *Mol. Cell. Endocrinol.* **24,** 85–101.

Hoffer, A. P., Hamilton, D. W., and Fawcett, D. W. (1973). The ultrastructure of the principal cells and intraepithelial leucocytes in the initial segment of the rat epididymis. *Anat. Rec.* **175,** 169–202.

Holtzman, E., Schacher, S., Evans, J., and Teichberg, S. (1977). Origin and fate of the membranes of secretion granules and synaptic vesicles: Membrane circulation in neurons, gland cells and retinal photoreceptors. *Cell Surf. Rev.* **4,** 165–246.

Jamieson, J. D., and Palade, G. E. (1968). Intracellular transport of secretory proteins in the pancreatic exocrine cell. IV. Metabolic requirements. *J. Cell Biol.* **39,** 589–603.

Jamieson, J. D., and Palade, G. E. (1971). Condensing vacuole conversion and zymogen granule discharge in pancreatic exocrine cells: Metabolic studies. *J. Cell Biol.* **48,** 503–523.

Jean-Faucher, C., Berger, M., De Turckheim, M., Veyssiére, G., and Jean, C. (1985). Permanent changes in the functional development of accessory sex organs and in fertility in male mice after exposure to cyproterone acetate. *J. Endocrinol.* **104,** 113–120.

Jean-Faucher, C., Berger, M., Gallon, C., De Turckheim, M., Veyssiére, G., and Jean, C. (1986). Imprinting of male sex tissues by neonatal endogenous androgens in mice. *Horm. Res.* **24,** 38–45.

Jean-Faucher, C., Berger, M., Gallon, C., De Turckheim, M., Veyssiére, G., and Jean, C. (1989). Long-term alterations on the male mouse genital tract associated with neonatal exposure to cyproterone acetate. Biochemical data. *J. Steroid Biochem.* **32,** 105–112.

Jonáková, V., Calvete, J. J., Mann, K., Schafer, W., Schmid, E. R., and Topfer-Petersen, E. (1992). The complete primary structure of three isoforms of a boar sperm-associated acrosin inhibitor. *FEBS Lett.* **297,** 147–150.

Kierszenbaum, A. L., DePhilip, R. M., Spruill, W. A., and Takenaka, I. (1983). Isolation and culture of rat seminal vesicle epithelial cells. The use of the secretory protein SVS IV as a functional probe. *Exp. Cell Res.* **145,** 293–304.

Kinghorn, E. M., Bate, A. S., and Higgins, S. J. (1987). Growth of rat seminal vesicle epithelial cells in culture: Neurotransmitters are required for androgen-regulated synthesis of tissue-specific secretory proteins. *Endocrinology (Baltimore)* **121,** 1678–1689.

Kistler, M. K., Ostrowsky, M., and Kistler, W. S. (1981). Developmental regulation of secretory protein synthesis in rat seminal vesicle. *Proc. Natl. Acad. Sci. U.S.A.* **78,** 737–741.

Koenig, H., Chung, Y., and Bakay, R. (1976). Testosterone and 6-N,2′-O-dibutyryladenosine 3′ : 5-cyclic monophosphate stimulate protein and lysosomal enzyme secretion in rat seminal vesicle. *Biochem. J.* **158,** 543–547.

Kraehenbuhl, J. P., Michetti, H. R., Perregaux, C., Mekalanus, J., and Neutra, M. (1992). Role of transepithelial transport in triggering a mucosal immune response and in delivery of mucosal antibodies into secretions. *NATO ASI Ser., Ser. H* **62,** 363–366.

Kwok, S. C. M., Soares, M. J., McMurtry, J. P., and Yurewicz, E. C. (1993). Binding characteristics and immunolocalization of porcine seminal protein, PSP-1. *Mol. Reprod. Dev.* **35**, 244–250.

Levine, N., and Kelly, H. (1980). Control of reabsorption of water and electrolytes by guinea-pig seminal vesicle in vivo. *J. Reprod. Fertil.* **58**, 421–427.

Lieber, M. M., Barham, S. S., and Veneziale, C. M. (1980). In vitro propagation of seminal vesicle epithelial cells. *Invest. Urol.* **17**, 348–355.

Luke, M. C., and Coffey, D. S. (1994). The male sex accessory tissues. Structure, androgen action, and physiology. *In* "The Physiology of Reproduction" (E. Knobil and J. D. Neil, eds.), 2nd ed., pp. 1435–1487. Plenum, New York.

Mainwaring, W. I., and Brandes, D. (1974). Functional and structural changes in accessory sex organs during aging. *In* "Male Accessory Sex Organs. Structure and Function in Mammals" (D. Brandes, ed.), pp. 469–500. Academic Press, London.

Manjunath, P., Chandonnet, L., Leblond, E., and Desnoyers, L. (1994). Major proteins of bovine seminal vesicles bind to spermatozoa. *Biol. Reprod.* **50**, 27–37.

Mariotti, A., Thornton, M., and Mawhinney, M. (1981). Actions of androgen and estrogen on collagen levels in male accessory sex organs. *Endocrinology (Baltimore)* **109**, 837–843.

Mariotti, A., Duremdes, G., and Mawhinney, M. (1983). Growth responses of the guinea pig seminal vesicle epithelium and muscle to long term treatment with dihydrotestosterone and/or estradiol benzoate. *Prostate (N.Y.)* **4**, 315–319.

Mata, L. R. (1976). Dynamics of HRPase absorption in the epithelial cells of the hamster seminal vesicle. *J. Microsc. Biol. Cell.* **25**, 127–132.

Mata, L. R. (1979). The secretory cycle of the hamster seminal vesicle epithelial cells as studied in vitro by electron microscopic autoradiography. *Biol. Cell.* **36**, 25–28.

Mata, L. R. (1981). Estudo morfo-funcional das células secretoras da vesícula seminal do Criceto (summary in English). Doctoral Thesis presented to the Faculty of Sciences, University of Lisbon.

Mata, L. R. (1994). Secretion and endocytosis in the seminal vesicles. *In* "Ultrastructure of Male Urogenital Glands" (A. Riva and P. M. Motta, eds.), pp. 51–60. Kluwer Academic Publishers, Boston, London.

Mata, L. R., and Christensen, E. I. (1990a). Membrane recycling in the epithelium of the seminal vesicle intersects the secretory pathway. *NATO ASI Ser., Ser. H* **62**, 161–168.

Mata, L. R., and Christensen, E. I. (1990b). Redistribution and recycling of internalized membrane in seminal vesicle secretory cells. *Biol. Cell.* **68**, 183–193.

Mata, L. R., and David-Ferreira, J. F. (1973). Transport of exogenous peroxidase to Golgi cisternae in the hamster seminal vesicle. *J. Microsc. (Paris)* **17**, 103–106.

Mata, L. R., and David-Ferreira, J. F. (1982). Testosterone interference with the intracellular transport of secretion in hamster seminal vesicle. *Biol. Cell.* **46**, 101–104.

Mata, L. R., and David-Ferreira, J. F. (1985). Secretory cell activity in the hamster seminal vesicle following castration. A morphometric ultrastructural study. *Biol. Cell.* **53**, 165–178.

Mata, L. R., and David-Ferreira, J. F. (1990). Testosterone interferes with the kinetics of endocytosis in the hamster seminal vesicle. *Biol. Cell.* **68**, 195–203.

Mata, L. R., and Maunsbach, A. B. (1982). Absorption of secretory protein by the epithelium of hamster seminal vesicle as studied by electron microscope autoradiography. *Biol. Cell.* **46**, 65–74.

Mata, L. R., Petersen, O. W., and van Deurs, B. (1986). Endocytosis in guinea- pig seminal vesicle epithelial cells cultivated in chemically defined medium. *Biol. Cell.* **58**, 211–220.

Mawhinney, M. G., and Neubauer, B. L. (1979). Actions of estrogen in the male. *Invest. Urol.* **16**, 409–420.

Mazzucchelli, L., Studer, U. E., and Zimmermann, A. (1992). Cystadenoma of the seminal vesicle: Case report and literature review. *J. Urol.* **147**, 1621–1624.

Meldolesi, J. (1974). Dynamics of cytoplasmic membranes in guinea pig pancreatic acinar cells. I. Synthesis and turnover of membrane proteins. *J. Cell Biol.* **61,** 1–13.

Meldolesi, J., Borgese, N., De Camilli, P., and Ceccarelli, B. (1978). Cytoplasmic membranes and the secretory process. *Cell Surf. Rev.* **5,** 509–627.

Metafora, S., Peluso, G., Persico, P., Ravagnan, G., Esposito, C., and Porta, R. (1989a). Immunosupressive and anti-inflammatory properties of a major protein secreted from the epithelium of the rat seminal vesicles. *Biochem. Pharmacol.* **38,** 121–131.

Metafora, S., Porta, R., Ravagnan, G., Peluso, G., Tufano, M. A., De Martino, L., Ianiello, R., and Galdiero, F. (1989b). Inhibitory effect of SV-IV, a major protein secreted from the rat seminal vesicle epithelium, on phagocytosis and chemotaxis of human polymorphonuclear leucocytes. *J. Leukocyte Biol.* **46,** 409–416.

Mifune, H., Noda, Y., Mohri, S., Suzuki, S., Nishinakagawa, H., and Otsuka, J. (1986). Fine structure of the seminal vesicle epithelium of the mouse and golden hamster. *Exp. Anim.* **35,** 149–158.

Morré, D. J., Kartenbeck, J., and Franke, W. W. (1979). Membrane flow and interconversions among endomembranes. *Biochim. Biophys. Acta* **559,** 71–152.

Negro-Vilar, A., Krulich, L., and McCann, S. M. (1973). Changes in serum prolactin and gonadotropins during sexual development of the male rat. *Endocrinology (Baltimore)* **93,** 660–664.

Negro-Vilar, A., Saad, W. A., and McCann, S. M. (1977). Evidence for a role of prolactin in prostate and seminal vesicle in immature male rats. *Endocrinology (Baltimore)* **100,** 729–737.

Neubauer, B., and Mawhinney, M. (1981). Actions of androgen and estrogen on guinea pig seminal vesicle epithelium and muscle. *Endocrinology (Baltimore)* **108,** 680–687.

Neubauer, B., Blume, C., Cricco, R., Greiner, J., and Mawhinney, M. (1981). Comparative effects and mechanisms of castration, estrogen anti-androgen, and anti-estrogen-induced regression of accessory sex organ epithelium and muscle. *Invest. Urol.* **18,** 229–234.

Neubauer, B. L., Biser, P., Jones, C. D., Mariotti, A., Hoover, D. M., Thornton, T., Thornton, M. O., and Goode, R. L. (1989). Antagonism of androgen and estrogen effects in guinea pig seminal vesicle epithelium and fibromuscular stroma by keoxifene (LY156758). *Prostate (N.Y.)* **15,** 189–273.

Norvitch, M. E., Harvey, S., Moore, J. T., and Wieben, E. D. (1988). Processing of two protein precursors yields four mature guinea pig seminal vesicle secretory proteins. *Biol. Reprod.* **38,** 1153–1164.

Norvitch, M. E., Harvey, S., Smith, S., Hagström, J. E., and Wieben, E. D. (1989). Androgens affect the processing of secretory protein precursors in the guinea pig seminal vesicle. I. Evidence for androgen-regulated proteolytic processing. *Mol. Endocrinol.* **3,** 1788–1796.

Okamoto, S., Ogasawara, Y., Yamane, T., Kitamura, Y., and Matsumoto, K. (1982). Proliferative response of mouse seminal vesicle epithelium to androgen and estrogen, assayed by incorporation of [^{125}I]iododeoxyuridine. *Endocrinology (Baltimore)* **110,** 1796–1803.

Ortiz, E. (1953). The effects of castration on the reproductive system of the golden hamster. *Anat. Rec.* **117,** 65–92.

Ortiz, H. E., and Cavicchia, J. C. (1990). Androgen-induced changes in nuclear pore number and in tight junctions in rat seminal vesicle epithelium. *Anat. Rec.* **226,** 129–134.

Ostrowsky, M. C., Kistler, M. K., and Kistler, W. S. (1979). Purification and cell-free synthesis of a major protein from rat seminal vesicle secretion. A potential marker for androgen action. *J. Biol. Chem.* **254,** 383–390.

Ostrowsky, M. C., Kistler, M. K., and Kistler, W. S. (1982). Effect of castration on the synthesis of seminal vesicle secretory protein IV in the rat. *Biochemistry* **21,** 3525–3530.

Palade, G. E. (1975). Intracellular aspects of the process of protein synthesis. *Science* **189**, 347–358.

Pinho, M. S., and Mata, L. R. (1992). Castration-resistant secretion in the hamster seminal vesicle does not depend on androgens. *Int. J. Androl.* **15**, 435–447.

Pinho, M. S., Sebastião, A. M., Rodrigues, G., Barroso, C. P., Ribeiro, J. A., Mata, L. R., and Gulbenkian, S. (1994). Vasoactive intestinal peptide in the hamster seminal vesicle: Distribution, binding sites and possible functions. *Neuroscience* **59**, 1083–1091.

Porta, R., Metafora, S., Esposito, C., Mariniello, L., Persico, P., Mancuso, F., and Peluso, G. (1993). Biological activities of a major protein secreted from the rat seminal vesicles after structural modification catalyzed by transglutaminase in vitro. *Immunopharmacology* **25**, 178–188.

Roth, T. F., and Porter, K. R. (1964). Yolk protein uptake in the oocyte of the mosquito *Aedes aegypti*, L. *J. Cell Biol.* **20**, 313–332.

Rumpler, Y. (1985). Seminal vesicles in evolution. *In* "Seminal Vesicles and Fertility. Biology and Pathology" (C. Bollack and A. Clavert, eds.), pp. 1–3. Karger, Basel.

Samuel, L. H., and Flickinger, C. J. (1986). Intracellular pathway and kinetics of protein secretion in the coagulating gland of mouse. *Biol. Reprod.* **34**, 107–117.

Samuel, L. H., and Flickinger, C. J. (1987). The relationship between the morphology of cell organelles and kinetics of the secretory glands of mice. *Cell Tissue Res.* **247**, 203–213.

Sanz, L., Calvete, J. J., Schafer, W., Mann, K., and Topfer-Petersen, E. (1992). Isolation and biochemical characterization of two isoforms of a boar sperm zona pellucida-binding protein. *Biochim. Biophys. Acta* **1119**, 127–132.

Schindelmeiser, J., Aumuller, G., Enderle-Schimitt, U., Bergmann, K., and Hoffmann, K. (1988). Photoperiodic influence on the morphology and the androgen receptor level of the ventral prostate gland and seminal vesicles of the djungarian hamster (*Phodopus sungorus*). *Andrologia* **20**, 105–113.

Schleicher, G., Stumpf, W. E., Drews, U., Thiedemann, K. U., and Sar, M. (1985). Differential distribution of ^{3}H-dihydrotestosterone and ^{3}H-estradiol nuclear binding sites in mouse male accessory sex glands. An autoradiographic study. *Histochemistry* **82**, 453–462.

Seitz, J., and Aumuller, G. (1991). Localization of intrinsic and extrinsic SVS II immunoreactivity in rat spermatozoa. *Andrologia* **24**, 27–31.

Senthilkumaran, B., Manimaran, R. R., Shavali, S. S., Aruldhas, M. M., and Govindarajulu, P. (1993). Effect of prepubertal hypothyroidism on some glycoprotein-associated monosaccharides in the seminal vesicles of Wistar rats. *Eur. Arch. Biol.* **103**, 251–255.

Soares-Pessoa, J. F., and David-Ferreira, J. F. (1980). Bidirectional transport of horseradish peroxidase by the rat Sertoli cells. An in vitro study. *Biol. Cell.* **39**, 301–304.

Sofikitis, N., Takahashi, C., Kadowaki, H., Okazaki, T., Shimamoto, T., Nakamura, I., and Miyagawa, I. (1992). The role of the seminal vesicles and coagulating glands in fertilization in the rat. *Int. J. Androl.* **15**, 54–61.

Spring-Mills, E., and Hafez, E. S. E. (1979). Functional anatomy and histology. *In* "Accessory Glands of Male Reproductive Tract" (E. S. E. Hafez and E. Spring-Mills, eds.), pp. 29–57. Ann Arbor Sci., Ann Arbor, MI.

Spring-Mills, E., Krall, A., and Apellaniz, M. (1982). Effects of bromocriptine and perphenazine on male accessory sex glands II. Ultrastructure of the seminal vesicles. *Arch. Androl.* **9**, 203–214.

Steinman, R. M., Mellman, I. S., Muller, W. A., and Cohn, Z. A. (1983). Endocytosis and the recycling of plasma membrane. *J. Cell Biol.* **96**, 1–28.

Stern, J. E., Gardner, S., Quirk, D., and Wira, C. R. (1992). Secretory immune system of the male reproductive tract: Effects of dihydrotestosterone and estradiol on IgA and secretory component levels. *J. Reprod. Immunol.* **22**, 73–85.

Szirmai, J. A., and Van Der Linde, P. (1965). Effect of castration on the endoplasmic reticulum of the seminal vesicle and other target epithelia in the rat. *J. Ultrastruct. Res.* **12,** 380–395.

Tajana, G. F., Locuratolo, P., Metafora, S., Abrescia, P., and Guardiola, J. (1984). Synthesis of a testosterone-dependent secretory protein by rat seminal vesicle-derived cell lines. *EMBO J.* **3,** 637–644.

Tam, C. C., and Wong, Y. C. (1993). Thiamine pyrophosphatase activity in secretory cells of the lateral prostate and seminal vesicle of normal and castrated guinea pigs and castrated treated with oestradiol. *Histochem. J.* **25,** 77–85.

Tam, C. C., Wong, Y. C., and Tang, F. (1985). Further regression of seminal vesicles of castrated guinea pig by administration of cyproterone acetate. *Acta Anat.* **124,** 65–73.

Tam, C. C., Wong, Y. C., and Tang, F. (1992). Ultrastructural and cytochemical studies of the effects of prolactin on the lateral prostate and the seminal vesicle of castrated guinea pig. *Cell Tissue Res.* **270,** 105–112.

Thaler, C. J., Knapp, P. M., McIntyre, J. A., Coulam, C. B, Critser, J. K., and Faulk, W. P. (1990). Congenital aplasia of seminal vesicles: Absence of trophoblast-lymphocyte cross-reactive antigens from seminal plasma. *Fertil. Steril.* **53,** 948–949.

Tomooka, Y., Harris, S. E., and McLachlan, J. A. (1985). Growth of seminal vesicle epithelial cells in serum-free collagen gel culture. *In Vitro Cell. Dev. Biol.* **21,** 237–244.

Toner, P. G., and Baillie, A. H. (1966). Biochemical, histochemical and ultrastructural changes in the mouse seminal vesicle after castration. *J. Anat.* **100,** 173–188.

Traish, A. M., Muller, R. E., Burns, M. E., and Woltz, H. H. (1981). Steroid binding proteins and human prostate. *In* "The Prostatic Cell: Structure and Function. Part A. Morphologic, Secretory and Biochemical Aspects" (G. P. Murphy, A. A. Sandberg, and J. P. Karr, eds.), pp. 417–434. Alan R. Liss, New York.

Tse, M. K. W., and Wong, Y. C. (1980). Structural study of the involution of the seminal vesicles of the guinea pig following orchiectomy. *Acta Anat.* **108,** 68–78.

van Deurs, B., Petersen, O. W., Olsnes, S., and Sandvig, K. (1989). The ways of endocytosis. *Int. Rev. Cytol.* **117,** 131–177.

Veneziale, C. M., Burns, J. M., Lewis, J. C., and Buchi, A. (1977). Specific protein synthesis in isolated epithelium of guinea-pig seminal vesicle. Effects of castration and androgen replacement. *Biochem. J.* **166,** 167–173.

Weinberger, M. J. (1984). Heterogeneity and distribution of estrogen binding sites in guinea pig seminal vesicle. *J. Steroid Biochem.* **20,** 1327–1332.

Williams-Ashman, H. G. (1983). Regulatory features of seminal vesicle development and function. *Curr. Top. Cell. Regul.* **22,** 201–275.

Yamane, T., Kitamura, Y., Terada, N., and Matsumoto, K. (1986). Proliferative response of seminal vesicle cells to androgen and estrogen in neonatally castrated mice. *J. Steroid Biochem.* **24,** 703–708.

Yamane, T., Ayata, M., Terada, N., Kitamura, Y., and Matsumoto, K. (1987). Roles of neonatal and prepubertal testicular androgens on androgen-induced proliferative response of seminal vesicle cells in adult mice. *J. Steroid Biochem.* **28,** 559–564.

Molecular Organization of Hepatocyte Peroxisomes

Takashi Makita

Department of Veterinary Anatomy, Yamaguchi University, Yoshida, Yamaguchi City 753, Japan

The matrix of peroxisomes has been considered to be homogeneous. However, a fine network of tubules is visible in electron micrographs at very high magnification. This substructure becomes more positive in a high-contrast photocopy and with an imaging-plate method.

Clofibrate, bezafibrate, and aspirin increase peroxisomes. In proliferated peroxisomes, the density of matrix is low and the fine network is more visible. The effect of proliferators is more significant in males than in females. This sex difference may involve the action of estrogen, growth hormone, cytochrome P-450 and thyroxine.

Mg-ATPase is localized on the limiting membrane of peroxisomes. Even on the membrane of irregular projections of proliferated peroxisomes, Mg-ATPase is evident cytochemically. Carnitine acetyltransferase is detectable in the matrix of proliferated peroxisomes.

Withdrawal of proliferators results in a rapid decrease of peroxisomes. This may indicate the existence of peroxisome suppressors. Alternatively, dynamic transformation of vesicular to tubular types in peroxisome reticulum may occur. Such transformation has been described in lysosomes and mitochondria.

KEY WORDS: Peroxisome, Hepatocyte, Peroxisome reticulum, Mg-ATPase, Carnitine acetyltransferase, Peroxisome proliferator, Tubular peroxisome, Peroxisome suppressor, Matrical fine network, Clofibrate.

I. Introduction

Rhodin (1954) first described in the renal tubule of the mouse the existence of previously unreported cell organelles, each surrounded by a single limiting membrane and containing a finely granular matrix. He named them microbodies. Later, Rouiller and Bernhard (1956) also found these bodies in rat hepatocytes and noted that they had a localized core or crystalloid structure.

De Duve and Baudhuin (1966) defined the peroxisome as an organelle containing at least one oxidase that formed H_2O_2 and catalase to decompose H_2O_2. Subsequently, Novikoff et al. (1973) observed small and coreless peroxisomes and named them microperoxisomes. These early studies and forms of nomenclature have been well documented in other reviews (Hruban and Rechcigal, 1969; Poole et al., 1969, 1970; de Duve, 1983; Hruban et al., 1972, 1974; Masters and Holmes, 1977; Hicks and Fahimi, 1977; Tolbert, 1981; Pais and Carrapico, 1982; Kindle and Lazarow, 1982; Kindle, 1982; Goldfischer and Reddy, 1984; Lazarow and Fujiki, 1985; Borst, 1986, 1989; Soto et al., 1993a; Gibson and Lake, 1993; Subramani, 1993; Sabatini and Adesnik, 1994; Lazarow, 1994).

As impressively summarized by Dr. H. D. Fahimi, the organizer of the International Symposium on Peroxisomes in Biology and Medicine (Heidelberg, 1986), there were three major conferences on peroxisomes held in 1969, 1981, and 1986. The first one (held in New York) was organized by Drs. C. de Duve and J. F. Hogg and was reported in the Annals of the New York Academy of Sciences (1970). The second meeting, also held in New York, was organized by Drs. H. Kindle and P. B. Lazarow (1982). The record of the Heidelberg Symposium organized by Drs. H. D. Fahimi and H. Sies was published as *Peroxisomes in Biology and Medicine* (1987) and as supplement No. 14 of the *European Journal of Cell Biology* (pp. 3–46, 1986). The list of participants and the areas covered at these three conferences include virtually all fields of peroxisome research.

From a survey of the proceedings of the three conferences, major topics related to the molecular structure of peroxisomes can be grouped into six categories;

Biogenesis of peroxisomes
Pathology of peroxisomes
Peroxisomal beta oxidation
Effects of peroxisome proliferators
Peroxisomes and lipid biosynthesis and metabolism
Peroxisomal membrane protein

Since the symposium in Heidelberg, an increasing number of researchers have used recombinant DNA technology to study the nature and location of the topogenic signals of peroxisomes. At the same time, advances in the immunocytochemical localization of enzymes and proteins which have not been identifiable previously with other staining procedures, and also super-high resolution electron microscopy, have made it possible to localize and distinguish the components of peroxisomes *in situ*.

In order to relate the general structure and substructure of peroxisomes at the ultrastructural level to their biogenesis, reference is made to the proposal of a fine network in the matrix or ground substance and to the tubulovesicular complex model of peroxisomes. A table of enzymes and proteins is also presented for further explanation of the immunohistochemical localization of enzymes. Mg-ATPase and carnitine acetyltransferase (CAT) activities are described in detail. Finally, experiments on proliferators and interesting data obtained by administration of clofibrate, bezafibrate, and aspirin are described.

Despite the availability of numerous references on the structure and function of peroxisomes in various organs of many animals ranging from humans to fireflies, the number of research groups focusing on the molecular organization of peroxisomes is surprisingly small. There are fewer than ten and they actively exchange information. The general subjects and trends of investigation are, therefore, encompassed by this small group.

This chapter focuses on the following points, which have not been fully dealt with in previous reviews on peroxisomes.

(1) The network substructure of the matrix and its potential as a source of matrix tubules and fibers.

(2) Modification of the peroxisome reticulum theory to include transformation from a vesicular to a tubular structure, analogous to that in lysosomes.

(3) Cytochemical localization of Mg-ATPase and carnitine acetyltransferase.

(4) Potential restriction factors which regulate the proliferation of peroxisomes, as suggested by the fact that withdrawal of proliferators results in a rapid decrease of induced peroxisomes without any recognizable residue of the limiting membrane.

Recent progress in molecular analysis has revealed several important molecular structures in peroxisomal proteins (Borst, 1989; Osumi and Fujiki, 1990; Gibson and Lake, 1993; Subramani, 1993; Soto et al., 1993b; Sabatini and Adesnik, 1994). These include peroxisome assembly factor-1 (PAF-1) (Tsukamoto et al., 1991; Shimozawa et al., 1992a), the Ser-Lys-Leu-(SKL) motif at the C-terminal as a peroxisome-targeting signal

(Miura *et al.*, 1992), nonspecific lipid transfer protein (nsLTP) (Fujiki *et al.*, 1989; Tsuneoka *et al.*, 1988), and species differences between mitochondrial and peroxisomal serine : pyruvate aminotransferase (SPT/AGT) activity (Noguchi, 1987; Ichiyama *et al.*, 1992).

Many further aspects also remain to be clarified, including the nature of the receptors on the limiting membrane of peroxisomes and molecular chaperones (Ellis and van der Vies, 1991). These topics are only briefly mentioned, since the author is not directly involved in these fields.

With the use of high-technology instruments, including an electron microscope (resolution 0.18 nm), electron cryomicroscope, imaging plate system, electron spectroscopic imaging, and laser-scanning confocal microscope, optimal high-contrast and high-magnification images of the fine network present in the matrix of peroxisomes have been obtained and are discussed here. The choice of topics was determined largely by a desire to complement other excellent reviews currently available (Gibson and Lake, 1993; Subramani, 1993; Soto *et al.*, 1993a; Sabatini and Adesnik, 1994; Lazarow, 1994).

II. Structure of Hepatic Peroxisomes

A. General Composition

Peroxisomes are characterized by marked variation in the number, size, enzyme composition, core structure, matrical plate, matrical tube, and other related structures with differences in species (Fahimi *et al.*, 1993), strain, sex, age, and cell and tissue type. However, the most typical image of a hepatic peroxisome is, as shown in an electron micrograph of a rat hepatocyte (Fig. 1), a round body surrounded by a single limiting membrane. The matrix, which appears finely granular, contains a core or crystalloid structure composed of multitubular crystal elements (Fig. 2). Very often the endoplasmic reticulum is closely apposed to the limiting membrane. The irregular shape of the limiting membrane reflects its fragility. When a specimen is prepared by rapid freezing and substitution, the outline of hepatic peroxisomes is round, with a smooth limiting membrane. As will be discussed later, the limiting membrane of newly induced peroxisomes is more irregular than that of mature peroxisomes, indicating either fragility of the membrane or lack of matrix substance in induced peroxisomes. The diameter of hepatocyte peroxisomes varies from 0.3 to 1.0 μm in normal rats.

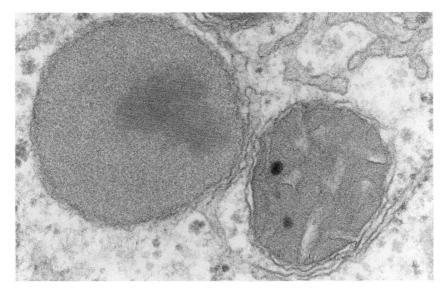

FIG. 1   A peroxisome in a rat hepatocyte. Limiting membrane, core structure, and matrix containing a complex network of tubules are discernible. Notice the close association of the endoplasmic reticulum (ER). Resemblance of the matrix to that of a mitochondrion is another noteworthy feature. ×77,500.

B. Matrix and Limiting Membrane

1. Peroxisome Matrix

Although the matrix of peroxisomes has been described as a homogeneous or fine-grained substance by electron microscopists, it is composed of many types of enzymes and proteins. A network of fine tubules or fibrils is discernible with careful observation (Makita and Hirose, 1990; Makita et al., 1990a,b). This substructural network contains variable sized small vesicles, dense grains, saccules and specific dense bodies (Figs. 3–9). They are more conspicuous in micrographs observed with an imaging-plate system, as will be discussed later (Makita and Ogawa, 1991; Makita et al., 1994). Components of the network may have Mg-ATPase activity (Fig. 10) and appear to be continuous, at least in part, with the limiting membrane.

Since the matrix of newly induced peroxisomes is paler than that of preexisting ones in general (Makita, 1990), the fine network is more visible in experimentally induced peroxisomes and a variety of matrical or mar-

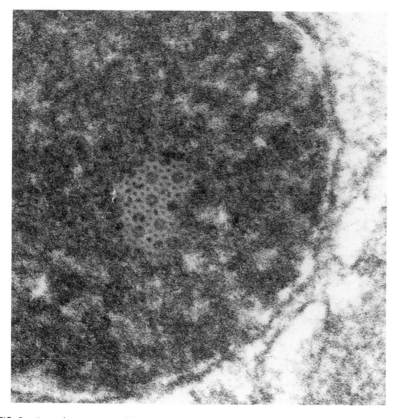

FIG. 2   An enlargement to illustrate the tubular substructure of the core. ×160,000.

ginal plates and other fibrillar structures are often seen. These substructures may originate from the matrical network substructure. However,
the marginal plate and other tubular or fibrillar structures are discernible
in the matrical network structure at high magnifications (Figs. 3 and 4).
At extremely high magnifications (Figs. 7 and 8), the dense spots in the
matrix network appear to be round or oval in shape and contain a crystalloi-
dal substructure.

In addition to the imaging-plate procedure (Fig. 8), photocopy of

---

FIG. 3   Limiting membrane, core structure, matrical plate and substructure of the matrix
stand out with the imaging plate method. ×92,500.

FIG. 4   Bundles of matrical plates are embedded in the fine network of matrix tubules in a
peroxisome after 2 weeks of clofibrate administration. ×92,500.

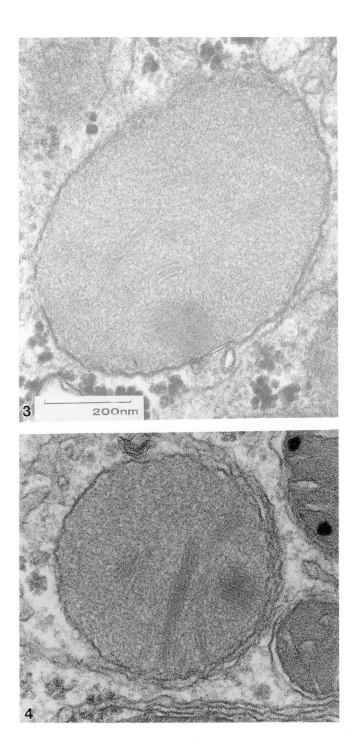

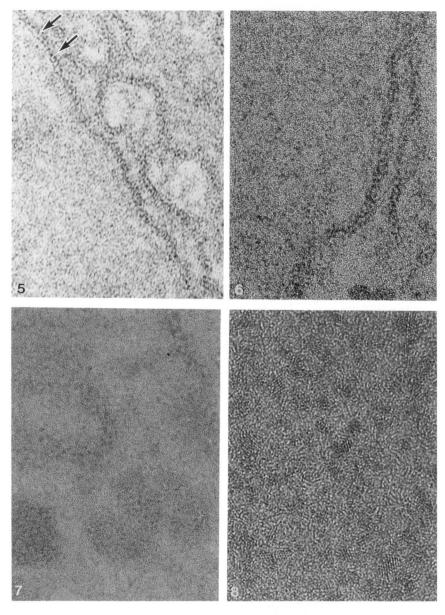

FIG. 5　A portion of the limiting membrane (arrows) of a peroxisome. The adjacent ER membrane resembles the limiting membrane in appearance. ×306,000.

FIG. 6　Porous appearance of the limiting membrane of the peroxisome at higher magnification. ×374,000.

FIG. 7　Dense spots in the matrix at high magnification. Upper right corner is a portion of the limiting membrane. ×578,000.

FIG. 8　Crystalline substructure of the dense spot in Fig. 7 at higher magnification. ×2,244,000.

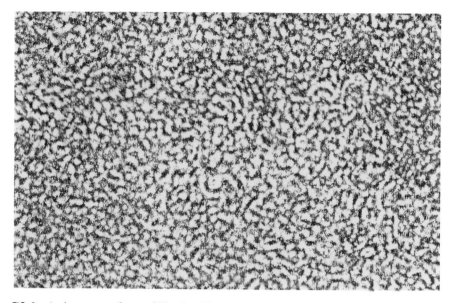

FIG. 9   A photocopy of part of Fig. 8 to illustrate the fine network in the matrix of a hepatic peroxisome. ×2,409,000.

electron micrographs visualizes the pattern of the matrical network (Fig. 9).

In mature peroxisomes, the matrical network of fine tubules appears to be embedded in protein or ground substance so that the network is not clearly visible. However, the network can be identified in a close examination of electron micrographs in other reports. The function of these networks in matrix remains to be investigated. From the localization

FIG. 10   Mg-ATPase activity in the substructural network of the matrix of a peroxisome in a beagle hepatocyte. ×134,900.

of platinum or cisplatin in the matrix, we speculate that peroxisomes may play a role in biocondensation. If so, the matrical network could be responsible (Makita *et al.,* 1985, 1986; Makita and Okawa, 1987).

2. Limiting Membrane

As reviewed by Hardeman *et al.* (1990) and Soto *et al.* (1993b) and Sabatini and Adesnik (1994), the peroxisomal membrane contains 70-kDa, 26-kDa, 22-kDa, and 15-kDa proteins as well as Mg-ATPase and Acyl-CoA synthetase (Table I). At very high magnification (Figs. 5 and 6), a micropore yor channel-like structure begins to be visible in the limiting membrane

TABLE I

Hepatic Peroxisomal Enzymes

Acetyl-CoA hydrolase

Acyl/Alkyl-DHAP reductase

Acyl-CoA synthetase

Alanine-glyoxylate aminotransferase

Aldehyde dehydrogenase

Alkyl-DHAP synthase

Allantoicase

Allantoinase

D-Amino acid oxidase

D-Aspartate oxidase

Bile acid-CoA-taurine *N*-acyltransferase

Catalase

Carnitine acetyltransferase (CAT)

Carnitine octanoyltransferase

Carnitine palmitoyltransferase

Cholesterol-7-α-hydroxylase

Dihydroalanosterol oxidase

Dihydroxyacetophosphate (DHAP) acyltransferase

Δ^3,Δ^2-Enol CoA isomerase

2-Enoyl-CoA hydratase

Epoxide hydrolase

Fatty acyl-CoA oxidase

Glucose-6-P-dehydrogenase

Glutaryl-CoA oxidase

(continued)

TABLE I (continued)

Glutathione-S-transferase

Glycerol-3-P-dehydrogenase

Glycolate oxidase

L-α-hydroxy acid oxidase

3-Hydroxy-3-methylglutaryl-CoA reductase

D-3-Hydroxyacyl-CoA dehydrogenase

Isocitrate dehydrogenase

3-Keto-acyl-CoA thiolase

Maleate synthase

NADPH cytochrome c (b5) reductase

NADPH cytochrome P-450 reductase

Palmitoyl-CoA hydrolase

Phosphogluconate dehydrogenase

L-Pipecolate oxidase

Polyamine oxidase

Prostaglandoyl-E_2/F_2-CoA oxidase

Serine-pyruvate-aminotransferase

Steroid-8 isomerase

Steroid-3 ketoreductase

Steroid-14 reductase

Superoide dismutase

Thiolase

Trihydroxycoprostanoyl-CoA oxidase

Ureidoglycolate lyase

Urate oxidase

Uricase (= urate oxidase)

Xanthine oxidase

(Makita et al., 1990b). It is not certain if this structure is comparable to that which has been postulated by biochemists (Labarca et al., 1987; Chen et al., 1988).

Together with the localization of Mg-ATPase on the limiting membrane, it is interesting that the membrane protein itself plays an important role in the transport of protein into the matrix (Wilson and King, 1991; Verheyden et al., 1992). Although receptors for protein transport must be integrated into the membrane, little, if any, information about such receptors is currently available.

Another item of interest from a structural point of view is the relationship between the limiting membrane and the fine network mentioned earlier. Catalase staining (Fig. 2) reveals a narrow space between the membrane and the matrix. However, matrix fibrils often appear to be continuous with the limiting membrane (Fig. 10). Catalase itself is known to cross the limiting membrane with or without the aid of a minipore structure.

Electron microscopists have reported that the endoplasmic reticulum (ER) running close to the peroxisome-limiting membrane often lacks ribosomes on the surface facing the peroxisome. Although the concept of peroxisome continuity with rough or smooth ER is not acceptable to biochemists (Hashimoto *et al.*, 1986), the topographic difference evident at the electron microscopic level should not be ignored.

C. Core or Crystalloid Structure

A dense core varying in size and shape is visible in the matrix of the peroxisome (Fig. 1). The shape is species specific. Peroxisomes in humans and primates are also devoid of a crystalloid structure, though such a structure has been reported occasionally. Microperoxisomes, which are distributed in practically all tissues, also lack a core structure. The core of renal peroxisomes is not as clearly crystal-like as that of hepatic peroxisomes, and often has a reticular structure. Only peroxisomes in the liver and kidney have compartments. The major component of the core is uricase, which has been histochemically localized (Hruban and Swift, 1964; Angermuller and Fahimi, 1986).

The structure of the crystalloid varies according to species, with the multiple tubules of hepatocyte peroxisomes (Fig. 2) being composed of ten subtubules. This 1 : 10 pattern is common in cats and rabbits, while that of guinea pigs is a hexagonal lattice structure (Tsukada *et al.*, 1971). Fahimi (1969) demonstrated catalase activity in the lumina of multitubules. The multitubules of hamster and mouse are smaller than those of rat. The outer diameters of a single tubule in rat, guinea pig, rabbit, cat, mouse, and hamster peroxisomes are approximately 45, 20, 15, 16.5, and 12 nm, respectively. On the whole they have a crystal form, but some are band-like.

It is interesting that crystallization of catalase results in the formation of tubules with monomolecular walls (Kiselev *et al.*, 1967). Calculation has shown that the crystal of ox liver catalase and catalase from erythrocytes has an external tubule diameter of about 31 nm and 42 nm, and a wall thickness of 6.5 nm and 6.5 nm, respectively. These lie within the range of single tubules in peroxisomes from different animals.

Despite the fact that the major component of the core crystalloid is uricase, other proteins are also present. The occasional appearance of a core in the human peroxisome which does not contain uricase indicates that the core is not exclusively crystalline uricase. Xanthine oxidase, D-amino acid oxidase, L-α-hydroxy acid oxidase, and some other enzymes are also localized in the crystalloid (Angermuller et al., 1987). Using an imaging plate, the author has found fibrils of the peroxisomal matrix inserted between multitubules of the core (Fig. 3) (Makita and Ogawa, 1991; Makita et al., 1994).

D. Marginal or Matrix Plates

Peroxisomes contain specific plates at the margin of the matrix. These structures are more frequent in renal than in hepatic peroxisomes. Zaar et al. (1991) isolated marginal plates from bovine renal peroxisomes and suggested that the plates might be aggregated L-α-hydroxy acid oxidase B.

Besides these intrinsic marginal plates, various types of matrix plates are induced in hepatic peroxisomes by peroxisome proliferators, and this will be discussed later. Gotoh et al. (1975) demonstrated the induction of matrix plates by feeding acetylsalicylic acid (aspirin), clofibrate, and dimethrin to rats. Interestingly, they were able to prevent aspirin from inducing matrix plates by administration of sodium citrate, but could not prevent induction by clofibrate or dimethrin. Feeding with propionate had a similar effect on rats fed aspirin but not on those fed clofibrate. Administration of aminotriazole prevented the formation of matrix plates in rats fed aspirin, although proliferation of peroxisomes occurred.

Hruban is of the opinion that matrix plates originate from coagulation of self-assembly of proteins in the matrix. In addition, they may be related to the fine tubule network in the matrix, as mentioned in Section II,B. Recently, we have been able to localize one to four matrix plates in renal peroxisomes of aspirin-fed Japanese monkeys (Wresdiyati and Makita, 1994, 1995).

E. Tubular Projections

Hepatic peroxisomes in aspirin-administered rats often have one or two prominent tubular projections in which matrix filaments or tubules are clearly visible. In the hepatic peroxisomes of rat (Makita et al., 1992; Hakoi et al., 1993; Makita and Hakoi, 1995) and dog (Hakoi and Makita, 1994), the formation of these tubular projections with aspirin administration appears to be dependent on dosage. Other peroxisome proliferators such as clofibrate, bezafibrate, and dimethrin also induce tubular projections on peroxisomes. Interestingly, the shape of these projections is specific to each hypolipidemic reagent to some extent, although the mechanism of projection formation has not been elucidated. It is possible that the alteration results from transformation and precipitation of protein in the matrix.

Another possibility is specific arrangement of peripheral circular profiles in the renal peroxisome (Beard and Novikoff, 1969; Barrett and Heideger, 1975). This type of circular profile may be a cross-section of a tubular projection. Though it is certain that these profiles are more frequent in renal than in hepatic peroxisomes, hepatocytes also contain a similar structure in which longitudinal and circular profiles of tubules coexist within a lipofuscin-like structure. This may indicate one aspect of the formation of tubular projections. Transformation of peroxisomes to tubular structures by clofibrate (Reddy and Svoboda, 1973) and formation of projections with a delicate helical fibrous coat by dimethrin (Hruban *et al.*, 1974) indicate the variety of mechanisms involved.

F. Bundles of Fibrils

Peculiar assemblies of short, curved, straight, circular, or long bundles of fibrils are found in peroxisomes as inclusion bodies (Figs. 3 and 4). They can be part of the matrix plate and tubules, but are distinguishable from the fine tubular network of the matrix.

Crystallization, precipitation, and assembly of fibers are typical modes of formation of inclusion bodies, but the formation of these fibrils in a specific form in response to a given peroxisomal proliferator remains to be investigated at the molecular level. As imaging plate views of these fibrils (Figs. 3 and 4) indicate, they are intermingled with the matrix network. Fibrils are also present in the projections discussed in Section II,E. Fibrils appear to protrude into the cytosol across the limiting membrane in some cases.

These variations in the form of matrix fibrils are reminiscent of much more variable types of inclusions evident in mitochondria in certain pathological and developmental states. For example, the large mitochondria in guinea pig oocytes contain long bundles of fibrils (Anderson *et al.*, 1970). Intramitochondrial crystals from the liver of patients with hepatitis and from other sources that resemble intraperoxisomal inclusions are only some examples of the many forms of these inclusions (Ghadially, 1988).

Needless to say, the substructures of peroxisomes described here are revealed by conventional electron microscopy. Advances in methodology have made it possible to visualize potential substructures. Immunohisto-chemistry has localized many enzymes and proteins which otherwise would not have been visualized in peroxisomes (Yokota, 1989; Usuda *et al.*, 1990). Application of quick-freezing and deep-etching methods (Ohno and Fujii, 1990), rapid-freezing and freeze-substitution fixation (Usuda *et al.*, 1990), freeze-fracture (Kryvi *et al.*, 1990) and imaging-plate methods (Mori *et al.*, 1990; Makita and Ogawa, 1991; Makita *et al.*, 1994) have also increased our understanding of peroxisome ultrastructure. The fine network of tubules in the matrix was visualized only after development of the high-resolution transmission electron microscope equipped with an imaging plate device. Recent progress in high-resolution cryo-electron microscopy (Kume, 1990; Fujiyoshi *et al.*, 1991; Goto *et al.*, 1994), which uses a high-resolution transmission electron microscope combined with a cryo-stage cooled with liquid helium, is expected to open a new field in the analysis of peroxisomal components. The atomic force microscope (AFM), one of the scanning probe microscopes (SPM), is another candi-date for advancing peroxisome research (Makita *et al.*, 1993). Instead of reconstructing the three-dimensional structure of peroxisomes from serial sections (Gorgas, 1987), between 16 and 25 optical focus sections can be obtained by laser-scanning confocal microscopy of a semithin epoxy sec-tion of hepatic tissue.

III. Enzymes in Peroxisomes

A. Main Enzymes

1. Catalase and Oxidases

As a marker enzyme of peroxisomes, catalase is the most abundant en-zyme. De Duve and Baudhuin (1966) illustrated the relationship between catalase and oxidases as follows;

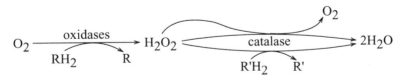

Peroxisomal oxidases (Hamilton, 1985) include D-amino acid oxidase (Angermuller and Fahimi, 1988), whose physiological significance is not clear, L-α-hydroxy acid oxidase, and urate oxidase or uricase, which is a major component of the core structure.

2. Enzymes Involved in Beta Oxidation of Fatty Acids

Acyl-CoA oxidase, enoyle-CoA hydratase: 3-hydroxyacyl-CoA dehydrogenase bifunctional enzyme, and 3-ketoacyl-CoA thiolase are enzymes of
fatty acid beta oxidation in peroxisomes (Lazarow and de Duve, 1976).
It is recognized that beta oxidation of fatty acids is the major function of
peroxisomes as well as mitochondria. However, the function of peroxisomes differs from that of mitochondria, as described in detail by Hashimoto (1982).

The reaction of acyl-CoA oxidase in mitochondria is coupled with the
electron transport system and can be inhibited by antimycin A or cyanide,
whereas in peroxisomes it yields H_2O_2 and it is not inhibited by cyanide.
The substrate of peroxisomal acyl-CoA oxidase is acyl-CoA longer than
8 carbons, and beta oxidation in peroxisomes plays a role in the decomposi-
tion of long-chain fatty acids to short ones.

3. Glyoxylic Acid Cycle

Glyoxisomes of plants and peroxisomes of yeasts contain enzymes related
to the glyoxylic acid cycle (Tolbert, 1981), but they are not present in the
peroxisomes of animals.

4. Enzymes for Plasmalogen Synthesis

Dihydroxyacetone phosphate acyltransferase and acyldihydroxyacetone
phosphate acyltransferase are enzymes involved in the formation of plas-
malogen. Both are known to be localized in the peroxisomal membrane
and in microsomes.

5. Serine : Pyruvate Aminotransferase (SPT) (EC 2.6.1.51) or Alanine : Glyoxylate Aminotransferase (SPT/AGT) (EC 2.6.1.44)

In rat liver, SPT is localized both in mitochondria (SPTm) and peroxisomes
(SPTp). SPTm is synthesized from a large mRNA as a precursor 45-kDa
translation product. On the other hand, SPTp is synthesized from a smaller
mRNA as a 43-kDa translation product. The smaller mRNA lacks a portion
of the 5-terminal sequence of the large mRNA which codes for the mitochondrial targeting the N-terminal extension peptide of the precursor.
Interestingly, mistargeting of SPTp to mitochondria causes peroxisomal
disease (primary hyperoxaluria type I). The ratio of SPTm to SPTp in rat
liver is about 1 : 1. Evolution of SPT has occurred rapidly and it is found
only in mammals, although there are considerable differences among spe-

cies. The hypolipidemic reagent, clofibrate, induces SPTp but not SPTm. Glucagon induces SPTm in rat, hamster, and mouse, but not SPTp in rat, hamster, mouse, and rabbit. Mori *et al.* (1991) mapped the rat SPT gene to the q34 to q36 region of the chromosome. Molecular aspects of species differences, and differences in induction by clofibrate, glucagon, and vitamin B_6 should be clarified in the near future.

6. 3-Hydroxy 3-Methylglutaryl CoA Reductase

The peroxisomal matrix contains this enzyme, which is identical to that in the endoplasmic reticulum membrane.

7. Enzymes and Proteins Localized in Peroxisomes

Major proteins and enzymes in peroxisomes investigated so far are listed in Table 1 (Tolbert, 1981; Lazarow and Fujiki, 1985; Goldfischer and Reddy, 1984; Soto *et al.*, 1993a; Guttierez *et al.*, 1988; Appelkvist *et al.*, 1990).

B. Cytochemical and Immunocytochemical Localization of Enzymes and Proteins in Peroxisomes

Compared with the biochemical identification of enzymes in peroxisomes (van den Bosch *et al.*, 1992), their cytochemical or immunocytochemical detection has been rather limited. Table II gives a summary of enzymes localized so far in peroxisomes. Typical enzymes are catalase and uricase. Yokota (1989), Yokota and Asayama (1992a,b), Usuda *et al.* (1990), Bendayan and Reddy (1982), and others have localized many other enzymes immunocytochemically in peroxisomes.

In this section, Mg^{2+}-ATPase as an example of an enzyme located on the limiting membrane, and carnitine acetyltransferase as an enzyme in the matrix, will be described in detail, since both were localized at the electron microscopic level in this laboratory (Makita *et al.*, 1990; Hakoi and Makita, 1994; Ohue and Makita, 1994).

1. Localization of Carnitine Acetyltransferase in the Matrix of the Rat Hepatic Peroxisome

CAT is one of the enzymes involved in the beta oxidation of fatty acids. Histochemically, it was first localized in the mitochondria of cardiac muscle (Higgins and Barrnett, 1970) and then in the mitochondria of liver (Makita and Sandborn, 1971) and skeletal muscle (Makita *et al.*, 1973).

TABLE II

Cytochemical and Immunocytochemical Localization of Enzymes in the Rat Peroxisomes

Enzymes	Hepatocyte			Renal tubule		
	Memb.	Core	Matrix	Memb.	Center	Periphery
Catalase	−	−	+	−	−	+
Mg-ATPase	+	−	±	+	±	±
CAT	−	−	±	−	−	−
Uricase	−	+	−	−	−	−
α-Hydroxy acid oxidase	−	−	+	−	±	+
D-Amino acid oxidase	−	−	+	−	±	+
SPT	−	−	+	−	±	+
Acyl-CoA oxidase	−	−	+	−	±	+
Bifunctional protein	−	−	+	−	±	+
Thiolase	−	−	+	−	±	+
Acyl-CoA synthetase	±	−	−	+	−	−
2,4-Dienoyl-CoA reductase	−	−	+			
Δ^3,Δ^2,-Enoyl-CoA isomerase	−	−	+			
70 kDa, 26 kDa, 22 kDa,						
15 kDa	±	−	−			

The main site of reaction is the outer surface of the inner mitochondrial membrane. The components of the incubation medium and inhibitors are shown in Table III.

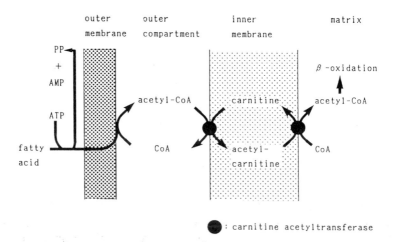

FIG. 11   Diagram of localization of carnitine acetyltransferase in mitochrondria.

TABLE III

Incubation Medium for Cytochemical Detection of
Carnitine Acetyltransferase[a]

Potassium ferricyanide	5.0 mg
Uranyl acetate	2.5 mg
Acetyl CoA	2.0 mg
Carnitine	4.0 mg
Dextrose	100.0 mg
0.05 M Maleate buffer (pH 7.0)	2.5 ml

[a] Inhibitor; $HgCl_2$ $4 \times 10^{-4}M$. (From Higgins and Barnett, 1970.)

CAT catalyses the following reaction;

$$\text{carnitine} + \text{acetyl CoA} \rightleftharpoons \text{acetylcarnitine} + \text{CoA}$$

CAT I is detectable on the outer surface of the inner mitochondrial membrane, while CAT II is localized on the matrix side of the inner mitochondrial membrane (Brosnan et al., 1973; Atkins and Claudinin, 1990).

Brdiczka et al. (1969) and Colucci and Gandour (1988) reported the presence of CAT in the mitochondrial matrix. A study by Markwell et al. (1973) showed that approximately 53% was present in mitochondria, 14% in peroxisomes Bieber et al. (1982), and 34% in the lipid-rich membrane fraction of mammalian liver and kidney. CAT has been extracted from chemically induced peroxisomes (Miyazawa et al., 1983). CAT activity can be enhanced up to 90-fold by clofibrate (Tosh et al., 1989), 35.8-fold by bezafibrate (Watanabe et al., 1989), and 17-fold by fenofibrate (Henninger et al., 1987). The enhancement is more significant in males than in females (Watanabe et al., 1989). In spite of this significant activation by hypolipidemic reagents, CAT has been difficult to localize in peroxisomes. Only after oral administration of bezafibrate for 2 weeks, was CAT found to be localized in the peroxisome matrix in rat hepatocytes (Ohue and Makita, 1994). CAT activity in hepatic homogenates from male, female, and castrated male and female rats is shown in Table IV.

2. Cytochemical Localization of Mg-ATPase and Ca-ATPase in Peroxisomes of Rat Hepatocytes

Cytochemical localization of ATPase by cerium method (van Norden and Frederinks, 1993) on the limiting membrane of peroxisomes was first described by Douma et al. (1987) in yeasts. ATPase in rat hepatocyte peroxisomes (Makita, 1988; Makita et al., 1990a, 1992) and canine hepatocytes (Hakoi and Makita, 1994) was found exclusively on the limiting

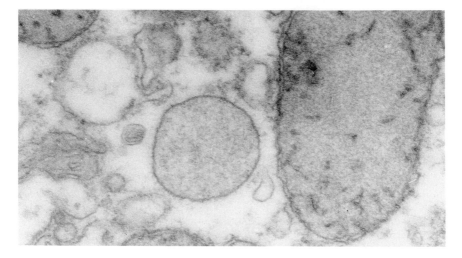

FIG. 12 Carnitine acetyltransferase (CAT) activity in a peroxisome (P) and in mitochondria. ×61,215.

membrane (Figs. 12, 13). Ca-ATPase was also localized at the same site but in smaller amounts compared with Mg-ATPase.

The reaction medium is given in Table V. This was a modification of

TABLE IV

Effects of Bezafibrate on CAT Activity

Animal	Liver weight (%)[a] Body weight (%)	CAT activity (nmol/min/g liver)[a] Liver	Brown adipose tissue
Control			
Male	4.69 ± 0.36	125.64 ± 19.92	129.16 ± 27.46
Female	4.04 ± 0.43	174.58 ± 47.64	122.34 ± 33.14
Castrated male	4.57 ± 0.20	231.88 ± 63.58	428.14 ± 118.77
Castrated female	5.06 ± 0.23	224.68 ± 45.58	384.55 ± 111.14
1 week			
Male	7.30 ± 0.21**	1381.18 ± 341.17*	696.92 ± 132.36*
Female	5.12 ± 0.27**	800.46 ± 145.50**	534.16 ± 163.28*
Castrated male	6.50 ± 0.33**	1368.24 ± 263.51**	467.56 ± 150.87
Castrated female	6.29 ± 0.26**	913.88 ± 122.81**	366.59 ± 116.62
2 weeks			
Male	8.39 ± 0.58**	1503.28 ± 429.59*	772.20 ± 194.94*
Female	5.53 ± 0.32**	670.82 ± 190.24*	249.38 ± 67.76***
Castrated male	6.82 ± 0.57**	1942.59 ± 148.81**	395.77 ± 80.44
Castrated female	7.28 ± 0.23**	1241.88 ± 455.82*	366.59 ± 107.52

[a] Significantly different from control group. *$P < .01$, **$P < .001$, ***$P < .05$.

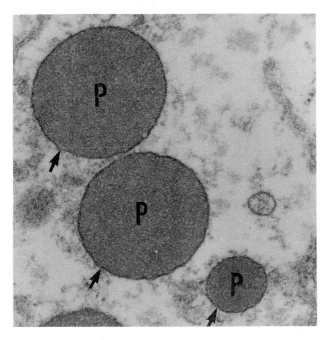

FIG. 13   Mg-ATPase activity on peroxisomes in a rat hepatocyte after 2 weeks' administration of clofibrate, a peroxisome proliferator. ×25,000.

TABLE V

Incubation Medium for Cytochemical Localization of Mg-ATPase and Ca-ATPase

0.2 M Tris–HCl buffer at pH 9.0	13 ml	(130 mM)
(glycine–NaOH buffer is also applicable)		
$MgSO_4 7H_2O$ (for Mg-ATPase)	49 mg	(Mg: 10 mM)
$CaCl_2$ (for Ca-ATPase)	23 mg	(Ca: 10 mg)
ATP2Na (substrate)	30 mg	(2.5 mM)
0.5% lead citrate dissolved in 50 mM NaOH	5 ml	(Pb: 2.5 mM)
Dimethylsulfoxide (DMSO)	1 ml	(5% v/v)
2 mM 2,4-dinitrophenol (DNP)	1 ml	(0.1 mM)
Total (final pH 8.8 to 9.0)	20 ml	

Inhibitors:
    Ouabain for Na-K-ATPase   10 mM
    Levamisole for alkaline phosphatase   2.5 mM
    p-chloromercuric benzoate (PCMB) for Mg-   10 mM
        ATPase
    Oligomycin for mitochondrial ATPase   0.1 mM
Control medium
    1. Substrate-free medium
    2. Medium in which ATP2Na was replaced with β-glycerophosphate
    3. Medium in which $MgSO_4$ was omitted

the medium used for the creatine phosphokinase reaction, and a stable reaction was demonstrated on the inner surface (matrix side) of the inner mitochondrial membrane (Makita *et al.*, 1990b; Makita, 1993).

Although the reaction product of either Mg-ATPase or Ca-ATPase in peroxisomes is most conspicuous on the limiting membrane, subtle reaction products are also distributed along the tubular network in the matrix (Fig. 10). As mentioned before, the requirement for, and presence of ATP on the limiting membrane as well as in the matrix have been postulated (Whitney and Bellion, 1991), and thus localization of ATPase at these two sites is not illogical.

NADPH, Mg, and ATP-dependent hydroxylase is known to exist in rat liver peroxisomes (Thompson and Krisans, 1985). Del Valle *et al.* (1988) detected ATPase in rat liver peroxisomes and suggested that it created a pH gradient (pH 5.8 to 6.0 in the peroxisome and pH 7.1 in the cytosol) across the limiting membrane. In isolated peroxisomes, enhancement of peroxisomal acyl-CoA oxidation by ATP is abolished when the limiting membrane is disrupted, indicating that an ATP-related complex is localized in the membrane (Leighton *et al.*, 1982).

In bile acid formation, liver peroxisomes play a role in the side-chain oxidation of 26-hydroxycholesterol and other cholesterol derivatives. The hydroxylase that catalyzes this 26-hydroxylation, probably a P-450 cytochrome, requires the NADPH-generating system, Mg^{2+} and also ATP. Translocation of acyl-CoA oxidase into peroxisomes requires ATP hydrolysis (Imanaka *et al.*, 1987). The 701-kDa peroxisomal membrane protein is a member of the ATP-binding protein family (Kamijyo *et al.*, 1990). All these data support the presence of ATPase in the limiting membrane and the matrix of liver peroxisomes.

IV. Important Clues Postulated by Various Experiments with Peroxisome Proliferators

It is well known that peroxisomes in a variety of species, including humans and nonhuman primates, can be increased in number using many kinds of hypolipidemic reagents, phthalate ester plasticizer, and herbicide. There are several review articles on the effects of hypolipidemic reagents on hepatic peroxisomes (Svoboda *et al.*, 1967; Baumgart *et al.*, 1987; Alexson *et al.*, 1985; Kolde *et al.*, 1976; Cohen and Grasso, 1981; Grasso, 1993; Reddy and Lalwani, 1983; Moody, 1974; Moody *et al.*, 1992). A possible carcinogenic effect was one of the areas of intensive research on these chemicals in clinical use.

Apart from their clinical significance (Moser *et al.*, 1991; Santos *et al.*,

1988b; Simozawa *et al.*, 1992b; Small *et al.*, 1982, 1988b), the hypothesis on the molecular kinetics of induction of hepatic peroxisomes postulated important factors concerning morphogenesis of this cell organelle (Flatmark *et al.*, 1982); differences with sex, age, species and strains; influence of steroid hormones (Dahl, 1971a,b), thyroid hormones, and cytochrome P-450; and induction of enzymes involved in beta oxidation of fatty acids. Since experiments in my laboratory have been limited to rats, dogs, and monkeys using clofibrate, bezafibrat (Fahimi *et al.*, 1982; Ohue and Makita, 1994), and aspirin, the following discussion is based on the data obtained from these three peroxisome proliferators.

A. Rapid Disappearance or Decrease of Peroxisomes after Withdrawal of Proliferators

Despite many attempts to estimate the increase of peroxisomes, little attention has been focused on why withdrawal of a proliferator results in rapid disappearance of proliferated peroxisomes without any remarkable membrane debris remaining in the cytosol. As shown in Fig. 14, the number of peroxisomes is reduced to almost a normal level only 1 week after withdrawal of aspirin. Other peroxisome proliferators such as clofibrate and bezafibrate are similar to aspirin in this respect.

1. Possible Existence of a Substance That Suppresses the Proliferation of Peroxisomes

A suppressor present in intact hepatocytes may prevent peroxisomes from proliferating. If such a substance exists, its characteristics remain to be elucidated. It is also important to consider the mechanism by which the putative suppressor can distinguish the newly induced peroxisomes from existing mature ones.

In general, induced peroxisomes are irregular in outline, with characteristic projections, as seen in rats after aspirin administration. The limiting membrane of proliferated peroxisomes appears to be more permeable to substances in the cytosol and also in the matrix, as judged by leakage of catalase revealed by cytochemical staining. Biochemical data also support this concept. Subsequently, the matrix of induced peroxisomes is less dark than preexisting mature ones, and thus the matrix plates, fibrils, and fine tubular network are much more conspicuous. The core or crystalloid structure of proliferated peroxisomes in rat hepatocytes is not conspicuous or is absent. All these morphological differences seen in peroxisomes induced by a proliferator imply a difference in molecular organization of the membrane, core, and matrix between mature and newly

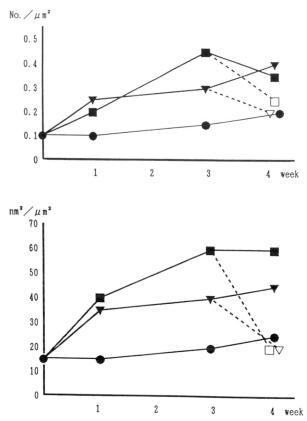

FIG. 14   Change of number (upper) and size (lower) of hepatic peroxisomes per square micrometer of cytoplasm with oral administration of clofibrate or asiprin and withdrawal of these peroxisome proliferators. ●—●, Basal diet; ■—■, 0.25% clofibrate; ■---□, 0.25% clofibrate → basal diet; ▼—▼, 1.0% aspirin; ▼---▽, 1.0% aspirin → basal diet.

induced peroxisomes (Crane *et al.*, 1985, 1990; Crane and Masters, 1986).

a. Phenobarbital as an Inhibitor or Suppressor of Peroxisome Proliferation   In the search for a suppressor of peroxisome proliferation, we focused on phenobarbital (PB), which is a well-known proliferator of smooth endoplasmic reticulum (s-ER). In fact, this reagent was the first chemical found to induce proliferation of cell organelles (Jirtle *et al.*, 1991). The inhibitory effect of PB on peroxisome biogenesis has been reported in rat (Staubli *et al.*, 1969) and mouse (Kanamura *et al.*, 1988) liver. The effect of PB on peroxisomes induced by clofibrate has not been reported.

Since clofibrate can induce proliferation of s-ER under certain conditions, it was a critical factor in demonstrating the suppression effect on peroxisome proliferation to select the timing of PB administration (Hakoi *et al.*, 1993; Hakoi and Makita, 1994).

Male rats were fed a diet containing 0.25% clofibrate for 14 days, and simultaneously 50 mg/kg body weight sodium PB was administered daily by gavage. The number and size of peroxisomes per square micrometer of cytoplasm were as follows;

	number/μm^2	area/μm^2
Control	0.14 ± 0.03	14.31 ± 2.92
PB	0.14 ± 0.04	14.30 ± 1.98
Clofibrate	0.55 ± 0.07	87.56 ± 13.90
Clofibrate + PB	0.15 ± 0.04	10.52 ± 3.14

PB alone does not affect the number, size, and morphology of peroxisomes, but when it is combined with clofibrate it prohibits the proliferation of peroxisomes and also increases the number of microperoxisomes. A similar increase of microperoxisomes occurs after partial hepatectomy, administration of triiodothyronine, and starvation stress (Makita *et al.*, to be published). The experiment using PB + clofibrate revealed an increase of microperoxisomes and suppression of proliferation of ordinary peroxisomes. In this context, the involvement of the thyroid gland (McClain *et al.*, 1989) in the inhibition of peroxisome proliferation by PB may be worth considering here.

Mitochondrial glycero-3-phosphate dehydrogenase in the liver and in a number of other tissues is regulated by thyroid hormone *in vivo* and *in vitro* (Shoemaker and Yamazaki, 1991). With T3 treatment, an increase in translatable mRNA for glycero-3-phosphate dehydrogenase has been demonstrated. T3 causes hyperthyroidism and increases the number of peroxisomes in hepatocytes (Fringes and Reith, 1982). Clofibrate stimulates hepatic glycero-3-phosphate dehydrogenase, although this induction is independent of thyroid hormone (Shoemaker and Yamazaki, 1991). Thyroidectomy abolished the hypolipidemic effect of clofibrate (Svoboda *et al.*, 1969). The interrelationship between hypolipidemic reagents such as clofibrate, thyroid hormone, and PB is one of the key factors involved in the mechanism of inhibition of peroxisome proliferation. It has been reported that thyroid hormone and peroxisome proliferators do not cross-react at the level of their nuclear receptors (Evans, 1988; Castelein *et al.*, 1993).

In cultured hepatocytes, a peroxisome proliferator, nafenopin, suppresses apoptosis (Bayly *et al.*, 1994), and this may explain the hepatocarcinogenicity of peroxisome proliferators. A number of explanations for the hepatocarcinogenicity of peroxisome proliferators have been postulated (Moody *et al.*, 1992; Rao and Reddy, 1991; Michaelopoulos, 1991; Green *et al.*, 1992). Complementary to sustained proliferation, the suppression of hepatocyte apoptosis would allow the survival of tumorgenic cells that would otherwise be eliminated (Bursh *et al.*, 1992). Nongenotoxic carcinogens such as PB and cyproterone acetate induce hyperplasia of rat hepatocytes, and their withdrawal causes regression of the hyperplastic liver with an increase of apoptosis. Nafenoprin, a peroxisome proliferator, as mentioned before, inhibits the apoptosis that follows withdrawal of cyproterone acetate. The liver growth regulator transforming growth factor (TGF b_1) induces hepatocyte apoptosis (Lin and Chou, 1992; Oberhammer *et al.*, 1992). Nefenopin protects hepatocytes from TGF b_1-induced apoptosis (Bayly *et al.*, 1994).

b. Other Inhibitors of Peroxisome Proliferation Induced by Hypolipidemic Reagents   Although they may not be directly related to the suppression of peroxisome proliferation, shortly after the withdrawal of peroxisome proliferators such as clofibrate, bezafibrate, and aspirin, some reagents have been recognized to suppress the effect of proliferators. Nicardipine, a well-known calcium antagonist, suppresses microsomal omega oxidation of laurate in clofibrate-administered rats (Itoga *et al.*, 1990). Although 2 weeks of clofibrate administration (0.25% in diet) alone induced a 15.3-fold increase of peroxisomes, coadministration of clofibrate and nicardipine (100 mg/kg body weight) resulted in a sixfold induction. An induction of beta oxidation by more than 50% in peroxisomes and carnitine acetyltransferase activity were also suppressed by nicardipine. Clofibrate reduces NADPH-cytochrome c reductase, whereas nicardipine increases it. Furthermore, clofibrate increases hepatic cytochrome P-450 and nicardipine increases both P-450 and b_5 (Watanabe *et al.*, 1991). The suppression of clofibrate-induced peroxisome proliferation and omega oxidation by nicardipine, therefore, indicates the involvement of microsomes in induction of enzymes in peroxisomes. Watanabe *et al.* (1992) have also reported that both trifluoperazine, a calmodulin antagonist, and H-7, a potent inhibitor of protein kinase C, suppressed the induction by clofibrate of carnitine acetyltransferase and the cyanide-insensitive fatty acyl-CoA oxidizing system in rat hepatocytes. According to them, there were two phases in the induction: 0 to 3 days after administration of clofibrate, and 3 to 5 days after the treatment. The suppression by these two antagonists was significant only in the later phase. Induction of catalase, a bifunctional enzyme, and the 69-kDa integral membrane protein of the peroxisome, may not occur through similar processes. Compared with the bifunctional enzyme, which is inhibited by nicardipine, the induction of the 69-kDa membrane protein by clofibrate is delayed.

All of these studies on inhibitors or suppressors of peroxisome proliferation should provide clues to the molecular mechanism of peroxisome induction, including the existence of receptors (Dautry-Varsat and Lodish, 1984), second messengers, hepatic coenzymes, and still-unknown factors.

How the putative depressor substance detects the difference between newly induced and existing peroxisomes after withdrawal of the proliferator is the key reaction explaining the sudden decrease in number of peroxisomes to a normal level.

2. Possible Explanation of the Sudden Disappearance of Round Peroxisomes

In addition to the existence of a suppressor as discussed above, the concept of the peroxisome reticulum (Lazarow *et al.*, 1982) would facilitate interpretation of the sudden disappearance of peroxisomes after withdrawal of a proliferator. The transformation of a tubular to a round type will be discussed in Section VI. The substance or factor which initiates and controls the transformation may or may not be identical to a peroxisome proliferator. The slight difference between the peroxisome reticulum postulated by Lazarow *et al.* (1982) and the concept presented here is in the tubular part. The tubular parts often induced by aspirin, clofibrate, and gemfibrozil are somewhat irregular and thick in shape whereas the tubules in the reticulum under consideration are rather uniform and very thin, as can be seen by scanning electron microscopy (Fig. 15). A subtle thin tubular component, which is often invisible under normal conditions and also after withdrawal of proliferators, would explain the dynamic change of peroxisome from round to tubular.

B. Factors That May Be Involved in Sex Differences in the Response of Peroxisomes to Proliferators

1. Steroid Hormones

As indicated by the set of data from experiments including bezafibrate administration to male and female rats (Table IV), males are more sensitive to peroxisome proliferators, although the extent of the increase in peroxisomes varies with the proliferator used and the amount administered. Naturally, steroid hormones such as androgen and estrogen will interfere with the response of peroxisomes to proliferators.

Our preliminary experimental data obtained using castrated male and

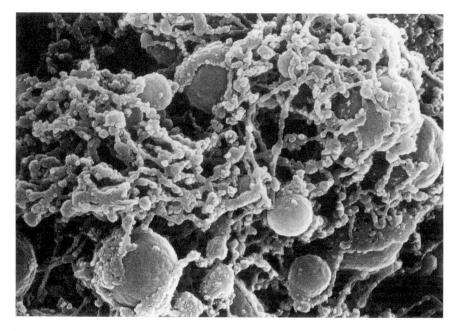

FIG. 15   A scanning electron micrograph of a network of tubules and vesicular peroxisomes in rat hepatocyte. ×40,150 (Courtesy of Dr. M. Hamasaki, Kurume, Japan.)

castrated female rats (Table IV) indicate that even after castration males are more responsive to bezafibrate. The preexistence of testosterone, for example, may preserve the ability of males to react to this peroxisome proliferator.

2. Cytochrome P-450

A tentative candidate for the sex difference in hepatic cytosol is cytochrome P-450$_{male}$ and P-450$_{female}$ (Kato and Yamazoe, 1992) in rats and mice. Both increase at 25 to 30 days of age and reach a plateau at 40 to 45 days (Maeda *et al.*, 1984). In castrated males, P-450$_{male}$ decreases and P-450$_{female}$ is not detectable. With testosterone administration, the P-450$_{male}$ level recovers. On the other hand, estradiol administration induces P-450$_{female}$ and abolishes P-450$_{male}$. In castrated females, testosterone induces P-450$_{male}$ and abolishes P-450$_{female}$ (Kamataki *et al.*, 1983).

As reviewed by Gibson *et al.* (1993) and Green *et al.* (1993), microsomal P-450 is responsible for fatty acid hydroxylation. Milton *et al.* (1990) *in vivo* and Bieri *et al.* (1991) *in vitro* observed that induction of hepatic cytochrome P-450 by peroxisome proliferators preceded the induction of peroxisomal enzymes. Many cytochrome P-450s are induced by several xenobiotics, including barbiturates (Porter and Coon, 1991; Guengerich, 1991). Among the P-450s, genes of the P-4504 family are most susceptible to induction by peroxisome proliferators, and individual genes are associated with endogenous lipid metabolism (Gibson, 1989).

P-4504A1 induction and peroxisome proliferation are closely related. Species differences in the response to P4504A1 induction by peroxisome proliferators are closely related to increases in peroxisomal beta oxidation enzymes. The rat and mouse are most responsive; the hamster and rabbit show an intermediate response; and the guinea pig, monkey, and marmoset are almost nonresponsive. According to Lock et al. (1989), peroxisome proliferators initially inhibit fatty acid oxidation by a dual mechanism involving inhibition of carnitine acetyltransferase activity in mitochondria or sequestration of essential CoA by the proliferator itself. Then medium and long-chain fatty acids accumulate in the cytoplasm and induction of P-4504A1 and associated fatty acid omega-hydroxylase occur, followed by accumulation of long-chain dicarboxylic acids. As mitochondria cannot efficient beta oxidize these, peroxisomal beta oxidation would be induced.

Coregulation of the P-4504A1 and acyl-CoA oxidase genes may be mediated through the mouse-derived peroxisome-proliferator-activated receptor (PPAR), which consists of a DNA-binding domain of about 70 amino acids and a ligand-binding domain of approximately 200–250 amino acids (Issemann and Green, 1990; Green, 1992). Inspection of the computer-modeled PPAR ligand-binding domain has revealed a likely binding site containing amino acid residues complementary with peroxisome proliferators (Lewis and Lake, 1993). Osumi et al. (1991a) have indicated the presence of multiple regulatory elements in the acyl-CoA gene.

PPAR is located in the nucleus and recognizes the TGACCT motif (to be discussed later) located upstream from the target genes. The activated PPAR, the receptor, would enhance or repress gene transcription to allow peroxisome proliferation. A three-dimensional model of PPAR (Lake and Lewis, 1993) has suggested the interaction of clofibrate with a putative binding site in a portion of the PPAR ligand-binding domain and the putative DNA-binding domain containing characteristic zinc fingers.

A peroxisome proliferator-bindng protein in rat liver (Lalwani et al., 1983, 1987) has been identified as a protein related to heat shock protein (hsp). The presence of this protein and the fact that peroxisome proliferators activate gene transcription suggest that the action of proliferator is similar to that of a steroid hormone (Reddy and Rao, 1986).

Taking this evidence together with the fact that cytochrome P-450 4AL is elevated 10- to 30-fold in response to a peroxisome proliferator (Orton and Parker, 1982; Gibson et al., 1982), clarification of the difference in response between males and females is important in order to solve the molecular mechanism of peroxisome proliferation.

The hsp70 heat-shock or stress protein is located on the outside of the peroxisomal membrane (Walton *et al.*, 1994). As reviewed by Welch (1991), hsp70 is intimately involved in protein folding and transport.

The nuclear hormone receptor superfamily comprises more than 25 mammalian genes that encode receptors for steroid hormones, thyroid hormones, vitamin D_3 and retinoic acid (Carson-Jurica *et al.*, 1990). Nuclear hormone receptors are DNA-binding proteins that recognize short DNA motifs (hormone response elements). One of the hormone receptor family, PPAR, can be activated by hypolipidemic agents, as mentioned earlier. PPAR is found not only in liver but also in brown adipose tissue, kidney, and cardiac muscle. It is weakly expressed in skeletal muscle, small intestine, testis, and thymus (Issemann and Green, 1990). PPAR would be activated by binding of a peroxisome proliferator, thereafter recognizing specific DNA sequence motifs and activating transcription of specific genes.

Recent evidence has shown that members of the hsp70 family bind to and stabilize nascent polypeptides when they emerge from the ribosome. The cytoplasmic hsp70 family is required for post-translational transport of protein across the membrane of the ER, mitochondria, nucleus, and lysosomes (Mackiewicz *et al.*, 1993).

Another class of stress proteins, hsp60, is located in the cytoplasm and mitochondria. These proteins are called "chaperonins" (Ellis, 1987), as will be discussed later. Chaperonin is a member of chaperones. Chaperonins participate directly in protein folding and the assembly of oligomeric protein complexes (Koll *et al.*, 1992). Both hsp60 and hsp70 are ATP-binding proteins.

The import of proteins into the peroxisome is an ATP-dependent process (Imanaka *et al.*, 1987) and requires N-ethylmaleimide (NEM)-insensitive cytosolic factors (Wendland and Subramani, 1993). It is not known whether members of the hsp70 family participate in the import of protein into peroxisomes or hsp60-like proteins are involved in the assembly of protein oligomers. Walton *et al.* (1994) have postulated two models for the role of hsp70 molecules in stimulating peroxisomal protein import. In the first model, hsp70 interacts directly with the peroxisome targeting signal and inhibits its folding to allow access of the signal to the targeting signal receptor and translocation machinery. This is, hsp70 does not unfold the transported protein or its peroxisomal targeting signal but simply stabilizes the signal when it is exposed during thermal fluctuations of the protein structure. In the second model, hsp70 mediates stimulation of peroxisomal protein import without interacting directly with the transported protein.

As reviewed by Ellis and van der Vies (1991), molecular chaperones are defined as a family of unrelated classes of protein that mediate the correct assembly of other polypeptides, but are not themselves components of the final functional structures. Chaperones convey no strict information for protein assembly, but kinetically assist self-assembly. Of the several types of chaperones, two types are sequence related to each other. The larger type is chaperonin 60 and the smaller type is chaperonin 10. Mitochondrial chaperonin 10 binds to chaperonin 60 in the presence of Mg-ATP and suppresses the ATPase activity of chaperonin 60. These responses to ATP are involved in the mechanism of action of mitochondrial chaperonins. Chaperonin 60 binds to unfolded forms of many polypeptides during processes such as protein synthesis and protein transport. This binding maintains the polypeptides in a state that (1) prevents them from misfolding, (2) assists them in arriving at their correct intra- and intermolecular folding pattern, (3) allows them to be transported more readily, and (4) permits proteases to degrade them. For some of these processes, chaperonin 10 and Mg-ATP are also required. In this context, localization of Mg-ATPase on the limiting membrane of peroxisomes may have functional significance, although no chaperone molecules of peroxisomes have been reported so far.

C. Age-Dependent Sensitivity of Peroxisomes to Proliferators

The age-dependent efficacy of clofibrate and other hypolipidemic reagents has been investigated in several laboratories, including ours. In male F344 rats aged 8, 52, and 117 weeks, susceptibility to clofibrate was as follows;

	8 weeks	52 weeks	117 weeks
Hepatomegaly	Significant	Significant	Slight
Total cholesterol	Decreased	Decreased	Decreased
Triglyceride	Decreased	Decreased	Decreased
Total cytochrome P-450	Increased	Increased	Slightly increased
Microsomal omega hydroxylation	Induced	Induced	Slightly induced
Peroxisomal beta oxidation	Induced	Induced	Slightly induced
Smooth ER	Proliferated	Proliferated	Slightly increased
Proliferation of peroxisomes	Significant	Significant	Slight

Table VI gives another set of data between 8 and 83 weeks (Yamoto *et al.,* 1995). As naturally expected, old animals showed a lower response to clofibrate. Current studies performed in this laboratory have indicated that the efficacy of clofibrate in mice is less evident than in rats, and senescence-accelerated mice (SAM) are more responsive to clofibrate

TABLE VI

Age-Dependent Sensitivity to Clofibrate

Age	Total cholesterol (mg/dl)	Triglyceride (mg/dl)	Cytochrome P-450 (nmol/mg protein)
8 weeks (control)	56.0	125.8	0.46
8 weeks (experimental)	43.8	56.8	0.66
83 weeks (control)	169.5	164.0	0.35
83 weeks (experimental)	182.0	123.5	0.29

	7-Alkoxycoumarin dealkylase (nmol/min/g liver)		Catalase (μ/mg protein)
	7-Methoxycoumarin	7-Ethoxycoumarin	
8 weeks (control)	44.44	46.80	162.8
8 weeks (experimental)	50.52	65.08	159.0
83 weeks (control)	25.45	20.83	170.0
83 weeks (experimental)	27.35	21.75	143.0

than ordinary aged mice. The interrelationship between aging and the proliferation of peroxisomes with hypolipidemic reagents requires further investigation.

D. Differences in Response to Clofibrate between Rats of Different Strains

Even within the same species of laboratory animal, such as the rat, different strains show different degrees of reaction to hypolipidemic reagents. An example of data using three different strains of rats—F344, Sprague-Dawley (SD), and Wistar-Lewis (WL)—is as follows (Yamoto et al., 1991).

Animals were fed clofibrate (150 mg/kg/day) for 14 days, and their blood and liver were analyzed on the 15th day.

	F344	SD	WL
Decrease of triglyceride	+	+	+ +
Increase of total cytochrome P-450	+	+	+
Hepatomegaly	+	+	+
Decrease of total cholesterol	+	+	+
7-Alkoxycoumarin O-dealkylase	+	+	+ +
Induction of carnitine acetyltransferase	+ +	+	+ +
Induction of KCN-insensitive fatty acid oxidation system (FAOS)	+ +	+	+ +
Proliferation of peroxisomes	+ +	+	+ +
Proliferation of S-ER	+ +	+	+ +

As shown by only the small number of examples in this section, experiments on peroxisome proliferators have provided a variety of important clues to the molecular mechanism of peroxisome morphogenesis. Many reviews have already described a number of postulated working hypotheses, but at present only tentative explanations are available. However, the increasing amount of molecular data on events involving peroxisomes suggests that we will soon be able to explain how they bind to receptors, if any, on the limiting membrane, followed by incorporation into the matrix. Although no studies along these lines have been conducted in this laboratory, some of them described in detail in other reviews will be briefly introduced in the following few sections as a basis for discussing current perspectives on peroxisome research.

V. Targeting, Binding to Receptor, Transmembrane, and Matrical Molecules

A. Peroxisomal Topogenic Signal

1. Ser-Lys-Leu (SKL) Motif

As the carboxyl-terminal Ser-Lys-Leu-related tripeptide of peroxisomal protein is regarded as a minimal peroxisome-targeting signal (Miura *et al.*, 1992), the peroxisome-targeting SKL motif sequences are present at the carboxyl-terminals of several proteins (Table VII). Not all SKL motif sequences are at the carboxyl-terminal; some are located internally (Subramani, 1992, 1993).

TABLE VII
SKL Motif

	⌐S—K—L¬
Acyl-CoA oxidase, rat	-Leu-Gly-Ser-Lys-Leu
Bifunctional protein, rat	-His-Gly-Ser-Lys-Leu
Luciferase, firefly	-Gly-Lys-Ser-Lys-Leu
Maltase synthase, cucumber	-Gly-Leu-Ser-Lys-Leu
Uricase, rat	-Leu-Pro-Ser-Arg-Leu
d-Amino acid oxidase, human, mouse, pig	-Pro-Pro-Ser-His-Leu
Nonspecific lipid transfer protein	
(sterol carrier protein) rat	-Asp-Lys-Ala-Lys-Leu
(sterol carrier protein) mouse	-Gly-Lys-Ala-Lys-Leu
(sterol carrier protein) human	-Gly-Asn-Ala-Lys-Leu
Peroxisomal membrane-associated 20-kDa protein (PMP20)	-Ile-Ile-Ala-Lys-Leu

2. Internal SKL Motif

The internal SKL motif (Ser-His-Leu) is found in the carboxyl-terminal 27-amino acid segment of human catalase, which directs the fusion protein to peroxisomes (Gould *et al.*, 1988). The internal SKL motif in rat catalase is Ser-His-Ile (Furuta, 1986).

3. Other Motifs

A cleavable peroxisomal targeting signal has been identified at the amino-terminal of rat 3-ketoacyl-CoA thiolase (Swinkles *et al.*, 1991). This is distinct from the previously identified tripeptide peroxisomal targeting signal (PTS) at the C-terminal. There are two types of thiolases in rat peroxisomes, synthesized as larger precursors with an amino-terminal prepiece of 36 (type A) or 26 (type B) amino acids. The thiolase B prepiece shows that the first 11 amino acids are sufficient for peroxisomal targeting. Osumi *et al.* (1991b) have also recognized the targeting signal at the amino-terminal of thiolase, and Hijikata *et al.* (1987) have isolated and analyzed the gene encoding rat peroxisomal 3-ketoacyl-CoA thiolase.

4. Targeting Signals for Peroxisomal Membrane Proteins (PMPs)

Peroxisomal membrane proteins must utilize signals other than peroxisomal targeting signals (PTS 1 and PTS 2) (Gould *et al.*, 1989; Miyazawa *et al.*, 1989). The genes for PMP22, PMP47, PMP70, PAF 1, and PAS 3 have been sequenced, and none have PTS 1 or PTS 2-like sequences at the C- or N-terminal. Keller *et al.* (1991) reported that antibodies against many peroxisomal proteins are located only in the matrix and never on the limiting membrane. Peroxisome membrane ghosts in fibroblasts from patients with Zellweger syndrome fail to import a majority of matrix polypeptides (Santos *et al.*, 1988a; Small *et al.*, 1988a). In PMP47, the C-terminal AKE and the internal SKL were each mutated by a K to A substitution. Single or double mutations did not prevent PMP47 from assembling into peroxisomes (Goodman *et al.*, 1992). The transmembrane domain of these proteins will be required in order to anchor them to the peroxisomal membrane.

5. Receptors for Peroxisomal Membrane Proteins (PTSs)

The PAS8 and PAS10 genes are homologous and encode members of the tetratricopeptide (TPR) protein family. In PAS8 protein, there are seven consecutive TPR domains in the C-terminal half of the protein, followed by a 52-amino acid tail. The PAS8 protein binds to the SKL peptide (McCollum *et al.*, 1993). The tight and selective binding of PAS8 to the peroxisomal membrane, the absence of an obvious transmembrane domain in the sequence of PAS8, and the localization of PAS8 on the cytosolic face of the peroxisomal membrane suggest that PAS8 is a tightly bound peripheral membrane protein that does not require its own PTS. The inducibility of PAS8 on peroxisome proliferation strongly suggests that PAS8 is part of a complex receptor that includes a specific peroxisomal membrane protein, to which membrane protein PAS8 would bind. The PAS7 protein is a candidate for the PTS2 receptor (Erdmann and Kunan, 1992).

Cytosolic SKL-binding factors are involved (Tsukamoto *et al.*, 1990; Wendland and Subramani, 1993) in the import of matrix proteins onto peroxisomes of permeabilized Chinese hamster ovary (CHO) cells. The purification of these factors will determine whether they are receptors or chaperonins.

The uptake of matrix proteins into peroxisomes can be divided into two stages. The binding of the protein to the peroxisome-limiting membrane occurs at 0°C and does not consume energy. Transport across the membrane occurs at 37°C and requires ATP. This step is not driven by a transmembrane potential because it occurs in the presence of proton ionophores such as carbonylcyanide-*m*-chlorophenylhydrazone (CCCP).

As reviewed by Small (1993), it is now established that more than one type of peroxisome targeting signal exists, although the receptors that recognize these signals have not been clearly identified. There will be more than one type of receptor, and a large amount of data implicating various signals, receptors, and carrier proteins have been reported from different laboratories (Keller *et al.*, 1989; Kragler *et al.*, 1993).

VI. Transformation of Tubular to Vesicular and Vesicular to Tubular Form of Peroxisomes Homologous to That in Mitochondria and Lysosomes

A. Transformation of Globoid to Tubular Mitochondria in Human Fibroblasts and Nude Mouse Hepatocytes

When stained with the vital mitochondrial fluorescent probe, dimethylaminostyryl methylpyridinium iodine (DASPM), nude mouse hepatocytes demonstrate normal globoid mitochondria. These change to a filamentous appearance in transformed cells. Similarly, mitochondria in human fibroblasts stained with rhodamine 134 display transformation from a normal shape to pseudo-fragments in Gaucher disease (Kohen *et al.*, 1992). These fluorescent probes are also applicable to lysosomes. Photobiology studies utilizing these probes can be combined with microspectrofluorometry and also with confocal optics for three-dimensional fluorescence determination. The tubular or filamentous configuration of mitochondria thus visualized provides a model for the formation of tubular peroxisomes.

B. Transformation of Vesicular Lysosomes to a Tubular Form

Lysosomes stained with vital fluorescent probes demonstrate transformation from a round or vesicular form to a tubular shape in human fibroblasts (Kohen *et al.*, 1992). This is also evident at the ultrastructural level. This is another model for the relationship between tubular and globular peroxisomes.

Oliver (1982, 1983); Oliver and Yuasa (1987) found unusually long lysosomes in the pancreas and salivary gland. As these long lysosomes were located in the basal portion of glandular cells, he named them "basal lysosomes." Swanson *et al.* (1987) also found tubular lysosomes in macrophages from the peritoneal cavity. Araki *et al.* (1992, 1993) used the term "nematolysosomes" for tubular lysosomes in macrophages, liver, and pancreas. Whatever the nomenclature, these long lysosomes are located basally, peripherally, or in the subsinusoidal space of the cell and are directed toward the Golgi complex. When microtubules are destroyed either by colchicine or low temperature, long lysosomes show a change in shape. Acid phosphatase and trimetaphosphatase are localized cytochemically in long lysosomes, but to lesser extent than in round lysosomes.

Hollenbeck and Swanson (1990) reported that dynein, a retrograde motor protein, and kinesins, orthodromic motor proteins of microtubules, are localized in long lysosomes. Electron microscopically, lysosomes often lie parallel to microtubules and are closely associated by a cross-linking structure (Knapp and Swanson, 1990). Physiological conditions that cause long lysosomes to change to the round type have been discussed by several investigators. When the cytosol is acidic, lysosomes move to the basal or peripheral portion of the cell, and when the cytoplasm recovers to neutral or alkaline pH, long lysosomes move to the center of the cell. Chloroquine increases the number of long lysosomes and NH_4Cl transforms long lysosomes to round ones (Young *et al.*, 1990). Knapp and Swanson (1990) demonstrated that large doses of sucrose or acridine orange accumulate in long lysosomes, which alter their shape to the large round type, and when sucrose is decomposed by invertase they revert to the long form. There are several pieces of evidence to suggest that actin and intermediate filaments are also involved in the formation of tubular lysosomes. The function of long, tubular, or nematolysosomes is not yet clear. Krstić (1988) has postulated that tubular lysosomes in rat and gerbil pinealocytes play a role in rapid cytoplasmic remodeling of the pinealocytes in response to various stimuli.

VII. Conclusion and Future Perspective

The fine network forming the intrinsic substructure of the peroxisomal matrix has been revealed by ultra-high-magnification electron microscopy. Currently, this is the most significant contribution of morphological studies to the molecular structure of peroxisomes. Although molecular studies have been carried out by chemists, observations made by morphologists may help corroborate certain concepts of molecular structure.

The fine network is not strictly specific to the matrix of peroxisomes. As illustrated in Fig. 16, a similar network can be discerned in the matrix of mitochondria, lysosomes, and lipofuscin granules as an intrinsic substructure. Although generalization of the matrix substructure in different cell organelles may oversimplify the subtle specifications of each cell organelle, it can provide clues to the process involved in the transformation

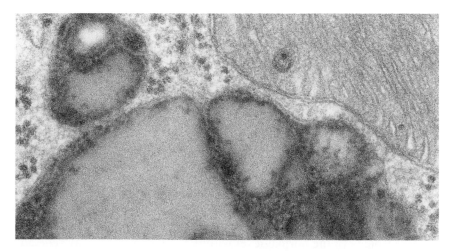

FIG. 16 Substructure of the matrix of a mitochondrion in the upper right-hand corner and that of lipofuscin granules. Note the fine network structure in the matrix. ×96,250.

of tubular mitochondria and lysosomes to vesicular or round forms by analogy with changes seen in peroxisomes.

Ignoring the distinct differences between peroxisomes and mitochondria, lysosomes, or lipofuscin granules in their morphogenesis, intracellular transport systems, targeting signals, assembly factors, receptors, potential functions, enzymic and protein contents, and many other features, a common mechanism of transformation of tubules to vesicles may apply to all these cell organelles. If so, the model of the peroxisome reticulum postulated by Lazarow and Fujiki (1985) can also be remodeled as a fundamental concept to interpret the rapid transformation of tubular peroxisomes to round ones upon application of proliferators and the rapid disappearance of proliferated peroxisomes after withdrawal of the proliferators without leaving debris of peroxisomal components.

The tubulovesicular complex model postulated in this article seems to be the most reasonable explanation for the mechanism of peroxisomal morphogenesis available at the moment. It seems rational to propose that peroxisomes arise from preexisting peroxisomes rather than from the endoplasmic reticulum. This is based mainly on biochemical evidence that newly synthesized catalase first appears in the cytosol and not in the microsomal fraction; that the protein composition of the peroxisomal membrane is totally different from that of the ER; and also that the peroxisomal proteins studied so far are synthesized on free ribosomes. Taking these facts into account, together with the findings of Gorgas (1987), who by electron microscopy and serial sections demonstrated that peroxisomes are independent structures, morphologists, including myself, cannot abandon the accepted concept that a tubular structure, most probably rough or smooth ER, is connected to round peroxisomes.

Topographic conformational differences do not necessarily prove discontinuity of membrane structures. The difference between apical and basolateral cell membranes, as well as between *cis* and *trans* Golgi sacs, for example, does not rule out the integrity of these structures. The reason I prefer the classic theory on the morphogenesis of peroxisomes is that it seems to have features essentially in common with that of other organelles.

The proliferation of peroxisomes in response to various hypolipidemic reagents and plasticizers is important, apart from its clinical significance, for understanding the morphogenesis of peroxisomes. In addition to the phenomena of proliferation and inhibition with other chemicals as discussed already, we have to pay attention to the evidence that simple withdrawal of a proliferator leads to rapid recovery, that is, a decrease in peroxisomes to the normal level within 7–10 days. This may reflect or indirectly indicate the existence of suppressor factors in normal cytosol.

It is beyond the scope of this chapter to speculate about the nature of the suppressor, whether the proliferator might interfere with the suppressor to increase or induce new peroxisomes, and whether removal of the proliferator allows the suppressor to reverse the transformation of tubular to vesicular peroxisomes.

In summary, this article supports the presence of a fine network in the matrix of peroxisomes, a lysosomal-type transformation of tubular to vesicular-type peroxisomes in a modified reticulum model, and the existence of a repressor factor that restricts the increase in the number of peroxisomes under normal conditions.

Acknowledgments

I owe a great deal to the following former and present graduate students of my laboratory: Drs. Kazuo Itakoi, Takashi Yamoto, Haruko Hirose, Kazushige Ogawa, Nobukazu Araki, Miho Ohue, and Takuya Ishida; Mr. Kazumusa Kondo, and Misses Eri Kanaya, Yan-Gai Yang, and Tutik Wresdiyati.

This review is based on their past and present experiments. Their data have been published separately as mentioned in the text, and some others are in preparation to be submitted elsewhere.

Thanks are also extended to the following researchers for sharing information and sending me reprints and valuable photographs:

Dr. Z. Hruban, Chicago; Dr. S. Yokota, Yamanashi Univ.; Dr. T. Watanabe, Tokyo College of Pharmacy; Dr. T. Orio, Gifu Univ.; Dr. Y. Fujiki, Kyushu University; Dr. T. Ohosumi, Himeji Engineering College; Dr. T. Noguchi, Kyushu Dental College; Dr. N. Usuda, Shinshu Univ.; Dr. M. Bendayan, Univ. Montreal; Dr. T. Morimoto, New York Univ.; and Dr. S. Ohono, Yamanashi Univ.

I am always grateful to the suppliers of experimental animals, and the students of my laboratory who are entrusted with the daily care of these animals.

Experiments included in this article were supported in part by grants from the Ministry of Education of Japan, New Energy and Industrial Technology Department Organization (NEDO), and by a supply of bezafibrate from Kissei Pharmaceutical Co.

References

Alexson, S. E. H., Fujiki, Y., Shio, H., and Lazarow, P. B. (1985). Partial disassembly of peroxisomes. *J. Cell Biol.* **101,** 294–305.

Anderson, E., Condon, W., and Sharp, D. (1970). A study of oogenesis and early embryogenesis in the rabbit, oryctolagus cuticles, with special reference to the structural changes of mitochondria. *J. Morphol.* **130,** 67–92.

Angermuller, S., and Fahimi, H. D. (1986). Ultrastructural cytochemical localization of uricase in peroxisomes of rat liver. *J. Histochem. Cytochem.* **34,** 159–165.

Angermuller, S., and Fahimi, H. D. (1988). Heterogeneous staining of D-amino acid oxidase in peroxisomes of rat liver and kidney. A light and electron microscopic study. *Histochemistry* **88**, 277–285.

Angermuller, S., Bruder, G., Voelkl, A., Wesch, H., and Fahimi, H. D. (1987). Localization of xanthine oxidase in crystalline cores of peroxisomes. A cytochemical and biochemical study. *Eur. J. Cell Biol.* **45**, 137–144.

Appelkvist, E.-L., Reinhart, M., Fischer, R., Billheimer, J., and Dallner, G. (1990). Presence of individual enzymes of cholesterol biosynthesis in rat liver peroxisomes. *Arch. Biochem. Biophys.* **282**, 318–325.

Araki, N., Takashima, Y., and Ogawa, K. (1992). Effect of colchicine on the three-dimensional organization of elongated lysosomes (nematolysosomes) in rat pancreatic exocrine cells. *Acta Histochem. Cytochem.* **25**, 105–112.

Araki, N., Ohono, J., Lee, T., Takashima, Y., and Ogawa, K. (1993). Nematolysosomes (elongate lysosome) in rat hepatocytes: Their distribution, microtubule dependence, and role in endocytotic transport pathways. *Exp. Cell Res.* **204**, 181–191.

Arias, I. M. (ed.) (1994). *In* "The Liver, Biology, and Pathobiology," 3rd ed., pp. 1–1591. Raven Press, New York.

Atkins, J., and Claudinin, M. T. (1990). Nutritional significance of factors affecting carnitine dependent transport of fatty acids in neonates: A review. *Nutr. Res.* **10**, 117–128.

Barrett, J. M., and Heideger, P. M., Jr. (1975). Microbodies of the rat renal peroxisomal tubule: Ultrastructural and cytochemical investigations. *Cell Tissue Res.* **157**, 283–305.

Baumgart, E., Stegmeir, K., Schmidt, F. H., and Fahimi, H. D. (1987). Proliferation of peroxisomes in pericentral hepatocytes of rat liver after administration of a new hypocholesterolemic agent BM-15766. *Lab. Invest.* **56**, 554–564.

Bayly, A. C., Roberts, R. A., and Dive, C. (1994). Suppression of liver cell apoptosis in vitro by the non-genotoxic hepatocarcinogen and peroxisome proliferator nafenopin. *J. Cell Biol.* **125**, 197–203.

Beard, M. E., and Novikoff, A. B. (1969). Distribution of peroxisomes (microbodies) in the nephrone of the rat. A cytochemical study. *J. Cell Biol.* **42**, 501–518.

Bendayan, M., and Reddy, J. K. (1982). Immunocytochemical localization of catalase and heat-labile enoyl-CoA hydratase in the livers of normal and peroxisome proliferator-treated rats. *Lab. Invest.* **47**, 364–369.

Bieber, L. L., Valkner, K., and Farrell, S. (1982). Carnitine acetyl transferase of liver peroxisomes. *Ann. N.Y. Acad. Sci.* **386**, 395–396.

Bieri, F., Meier, V., Staubli, W., Muakkassah-Kelly, S. F., Waechter, F., Sagelsdorff, P., and Bentley, P. (1991). Studies on the mechanism of induction of microsomal cytochrome P452 and peroxisomal bifunctional mRNAs by nafenopin in primary cultures of adult rat hepatocytes. *Biochem. Pharmacol.* **41**, 310–312.

Bonder, A. G., and Rachubinski, R. A. (1991). Characterization of the integral membrane polypeptides of rat liver peroxisomes isolated from untreated and clofibrate treated rats. *Biochem. Cell Biol.* **69**, 499–508.

Borst, P. (1986). How proteins get into microbodies (peroxisomes, glyoxysomes, glycosomes). *Biochim. Biophys. Acta* **866**, 179–203.

Borst, P. (1989). Peroxisomal biogenesis revisited. *Biochim. Biophys. Acta* **1008**, 1–13.

Brdiczka, D., Gerbitz, K., and Pette, D. (1969). Localization and function of external and internal carnitine acetyltransferases in mitochondria of rat liver and kidney. *Eur. J. Biochem.* **11**, 234–240.

Brosnan, T., Kopec, B., and Fritz, I. B. (1973). The localization of carnitine palmitoyltransferase on the inner membrane of bovine liver mitochondria. *J. Biol. Chem.* **248**, 4075–4082.

Bursh, W., Oberhammer, F., and Schulte-Hermann, R. (1992). Cell death by apoptosis and its protective role against disease. *Trends Pharmacol. Sci.* **13**, 245–251.

Carson-Jurica, M. A., Schrader, W. T., and O'Malley, B. W. (1990). Steroid receptor family: Structure and functions. *Endocrinol. Rev.* **11**, 201–220.

Castelein, H., Declercq, R. E., Mannaerts, G. P., and Baes, M. I. (1993). Peroxisome proliferators and T_3 operate by way of distinct receptors. *FEBS Lett.* **332**, 24–26.

Chen, N., Crane, D. I., and Masters, C. J. (1988). Analysis of the major integral membrane proteins of peroxisomes from mouse liver. *Biochim. Biophys. Acta* **945**, 135–144.

Cohen, A. J., and Grasso, P. (1981). Review of the hepatic response to hypolipidemic drugs in rodents and assessment of its toxicological significance to man. *Food Cosmet. Toxicol.* **19**, 585–605.

Colucci, W. J., and Gandour, R. D. (1988). Carnitine acetyltransferase: A review of bioorganic chemistry. *Bioorg. Chem.* **16**, 307–334.

Crane, D. I., and Masters, C. J. (1986). The effect of clofibrate on the phospholipid composition of the peroxisomal membranes in mouse liver. *Biochim. Biophys. Acta* **876**, 256–263.

Crane, D. I., Hemsley, A. C., and Masters, C. J. (1985). Purification of peroxisome from livers of normal and clofibrate-treated mice. *Anal. Biochem.* **148**, 436–445.

Crane, D. I., Zamattia, J., and Masters, C. J. (1990). Alterations in the integrity of peroxisomal membranes in livers of mice treated with peroxisome proliferators. *Mol. Cell. Biochem.* **96**, 153–162.

Dahl, E. (1971a). The effect of steroids on the granulosa cells in the domestic fowl. *Z. Zellforsch. Mikrosk. Anat.* **119**, 178–187.

Dahl, E. (1971b). The effect of clomiphene on the granulosa cells of the domestic fowl. *Z. Zellforsch. Mikrosk. Anat.* **119**, 188–194.

Dautry-Varsat, A., and Lodish, H. F. (1984). How receptors bring proteins and particles into cells. *In* "Cell Communication in Health and Disease" (H. Rasmussen, ed.), pp. 90–101. Freeman, New York.

de Duve, C. (1969). *Ann. N.Y. Acad. Sci.* **168**, 369–381.

de Duve, C. (1983). Microbodies in the living cell. *Sci. Am.* **248**, 74–84.

de Duve, C., and Baudhuin, P. (1966). Peroxisomes (microbodies and related particles). *Physiol. Rev.* **46**, 323–357.

del Valle, R., Soto, U., Necochea, C., and Leighton, F. (1988). Detection of an ATPase in rat liver peroxisomes. *Biochem. Biophys. Res. Commun.* **156**, 1353–1359.

Douma, A. C., Veenhuts, M., Sulter, G. J., and Harder, W. (1987). A proton-translocating ATPase is associated with the peroxisomal membrane of yeasts. *Arch. Microbiol.* **147**, 42–47.

Ellis, R. J. (1987). Proteins as molecular chaperones. *Nature (London)* **328**, 378–379.

Ellis, R. J., and van der Vies, S. M. (1991). Molecular chaperones. *Annu. Rev. Biochem.* **60**, 321–347.

Erdmann, R., and Kunan, W. H. (1992). A genetic approach to the biogenesis of peroxisomes in the yeast *Saccharomyces cervisial. Cell Biochem. Funct.* **10**, 167–174.

Evans, R. M. (1988). The steroid and thyroid hormone receptor superfamily. *Science* **240**, 889–895.

Fahimi, H. D. (1969). Cytochemical localization of peroxidatic activity of catalase in rat hepatic microbodies (peroxisomes). *J. Cell Biol.* **43**, 275–288.

Fahimi, H. D., and Sies, H. (1987). "Peroxisomes in Biology and Medicine." Springer-Verlag, Berlin.

Fahimi, H. D., Reinicke, A., Sujatta, M., Yokota, S., Özel, M., Hartig, F., and Stegmeier, K. (1982). The short- and long-term effects of bezafibrate in the rat. *Ann. N.Y. Acad. Sci.* **386**, 111–135.

Fahimi, H. D., Baumgart, E., Beier, K., Pill, J., Hartig, F., and Völkl, A. (1993). Ultrastructural and biochemical aspects of peroxisome proliferation and biogenesis in different mammalian species. *In* "Peroxisomes: Biology and Importance in Toxicology and Medicine" (G. Gibson and B. Lake, eds.), pp. 395–424. Taylor & Francis, London and Washington, DC.

Flatmark, T., Kryvi, H., and Christiansen, E. N. (1982). Polydispersity and biochemical heterogeneity of rat liver peroxisomes induced by clofibrate. A model for studying peroxisomal biogenesis. *Ann. N.Y. Acad. Sci.* **386,** 456–459.

Fringes, B., and Reith, A. (1982). Time course of peroxisome biogenesis during adaptation to mild hyperthyroidism in rat liver. *Lab. Invest.* **47,** 19–26.

Fujiki, T., Tsuneoka, M., and Tashiro, Y. (1989). Biosynthesis of nonspecific lipid transfer protein (sterol carrier protein 2) on free polyribosomes as a larger precursor in rat liver. *J. Biochem. (Tokyo)* **106,** 1126–1131.

Fujiyoshi, Y., Mizusaki, T., Morikawa, K., Yamagishi, H., Aoki, Y., Kihara, H., and Harada, Y. (1991). Development of a superfluid helium stage for high-resolution electron microscopy. *Ultramicroscopy* **38,** 241–251.

Furuta, S., Miyazawa, S., and Hashimoto, T. (1982). Biosynthesis of enzymes of peroxisomal b-oxidation. *J. Biochem. (Tokyo)* **92,** 319–326.

Ghadially, F. N. (1988). "Ultrastructural Pathology of the Cell and Matrix," 3rd ed., (vol. 1) pp. 191–328, (vol. 2) 767–786. Butterworth, London.

Gibson, G. G. (1989). Comparative aspects of the mammalian cytochrome P450 IV gene family. *Xenobiotica* **19,** 1175–1180.

Gibson, G. G., and Lake, B., eds. (1993). "Peroxisomes: Biology and Importance in Toxicology and Medicine." Taylor & Francis, London and Washington, DC.

Gibson, G. G., Orton, T. C., and Tamburini, P. P. (1982). Cytochrome P450 induction by clofibrate: Purification and properties of a hepatic cytochrome P450 relatively specific by the 12- and 11-hydroxylation of dodecanoic (lauric) acid. *Biochem. J.* **203,** 161–168.

Gibson, G. G., Chinje, E., Sabzevari, O., Kentish, P., and Lewis, D. F. V. (1993). Peroxisome proliferators as cytochrome P450 inducers. *In* "Peroxisomes: Biology and Importance in Toxicology and Medicine" (G. G. Gibson and B. Lake, eds.), pp. 110–136. Taylor & Francis, London and Washington, DC.

Goldfischer, S., and Reddy, J. K. (1984). Peroxisomes (microbodies) in cell pathology. *Int. Rev. Exp. Pathol.* **26,** 45–84.

Goodman, J. M., Garrard, L. J., and McCammon, M. T. (1992). Structure and assembly of peroxisomal membrane proteins. *In* "Membrane Biogenesis and Protein Targeting" (W. Newpart and R. Lill, eds.), pp. 221–229. Elsevier, Amsterdam.

Gorgas, K. (1987). Morphogenesis of peroxisomes in lipid-synthesizing epithelia. *In* "Peroxisomes in Biology and Medicine" (H. D. Fahimi and H. Sies, eds.), pp. 3–17. Springer-Verlag, Berlin.

Goto, T., Ashino, T., Fujiyoshi, Y., Kume, N., Yamagischi, H., and Nakai, M. (1994). Projection structures of human immunodeficiency virus type 1 (HIV-1) observed with high resolution electron cryo-microscopy. *J. Electron Microsc.* **43,** 16–19.

Gotoh, M., Griffin, C., and Hruban, Z. (1975). Effect of citrate and aminotriazole on matrical plates induced in hepatic microbodies. *Virchows Arch. B* **17,** 279–294.

Gould, S. J., Keller, G. A., and Subramani, S. (1988). Identification of peroxisomal targeting signals located at the carboxy terminus of four peroxisomal proteins. *J. Cell Biol.* **107,** 897–905.

Gould, S. J., Keller, G. A., Hosken, N., Wilkinson, J., and Subramani, S. (1989). A conserved tripeptide sorts proteins to peroxisomes. *J. Cell Biol.* **108,** 1657–1664.

Grasso, P. (1993). Hepatic changes associated with peroxisome proliferation. *In* "Peroxisomes: Biology and Importance in Toxicology and Medicine" (G. G. Gibson and B. Lake, eds.), pp. 639–652. Taylor & Francis, London and Washington, DC.

Green, S. (1992). Receptor-mediated mechanisms of peroxisome proliferators. *Biochem. Pharmacol.* **43,** 393–401.

Green, S., Tugwood, J. D., and Issemann, I. (1992). The molecular mechanism of peroxisome proliferator action: A model for species differences and mechanistic risk assessment. *Toxicol. Lett.* **64,** 131–139.

Green, S., Issemann, I., and Tugwood, J. D. (1993). The molecular mechanism of peroxisome proliferator action. *In* "Peroxisomes: Biology and Importance in Toxicology and Medicine" (G. G. Gibson and B. Lake, eds.), pp. 99–118. Taylor & Francis, London and Washington, DC.

Guengerich, F. P. (1991). Reactions and significances of cytochrome P450 enzymes. *J. Biol. Chem.* **266**, 10019–10023.

Guttierrez, C., Okita, R., and Krisans, S. (1988). Demonstration of cytochrome reductases in rat liver peroxisomes biochemical and immunochemical analyses. *J. Lipid Res.* **29**, 613–628.

Hakoi, K., and Makita, T. (1994). Ultrastructural localization of magnesium dependent adenosine triphosphatase in dog hepatic peroxisomes treated with a peroxisome proliferator, acetylsalicylic acid. *Biomed. Res.* **15**, 183–190.

Hakoi, K., Irimura, K., and Makita, T. (1993). Ultrastructural study of magnesium dependent adenosine triphosphatase (Mg-ATPase) activity in rat liver peroxisomes after administration of aspirin and clofibrate. *Med. Electron Microsc.* **26**, 161–168.

Hakoi, K., Ohsugi, Y., and Makita, T. (1995). Phenobarbitar suppressed proximosal proliferation by clofibrate in rat hepatocytes. *Toxicol. Lett.* in press.

Hamilton, G. A. (1985). Peroxisomal oxidases and suggestions for the mechanism of action of insulin and other hormones *Adv. Enzymol. Relat. Areas Mol. Biol.* **57**, 85–178.

Hardeman, D., Versantvoort, C., Van Den Brink, J. M., and Van Den Bosch, H. (1990). Studies on peroxisomal membranes. *Biochim. Biophys. Acta* **1027**, 149–154.

Hashimoto, T. (1982). Individual peroxisomal β-oxidation enzymes. *Ann. N.Y. Acad. Sci.* **386**, 5–12.

Hashimoto, T., Kuwabara, T., Usuda, N., and Nagata, T. (1986). Purification of membrane polypeptides of rat liver peroxisomes. *J. Biochem. (Tokyo)* **100**, 301–310.

Henninger, C., Clouet, P., Danh, H. C., Pascal, M., and Bezard, J. (1987). Effects of fenofibrate treatment on fatty acid oxidation in liver mitochondria of obese zucker rats. *Biochem. Pharmacol.* **36**, 3231–3236.

Hicks, L., and Fahimi, H. D. (1977). Peroxisomes (microbodies) in the myocardium of rodents and primates. *Cell Tissue Res.* **175**, 467–481.

Higgins, J. A., and Barnett, R. J. (1970). Cytochemical localization of transferases activities: Carnitine acetyltransferase. *J. Cell Sci.* **6**, 29–51.

Hijikata, M., Ishii, N., Kagamiyama, H., Osumi, T., and Hashimoto, T. (1987). Structural analysis of cDNA for rat peroxisomal 3-ketoacyl-CoA thiolase. *J. Biol. Chem.* **262**, 8151–8158.

Hollenbeck, P. J., and Swanson, J. A. (1990). Radical extension of macrophage tubular lysosomes supported by kinesin. *Nature (London)* **346**, 864–866.

Hruban, Z., and Rechcigal, M., Jr. (1969). Microbodies and related particles. Morphology, biochemistry and physiology. *Int. Rev. Cytol., Suppl.* **1**.

Hruban, Z., and Swift, H. (1964). Uricase: Localization in hepatic microbodies. *Science* **146**, 1316–1318.

Hruban, Z., Vigil, E. L., Slesers, A., and Hopkins, E. (1972). Microbodies, constituent organelles of animal cells. *Lab. Invest.* **27**, 184–191.

Hruban, Z., Gotoh, M., Slesers, A., and Chou, S.-F. (1974). Structure of hepatic microbodies in rats treated with acetylsalicylic acid, clofibrate, and dimethrin. *Lab. Invest.* **30**, 64–75.

Ichiyama, A., Miyagawa, T., Suzuki, T., Matsumoto, N., Sakata, M., and Uchida, C. (1992). Glyoxylate metabolism in peroxisomes. *In* "Frontiers and New Horizons in Amino Acid Research" (K. Takai, ed.), pp. 423–427. Elsevier, Amsterdam.

Imanaka, T., Small, G. M., and Lazarow, P. B. (1987). Translocation of acyl-CoA oxidase into peroxisomes require ATP hydrolysis but not a membrane potential. *J. Cell Biol.* **105**, 2915–2922.

Issemann, I., and Green, S. (1990). Activation of a member of the steroid hormone receptor superfamily by peroxisome proliferators. *Nature (London)* **347**, 645–650.

Itoga, H., Tamura, H., Watanabe, T., and Suga, T. (1990). Characteristics of the suppresive effect of nicardipine on peroxisome induction in rat liver. *Biochim. Biophys. Acta* **1051**, 21–28.

Jirtle, R. L., Meyer, S. A., and Brockenbrough, J. S. (1991). Liver tumor promotor phenobarbital: A biphasic modulator of hepatocyte proliferation. *In* "Chemically Induced Cell Proliferation: Implications for Risk Assessment" (B. E. Butterworth, T. J. Slaga, W. Farland, and M. McClain, eds.), pp. 209–216. Wiley-Liss, New York.

Kamataki, T., Maeda, K., Yamazoe, Y., Nagai, T., and Kato, R. (1983). Sex differences of cytochrome P-450 in the rat: Purification, characterization, and quantitation of constitutive forms of cytochrome P-450 from liver microsomes of male and female rats. *Arch. Biochem. Biophys.* **225**, 758–770.

Kamijyo, K., Taketani, S., Yokota, S., Osumi, T., and Hashimoto, T. (1990). The 70-kDa peroxisomal membrane protein is a member of the Mde(P-glycoprotein)-related ATP-binding protein superfamily. *J. Biol. Chem.* **265**, 4534–4540.

Kanamura, S., Kanai, K., Asaka, Y., and Watanabe, J. (1988). Inhibitory effect of phenobarbital on peroxisome biogenesis in mouse hepatocytes. *J. Ultrastruct. Mol. Struct. Res.* **100**, 269–277.

Kato, R., and Yamazoe, Y. (1992). Sex-specific cytochrome P_{450} as a cause of sex- and species-related differences in drug toxicity. *Toxicol. Lett.* 64/65: 661–667.

Keller, G. A., Scullen, T. J., Clarke, D., Maher, P. A., Krisana, S. K., and Singer, S. J. (1989). Subcellular localization of sterol carrier protein 2 in rat hepatocytes: Its primary localization to peroxisomes. *J. Cell Biol.* **108**, 1353–1361.

Keller, G. A., Krisans, S., Gould, S. J., Sommer, J. M., and Wang, C. C. (1991). Evolutionary conservation of a microbody targeting signal that targets proteins to peroxisomes, glyoxysomes, and glycosomes. *J. Cell Biol.* **114**, 893–904.

Kindle, H. (1982). The biogenesis of microbodies (peroxisomes, glyoxysomes). *Int. Rev. Cytol.* **90**, 193–229.

Kindle, H., and Lazarow, P. B. (1982). Peroxisomes and glyoxysomes. *Ann. N.Y. Acad. Sci.* **386**, 1–550.

Kiselev, N. A., Shpitzberg, C. L., and Vainshtein, B. K. (1967). Crystallization of catalase in the form of tubes with monomolecular walls. *J. Mol. Biol.* **25**, 433–441.

Knapp, P. E., and Swanson, J. A. (1990). Plasticity of the tubular lysosomal compartment in macrophages. *J. Cell Sci.* **95**, 433–439.

Kohen, E., Kohen, C., Hirschberg, J. G., Prince, J., Santus, R., Morliere, P., Schachtschabil, D. O., Shapiro, B. L., Mangel, W. F., and Grabowsky, G. A. (1992). The spatiotemporal organization of metabolism in living cells. *In* "Fundamentals of Medical Cell Biology, Vol. 3B. Chemistry of the Living Cell" (E. E. Bitton, ed.), Chap. 19, pp. 561–606.

Kolde, G., Roessner, A., and Themann, H. (1976). Effects of clofibrate (alpha-p-chlolophenoxy-isobutyryl-ethyl-ester) on male rat liver. *Virchows Arch. B* **22**, 78–87.

Koll, H., Guiard, B., Rassow, J., Ostermann, J., Horwich, A. L., Newpert, W., and Hartle, F.-U. (1992). Antifolding activity of hsp60 couples protein import into the mitochondrial matrix with export to the intermembrane space. *Cell (Cambridge, Mass.)* **68**, 1163–1175.

Kragler, F., Langeder, A., Raupachova, J., Binder, M., and Hartig, A. (1993). Two independent peroxisomal targeting signals in catalase A of saccharomyces cereviside. *J. Cell Biol.* **120**, 665–673.

Krstić, R. (1988). Tubular lysosomes in rat and gerbil pinealocytes. *Histochemistry* **88**, 203–206.

Kryvi, H., Kvannes, J., and Hatmark, T. (1990). Freeze-fracture study of rat liver peroxisomes evidence for an induction of intramembrane particles by agents stimulating peroxisome proliferation. *Eur. J. Cell Biol.* **53**, 227–233.

Kume, N. (1990). Methods in high resolution cryo-electron microscopy (3). From cryofixation into cryo-transfer. *J. Electron Microsc.* **25**, 58–63.

Labarca, P., Wolff, D., Soto, U., Uecocha, C., and Leighton, F. (1987). Large cation-selective pores from rat liver peroxisomal membranes incorporated to planar lipid bilayers. *J. Membr. Biol.* **94,** 285–292.

Lake, B. G., and Lewis, D. F. V. (1993). Structure-activity relationships for chemically induced peroxisome proliferation in mammalian liver. *In* "Peroxisomes: Biology and Importance in Toxicology and Medicine" (G. G. Gibson and B. Lake, eds.), pp. 313–342. Taylor & Francis, London and Washington, DC.

Lalwani, N. D., Reddy, M. K., Qureshi, S. A., Sirtori, C. R., Abiko, Y., and Reddy, I. K. (1983). Evaluation of selected hypolipidemic agents for the induction of peroxisomal enzymes and peroxisome proliferation in the rat liver. *Hum. Toxicol.* **2,** 27–48.

Lalwani, N. D., Alvares, K., Reddy, M. K., Reddy, M. N., Parikh, I., and Reddy, J. K. (1987). Peroxisome proliferator-binding protein: Identification and partial characterization of nafenopin-, clofibric acid- and ciprofibrate-binding proteins from rat liver. *Proc. Natl. Acad. Sci. U.S.A.* **84,** 5342–5346.

Lazarow, P. B., and de Duve, C. (1976). A fatty acyl-CoA oxidizing system in rat liver peroxisomes: Enhancement by clofibrate, a hypolipidemic drug. *Proc. Natl. Acad. Sci. U.S.A.* **73,** 2043–2046.

Lazarow, P. B., and Fujiki, Y. (1985). Biogenesis of peroxisomes. *Annu. Rev. Cell Biol.* **1,** 489–530.

Lazarow, P. B., Robbi, M., Fujiki, Y., and Wong, L. (1982). Biogenesis of peroxisomal proteins in vivo and in vitro. *Ann. N.Y. Acad. Sci.* **386,** 285–300.

Leighton, F., Brandau, E., Lazo, O., and Bronfman, M. (1982). Subcellular fractionation studies on the organization of fatty acid oxidation by liver peroxisomes. *Ann. N.Y. Acad. Sci.* **386,** 62–80.

Lewis, D. F. V., and Lake, B. G. (1993). The interaction of some peroxisome proliferators with the mouse liver peroxisome proliferator-activated receptor (PPAR): A molecular modeling and quantitative structure-activity relationship (QSAR) study. *Xenobiotica* **23,** 79–96.

Lin, J. K., and Chou, C. K. (1992). In vitro apoptosis in a human hematoma cell line induced by transforming growth factor β. *Cancer Res.* **52,** 385–388.

Lock, E. A., Mitchell, A. M., and Elcombe, C. R. (1989). Biochemical mechanisms of induction of hepatic peroxisome proliferation. *Ann. Rev. Pharmacol. Toxicol.* **29,** 145–163.

Mackiewicz, A., Kushner, I., and Baumann, H. (1993). Acute phase proteins. *In* "Molecular Biology, Biochemistry, and Clinical Applications," p. 686. CRC Press, Boca Raton, FL.

Maeda, K., Kamataki, T., Nagai, T., and Kato, R. (1984). Postnatal development of constitutive forms of cytochrome P-450 in liver microsomes of male and female rats. *Biochem. Pharmacol.* **33,** 509–512.

Makita, T. (1988). Cytochemical localization of Mg-ATPase activity in the limiting membrane of peroxisomes in hepatocytes of the rat. *Proc. Int. Congr. Cell Biol., 4th,* Montreal, August. p. 343.

Makita, T. (1990). Ultrastructure of peroxisomes in hepatocytes of clofibrate administered rats. *J. Clin. Electron Microsc.* **23,** 1044–1055.

Makita, T. (1993). Cell organelles. Mitochondria. *In* "Electron Microscopic Cytochemistry and Immunocytochemistry in Biomedicine" (K. Ogawa and T. Barka, eds.), pp. 505–517. CRC Press, Boca Raton, FL.

Makita, T., and Hakoi, K. (1994). Proliferation and alteration of hepatic peroxisomes and reduction of ATPase activity on their limiting membrane after oral administration of acetylsalicylic acid (aspirin) for four weeks to male rats. *Ann. N.Y. Acad. Sci.* (in press).

Makita, T., and Hirose, H. (1990). Substructural tubular network in the matrix of induced peroxisomes of rat hepatocyte. *Proc. Int. Congr. Electron Microsc., 12th, 1990,* pp. 556–557.

Makita, T., and Ogawa, K. (1991). Visualization of intramatrix tubular structure of peroxisome in hepatocytes by imaging plate (PIXsysTEM). *Anat. Rec.* **229**, 56A–57A.

Makita, T., and Okawa, T. (1987). Localization of platinum and iron in the peroxisomes of liver and kidney of rats after sucessional administration of cisplatin an anti-tumor drug. *Proc. Int. Congr. Inborn Errors Metab., 4th,* P-8.

Makita, T., and Sandborn, E. B. (1971). A comparison of ultrastructural localization of carnitine acetyltransferase activity in mouse liver mitochondria with that in cardiac muscle. *Experientia* **27**, 184–187.

Makita, T., Kiwaki, S., and Sandborn, E. B. (1973). Scanning electron microscopy and cytochemical localization of carnitine acetyltransferase activity in normal and dystrophic muscle of mice. *Histochem. J.* **5**, 335–342.

Makita, T., Itagaki, S., and Okawa, T. (1985). X-ray microanalysis and ultrastructural localization of cisplatin in liver and kidney of the rat. *Jpn. J. Cancer Res.* **76**, 895–901.

Makita, T., Hakoi, K., and Okawa, T. (1986). X-ray microanalysis and electron microscopy of platinum complex in the epithelium of proximal renal tubules of the cisplatin administered rabbits. *Cell Biol. Int. Rep.* **10**, 447–454.

Makita, T., Hakoi, K., and Araki, N. (1990a). Cytochemical localization of Mg-ATPase and Ca-ATPase on the limiting membrane of rat liver peroxisomes. *Acta Histochem. Cytochem.* **23**, 601–611.

Makita, T., Ishida, T., and Yamoto, T. (1990b). Substructure of rat hepatocyte peroxisomes: Micropores in the limiting membrane and tubules in the matrix. *J. Electron Microsc.* **39**, P-17-F6 (abstr.).

Makita, T., Hakoi, K., and Irimura, K. (1992). Effect of oral administration of aspirin on the ultrastructure and Mg-ATPase localization of peroxisomes and mitochondria in rat hepatocytes. *J. Clin. Electron Microsc.* **25**, 543–544.

Makita, T., Ohue, M., Yamoto, T., and Hakoi, K. (1993). Atomic force microscopy (AFM) of the cuticular pigment globules of the quail egg shell. *J. Electron. Microsc.* **42**, 189–192.

Makita, T., Hakoi, K., Ogawa, K., and Hirose, H. (1994). Imaging-plate as a tool to reveal substructural tubular network in the ground substance of peroxisomes, lipofuscin granules and mitochondria. *Proc. Int. Congr. Electron Microsc., 13th,* **3B**, pp. 965–966. Paris, July 17–22, 1994.

Markwell, M. A. K., Mcgroarty, E. J., Bieber, L. L., and Tolbert, N. E. (1973). The subcellular distribution of carnitine acyltransferases in mammalian liver and kidney. *J. Biol. Chem.* **248**, 3426–3432.

Masters, C., and Holmes, R. (1977). Peroxisomes: New aspects of cell physiology and biochemistry. Physiology and biochemistry. *Physiol. Rev.* **57**, 816–882.

McClain, R. M., Levin, A. A., Posch, R., and Downing, J. C. (1989). The effect of phenobarbital on the metabolism and excretion of thyroxine in rats. *Toxicol. Appl. Pharmacol.* **99**, 216–228.

McCollum, D., Monosov, E., and Subramani, S. (1993). The pas8 mutant of Pichia pastoris exhibits the peroxisomal protein import deficiencies of Zellweger syndrome cells. The PAS8 protein binds to the COOH-terminal tripeptide peroxisomal targeting signal, and is a membrane of the TPR protein family. *J. Cell Biol.* **121**, 761–774.

Michaelopoulos, G. (1991). Control of hepatocyte proliferation in regenerating, augmentative hepatomegaly, and neoplasia. *In* "Chemically Induced Cell Proliferation: Implication for Risk Assessment" (B. E. Butterworth, T. J. Slaga, W. Farland, and M. McClain, eds.), pp. 227–236. Wiley-Liss, New York.

Milton, M. N., Elcombe, C. R., and Gibson, G. G. (1990). On the mechanism of induction of cytochrome P450IVA1 and peroxisome proliferation in rat liver by clofibrate. *Biochem. Pharmacol.* **40**, 2727–2732.

Miura, S., Kasuya-Arai, I., Mori, H., Miyazawa, S., Osumi, T., Hashimoto, T., and Fujiki, Y. (1992). Carboxyl-terminal consensus Ser-Lys-Leu-related tripeptide of peroxisomal proteins fractions in vitro as a minimal peroxisome-targeting signal. *J. Biol. Chem.* **267**, 14405–14411.

Miyazawa, S., Ozasa, H., Furuta, S., Osumi, T., and Hashimoto, T. (1983). Purification and properties of carnitine acetyltransferase from rat liver. *J. Biochem.* (*Tokyo*) **93**, 439–451.

Miyazawa, S., Osumi, T., Hashimoto, T., Ohno, K., Miura, S., and Fujiki, Y. (1989). Peroxisome targeting signal of rat liver acyl-Coenzyme A oxidase resides at the carboxy terminus. *Mol. Cell. Biol.* **9**, 83–91.

Moody, D. E., Gibson, G. G., Grant, D. F., Magdalon, J., and Sambasiva, M. (1992). Peroxisome proliferators, a unique set of drug-metabolizing enzyme inducers: Commentary on a symposium. *Drug Metab. Dispos.* **20**, 779–791.

Moody, D. E. (ed.). (1994). *In* "Peroxisome Proliferators: Unique Inducers of Drug-Metabolizing Enzymes," pp. 1–187. CRC Press, Boco Raton, FL.

Mori, N., Oikawa, T., Harada, Y., and Miyahara, J. (1990). Development of the imaging plate for the transmission electron microscope and its characteristics. *J. Electron Microsc.* **39**, 433–436.

Mori, T., Tsukamoto, T., Mori, H., Tashiro, Y., and Fujiki, Y. (1991). Molecular cloning and deduced amino acid sequence of nonspecific lipid transfer protein (sterol carrier protein 2) of rat liver: A higher molecular mass (60kDa) protein contains the primary sequence of nonspecific lipid transfer protein as its C-termin 1 part. *Proc. Natl. Acad. Sci. U.S.A.* **88**, 4338–4342.

Moser, H. W., Bergin, A., and Cornblath, D. (1991). Peroxisomal disorders. *Biochem. Cell Biol.* **69**, 463–474.

Noguchi, T. (1987). Amino acid metabolism in animal peroxisomes. *In* "Peroxisomes in Biology and Medicine" (H. D. Fahimi and H. Sies, eds.), pp. 234–243. Springer-Verlag, Berlin.

Novikoff, P. M., Novikoff, A. B., Quintana, N., and Davis, C. (1973). Studies on microperoxisomes. III. Observations on human and rat hepatocytes. *J. Histochem. Cytochem.* **21**, 540–558.

Oberhammer, F. A., Pavelka, M., Sharma, S., Tiefenbacher, R., Purchio, A. F., Bursch, W., and Schulte-Hermann, R. (1992). Induction of apoptosis in cultured hepatocytes and in regressing liver by transforming growth factor β_1. *Proc. Natl. Acad. Sci. U.S.A.* **89**, 5408–5412.

Ohno, S., and Fujii, Y. (1990). Three-dimensional and histochemical studies of peroxisomes in cultured hepatocytes by quick-freezing and deep-etching method. *Histochem. J.* **22**, 143–154.

Ohue, M., and Makita, T. (1994). Effect of oral administration of bezafibrate on peroxisomes and carnitine acetyltransferase in rat hepatocytes. *J. Vet. Med. Sci.* **56**, 541–546.

Oliver, C. (1982). Endocytic pathways at the lateral and basal cell surface of endocrine acinar cells. *J. Cell Biol.* **95**, 154–161.

Oliver, C. (1983). Characterization of basal lysosomes in endocrine acinar cells. *J. Histochem. Cytochem.* **31**, 1209–1216.

Oliver, C., and Yuasa, Y. (1987). Distribution of basal lysosomes in exocrine acinar cells. *J. Histochem. Cytochem.* **35**, 565–570.

Orton, T. C., and Parker, G. L. (1982). The effect of hypolipidemic agents on the hepatic microsomal drug-metabolizing enzyme system of the rat. Induction of cytochrome (s)P-450 with specificity toward terminal hydroxylation of lauric acid. *Drug Metab. Dispos.* **10**, 110–115.

Osumi, T., and Fujiki, T. (1990). Topogenesis of peroxisomal proteins. *BioEssays* **12**, 217–222.

Osumi, T., Wen, J.-K., and Hashimoto, T. (1991a). Two cis-acting regulatory sequences in the peroxisome proliferator-responsive enhancer region of rat acyl-CoA oxidase gene. *Biochem. Biophys. Res. Commun.* **175**, 866–871.

Osumi, T., Tsukamoto, T., Hata, S., Yokota, S., Miura, S., Fujiki, Y., Hijikata, M., Miyazawa, S., and Hashimoto, T. (1991b). Amino-terminal presequence of the precursor of peroxisomal 3-ketoacyl-CoA thiolase is a cleavable signal peptide for peroxisomal targeting. *Biochem. Biophys. Res. Commun.* **181**, 947–954.

Pais, M. S. S., and Carrapico, F. (1982). Micobodies—a membrane compartment. *Ann. N.Y. Acad. Sci.* **386**, 510–513.

Poole, B., Higashi, T., and de Duve, C. (1977). The synthesis and turnover of rat liver peroxisome. III. The size distribution of peroxisomes and the incorporation of new catalase. *J. Cell Biol.* **45**, 408–415.

Porter, T. D., and Coon, M. J. (1991). Cytochrome P450 multiplicity: Isoforms, substrate and catalytic and regulatory mechanisms. *J. Biol. Chem.* **266**, 13469–13472.

Rao, M. S., and Reddy, J. K. (1991). An overview of peroxisome proliferator-induced hepato carcinogenesis. *Environ. Health Perspect.* **93**, 205–209.

Reddy, J., and Svoboda, D. (1973). Further evidence to suggest that microbodies do not exist as individual entities. *Am. J. Pathol.* **70**, 421–432.

Reddy, J. K., and Lalwani, N. D. (1983). Carcinogenesis by hepatic peroxisome proliferators: Evaluation of the risk of hypolipidemic drugs and industrial plasticizers to humans. *CRC Crit. Rev. Toxicol.* **12**, 1–58.

Reddy, J. K., and Rao, M. S. (1986). Peroxisome proliferators and cancer: Mechanisms and Implications. *Trends Pharmacol. Sci.* **7**, 438–443.

Rhodin, J. (1954). Correlation of ultrastructural organization and function in normal and experimentally changed proximal convoluted tubule cells of the mouse kidney. Ph.D. Thesis, Karolinska Inst. Aktiebolaget Godvil, Stockholm.

Rouiller, C., and Bernhard, W. (1956). "Microbodies" and the problem of mitochondria regeneration in liver cells. *J. Biophys. Biochem. Cytol., Suppl.* **2**, 355–359.

Sabatini, D. D., and Adesnik, M. (1994). Biogenesis of peroxisomes. *In* "The Metabolic Basis of Inherited Diseases" (C. R. Scriver, ed.), 3rd ed. McGraw-Hill, New York (in press).

Santos, M. J., Imanaka, T., Shio, H., Small, G. M., and Lazarow, P. B. (1988a). Peroxisomal membrane ghosts in Zellweger syndrome-aberrant organelle assembly. *Science* **239**, 1536–1538.

Santos, M. J., Imanaka, T., Shio, H., and Lazarow, P. B. (1988b). Peroxisomal integral membrane proteins in control and Zellweger fibroblasts. *J. Biol. Chem.* **263**, 10502–10509.

Shimozawa, N., Tsukamoto, T., Suzuki, Y., Orii, T., Shirayoshi, Y., Mori, T., and Fujiki, T. (1992a). A human gene responsible for Zellweger syndrome that affects peroxisome assembly. *Science* **255**, 1132–1134.

Shimozawa, N., Tsukamoto, T., Suzuki, Y., Orii, T., and Fujiki, Y. (1992b). Animal cell mutants represent two complementation groups of peroxisome-defective Zellweger syndrome. *J. Clin. Invest.* **90**, 1864–1870.

Shoemaker, R. L., and Yamazaki, R. K. (1991). Thyroid hormone-independent regulation of mitochondrial glycero-3-phosphate dehydrogenase by the peroxisome proliferator clofiric acid. *Biochem. Pharmacol.* **41**, 652–655.

Small, G. M. (1993). Peroxisome biogenesis. *In* "Peroxisomes: Biology and Importance in Toxicology and Medicine" (G. G. Gibson and B. Lake, eds.), pp. 1–7. Taylor & Francis, London and Washington, DC.

Small, G. M., Burdett, K., and Connock, M. J. (1982). Clofibrate-induced changes in enzyme activities in liver, kidney and small intestine of male mice. *Ann. N.Y. Acad. Sci.* **386**, 460–463.

Small, G. M., Szabo, L. J., and Lazarow, P. B. (1988a). Acyl-CoA oxidase contains two targeting sequences each of which can mediate protein import into peroxisomes. *EMBO J.* **7**, 1167–1173.

Small, G. M., Santos, M. J., Imanaka, T., Poulos, A., Danks, D. M., Moser, H. W., and

Lazarow, P. B. (1988b). Peroxisomal integral membrane proteins in livers of patients with Zellweger syndrome, infantile Resum's disease and X-linked adrenoleukodystrophy. *J. Inherited Metab. Dis.* **11**, 358–371.

Soto, U., Rapp, S., Gorgas, K., and Just, W. W. (1993a). Peroxisomes and lysosomes. In "Molecular and Cell Biology of the Liver" (A. V. LeBouton, ed.), pp. 181–262. CRC Press, Boca Raton, FL.

Soto, U., Repperkoh, R., Ansorge, W., and Just, W. W. (1993b). Import firefly luciferase into mammalian peroxisomes in vivo requires nucleoside triphosphates. *Exp. Cell Res.* **205**, 66–75.

Staubli, W., Hess, R., and Weibel, E. R. (1969). Microbodies in experimentally altered cells. *J. Cell Biol.* **42**, 92–111.

Subramani, S. (1992). Targeting of proteins into the peroxisomal matrix. *J. Membr. Biol.* **125**, 99–106.

Subramani, S. (1993). Protein import into peroxisomes and biogenesis of the organelle. *Annu. Rev. Cell Biol.* **9**, 445–478.

Svoboda, D., Grady, H., and Azarnoff, D. (1967). Microbodies in experimentally altered cells. I. *Cell Biol.* **35**, 127–152.

Svoboda, D., Azarnoff, D., and Reddy, J. (1969). Microbodies in experimentally altered cells. II. The relationship of microbody proliferation to endocrine gland. *J. Cell Biol.* **40**, 734–746.

Swanson, J., Burke, E., and Silverstein, S. C. (1987). Tubular lysosomes accompany stimulated pinocytosis in macrophages. *J. Cell Biol.* **104**, 1217–1222.

Swinkles, B. W., Gould, S. J., Bodner, A. G., Rachubinski, R. A., and Subramani, S. (1991). A novel, cleavable peroxisomal targeting signal at the amino-terminus of the rat 3-ketoacyl-CoA thiolase. *EMBO J.* **10**, 3255–3262.

Thompson, S. L., and Krisans, S. K. (1985). Evidence for peroxisomal hydroxylase activity in rat liver. *Biochem. Biophys. Res. Commun.* **130**, 708–716.

Tolbert, N. E. (1981). Metabolic pathways in peroxisomes and glyoxysomes. *Annu. Rev. Biochem.* **50**, 133–157.

Tosh, D., George, K., Alberti, M. M., and Agius, L. (1989). Clofibrate induces carnitine acetyltransferases in periportal and perivenous zones of rat liver and does not disturb the acinar zonation of gluconeogenesis. *Biochim. Biophys. Acta* **992**, 245–250.

Tsukada, H., Koyama, S., Gotoh, M., and Tadano, H. (1971). Fine structure of crystalloid nucleoids of compact type in hepatocyte microbodies of guinea pigs, cats and rabbits. *J. Ultrastruct. Res.* **36**, 159–175.

Tsukamoto, T., Yokota, S., and Fujiki, Y. (1990). Isolation and characterization of Chinese hamster ovary cell mutants defective in assembly of peroxisomes. *J. Cell Biol.* **110**, 651–660.

Tsukamoto, T., Miura, S., and Fujiki, Y. (1991). Restoration by a 35k membrane protein of peroxisome assembly in a peroxisome-deficient mammalian cell mutant. *Nature (London)* **350**, 77–81.

Tsuneoka, M., Yamamoto, A., Fujiki, Y., and Tashiro, Y. (1988). Nonspecific lipid transfer protein (sterol carrier protein-2) is located in rat liver peroxisomes. *J. Biochem.* **104**, 560–564.

Usuda, N., Ma, H., Hanai, T., Yokota, S., Hashimoto, T., and Nagata, T. (1990). Immunoelectron microscopy of tissues processed by rapid freezing and freeze-substitution without chemical fixatives: Application to catalase in rat liver hepatocytes. *J. Histochem. Cytochem.* **38**, 617–623.

van den Bosch, H., Schutagens, R. B. H., Wanders, R. J. A., and Tager, J. M. (1992). Biochemistry of peroxisomes. *Annu. Rev. Biochem.* **61**, 157–197.

Van Noorden, C. J. F., and Frederinks, W. M. (1993). Cerium methods for light and electron microscopical histochemistry. *J. Microsc. (Oxford)* **171**, 3–16.

Verheyden, K., Fransen, M., Van Veldhoven, P. P., and Mannaerts, G. P. (1992). Presence of small GTP-binding proteins in the peroxisomal membrane. *Biochim. Biophys. Acta* **1109**, 48–54.

Walton, P. A., Wendland, M., Subramani, S., Rachubinski, R. A., and Welch, W. J. (1994). Involvement of 70-kD heat-shock proteins in peroxisomal import. *J. Cell Biol.* **125**, 1037–1046.

Watanabe, T., Horie, S., Yamada, J., Isaji, M., Nishigaki, T., Naito, J., and Suga, T. (1989). Species differences in the effects of bezafibrate, a hypolipidemic agent, on hepatic peroxisome-associated enzymes. *Biochem. Pharmacol.* **38**, 367–371.

Watanabe, T., Itoga, H., Okawa, S., Tamura, H., and Suga, T. (1991). Co-suppression by nicardipine, a calcium antagonist, of induction of microsomal lauric acid hydroxylation with peroxisome proliferation in clofibrate-treated rat liver. *Chem. Pharmacol. Bull.* **39**, 1320–1322.

Watanabe, T., Okawa, S., Itoga, H., Imanake, T., and Suga, T. (1992). Involvement of calmodulin- and protein kinase C-related mechanism in an induction process of peroxisomal fatty acid oxidation-related enzyme by hypolipidemic peroxisome proliferators. *Biochim. Biophys. Acta* **1135**, 84–90.

Welch, W. J. (1991). The role of heat-shock proteins as molecular chaperones. *Curr. Opin. Cell Biol.* **3**, 1033–1038.

Wendland, M., and Subramani, S. (1993). Cytosol-dependent peroxisomal protein import in a permeabilized cell system. *J. Cell Biol.* **120**, 675–685.

Whitney, A. B., and Bellion, E. (1991). ATPase activities in peroxisome-profile rating yeast. *Biochim. Biophys. Acta* **1058**, 345–355.

Wilson, G. N., and King, T. E. (1991). Structure and variability of mammalian peroxisomal membrane proteins. *Biochem. Med. Metab. Biol.* **46**, 235–245.

Wresdiyati, T., and Makita, T. (1994). Remarkable increase of peroxisomes in the renal tubule cells of Japanese monkey under fasting stress. *Pathophysiology* **1** (Supple), **386** (abst.).

Yamoto, T., Ohashi, Y., Kimura, K., and Matsunami, N. (1991). Effect of clofibrate on liver peroxisomes in several rat strains. *Proc.* 18th *Jpn. Soc. Toxicol. Sci.* **18**, 269.

Yamoto, T., Ohashi, Y., Teranishi, M., Takaoka, M., Manabe, S., Matsunuma, N., and Makita, T. (1995). Age-related changes in the susceptibility to clofibrate acid, a hypolipidemic agent, of male rat liver. *Toxicol. Lett.* (in press).

Yokota, S. (1989). Immunocytochemical localization of peroxisomal enzymes in rat liver and kidney revealed by immunoenzyme and immunogold techniques. *Rev. Immunoassay Technol.* **3**, 145–166.

Yokota, S., and Asayama, K. (1992a). Induction of acyl-CoA oxidase in rat heart and soleus muscles after starvation. *Biomed. Res.* **13**, 87–94.

Yokota, S., and Asayama, K. (1992b). Proliferation of myocardial peroxisomes in experimental rat diabetes: A biochemical and immunocytochemical study. *Virchows Arch. B* **63**, 43–49.

Young, M. R., Gordon, A. H., and Hart, P. D. (1990). Tubular lysosomes and their drug reactivity in cultured resident macrophages and in cell-free medium. *Exp. Cell Res.* **190**, 283–289.

Zaar, K., Völkl, A., and Fahimi, H. D. (1991). Purification of marginal plates from bovine renal peroxisomes: Identification with L-α-hydroxy acid oxidase B. *J. Cell Biol.* **113(1)**, 113.

INDEX